Quadratic Formula

If $ax^2 + bx + c = 0$, for $a \neq 0$, then

$$x = \frac{-b \pm \sqrt{b^2 - 4ac}}{2a}$$

Cartesian Coordinates

Let $P_1 = (x_1, y_1)$ and $P_2 = (x_2, y_2)$; then:

1 The distance d between P_1 and P_2 is
$$d = \sqrt{(x_2 - x_1)^2 + (y_2 - y_1)^2}$$

2 The slope m of the line segment $\overline{P_1 P_2}$ is
$$m = \frac{y_2 - y_1}{x_2 - x_1}$$

3 The slope intercept form of a line is
$$y = mx + b$$

4 The point slope form of a line is
$$y - y_1 = m(x - x_1)$$

Logarithm Properties

1 $\log_b MN = \log_b M + \log_b N$

2 $\log \dfrac{M}{N} = \log_b M - \log_b N$

3 $\log_b M^k = k \log_b M$

Algebra for College Students

Algebra for College Students

M.A. MUNEM

W. TSCHIRHART

MACOMB COMMUNITY COLLEGE

WORTH PUBLISHERS, INC.

ALGEBRA FOR COLLEGE STUDENTS

EDITOR: ROSALIND LIPPEL

PRODUCTION: JOSE FONFRIAS

DESIGN: MALCOLM GREAR DESIGNERS

ILLUSTRATIONS: REPRODUCTION DRAWINGS LIMITED

TYPOGRAPHER: TYPOTHETAE BOOK COMPOSITION

PRINTING AND BINDING: VON HOFFMANN PRESS

COVER: COMPUTER GRAPHICS BY TOM NORTON

WORTH PUBLISHERS, INC.

444 PARK AVENUE SOUTH

NEW YORK, NEW YORK 10016

Preface

Purpose This textbook was written to provide students who have had a first course in algebra with the information and skills needed to succeed in precalculus and other mathematics courses. Basic concepts are presented in a clear, straightforward manner.

Prerequisites One year of high-school algebra or an equivalent course in beginning algebra is sufficient. Conscientious students with less preparation may, however, be able to master the contents of this book, particularly if they use the accompanying *Study Guide* as an aid.

Features We have taken great care to enable the average student to learn to solve routine problems, to develop computational skills, and to build confidence in his or her mastery of important concepts directly from this textbook. Several features of this book contribute to meeting these objectives.

1. *Pedagogy* Topics are presented in brief sections that progress logically from basic algebraic concepts and skills to more complex material. Each topic is illustrated with worked-out examples, problems for students to work themselves, and applications. This arrangement, together with the logical ordering of topics, enables students to build continually upon what they have learned. Many problems and examples are provided to ensure that students have plenty of opportunity to test their understanding and develop problem-solving skills.

2. *Examples* Examples clearly illustrate the important concepts and procedures. The 653 examples in this textbook are worked in detail, with all the substitutions shown.

3. *Procedural Guides* Concise, step-by-step guides are given to explain the basic algebraic operations.

4. *Problem Sets* Problem sets at the end of each section contain 2,891 problems, most of which correspond to the examples in the text. Each set of problems is graded in difficulty, building from very simple drill problems to those that require a slightly higher level of skill. Occasionally—in the even-numbered problems towards the end of a problem set—there are more challenging problems. The answers to the odd-numbered problems are given in an appendix.

5. *Review Problem Sets*   Review problem sets at the end of each chapter contain an additional 1,066 problems. These review problems can be used in several ways: Instructors may wish to use them for supplementary assignments or for quizzes and exams; students may use them to review essential material from the entire chapter and determine areas where additional study is needed.

6. *Applications*   There are many applications—to business, science, economics, and engineering—in this textbook, both in examples and in problem sets. The frequent use of applications shows students the relevance of algebra in their personal lives and in their other studies. Sections devoted entirely to applications appear in Chapters 4, 6, 7, and 8.

7. *Common Errors*   Students are warned about common errors in such a way that there will be no confusion about what is correct and what is not. (For example, see pages 43 and 72.)

8. *Use of Calculators*   Some professors may want to emphasize calculator use; others may not. The material in this textbook has been written so that both options are possible. Problems that can be more appropriately solved with a calculator (see Chapter 8, for example) are labeled with the symbol ▣.

9. *Design*   An open design makes the book easy to read and understand. Important elements are accented with color, ruled boxes, and boldface type. Figures are used to enhance and supplement explanations whenever possible.

Student Aids

The *Study Guide* is a supplementary learning resource, offering tutorial aid on each topic in the textbook. It is a useful source of drill problems for students who lack confidence, who need additional reinforcement and practice, who have missed some classes and need to catch up, or who need assurance that they are ready for exams. The *Study Guide* contains study objectives, fill-in statements and problems (broken down into simple units), and a self-test for each chapter. Answers to all problems and tests in the *Study Guide* are included, to provide immediate reinforcement for correct responses.

Instructor Aids

An *Instructor's Resource Manual* is available, which contains (1) a unique Algebra Placement Test, a readily scored, branching test that gives immediate information for placing students in the appropriate algebra course; (2) a syllabus; (3) suggestions concerning the appropriate lecture schedule for one-quarter and one-semester courses; (4) two tests for each chapter, one of which is multiple choice; (5) a multiple-choice final examination; and (6) answers to the even-numbered problems in the textbook.

Acknowledgments   A great many people have contributed to the development of this textbook. We wish to thank the following people, who reviewed this manuscript and offered many helpful suggestions: William Coppage, *Wright State University*, Linda Exley, *DeKalb Community College*, Merle Friel, *Humboldt State College*, Gus Pekara, *South Oklahoma City Junior College*, Howard Taylor, *West Georgia College*, Stuart Thomas, *Oregon State University*, Roger Willig, *Montgomery County Community College*, and Jim Wolfe, *Portland Community College*.

Special thanks are due to our colleague William Hart for solving all the problems in the book and for assisting in the proofreading, and to Rosemarie Zidzik for help in the proofreading. We are also grateful to the staff of Worth Publishers for their continuous cooperation and assistance throughout this project. In particular, we wish to thank our editor, Rosalind Lippel, and, once again, Robert Andrews.

M. A. Munem
W. Tschirhart

Contents

1

Numbers and Their Properties

Algebra is a generalization of arithmetic in which letter symbols, such as x or y, are used to represent numbers. An understanding of arithmetic is therefore an essential foundation for the study of algebra. This will become clear as you review the basic concepts in this chapter.

1.1 Sets of Numbers

The idea of a **set** allows us to classify numbers. A set may be thought of as a collection of objects. Any one of the objects in a set is called an **element** or a **member** of the set. Capital letters, such as A, B, C, and D, or braces { } enclosing the elements in a set are often used to denote sets. Thus, if we write {1, 2, 3, 4, 5}, we mean the set whose elements are the numbers 1, 2, 3, 4, and 5.

Sets of numbers can be visualized by using a **number line** or a **coordinate axis.** To construct a number line, we draw a horizontal line L, with arrowheads at both ends to show that the line extends endlessly in both directions. Then we choose a point O on this line, and we associate it with the number 0. This point is called the **origin.** Next, we select another point U on the number line, called the **unit point.** We associate the number 1 with the point U. The distance between the origin O and the unit point U is called the **unit distance.** This distance may be 1 inch, 1 centimeter, or one unit of whatever measure you choose (Figure 1a).

By measuring one unit distance to the right of the point U, we find a point that we associate with the number 2. Repeating this process, we can find points to associate with 3, 4, 5, . . . , (the three dots shown here and in Figure 1b mean "and so on").

Figure 1

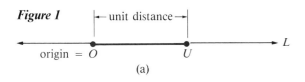

(a)

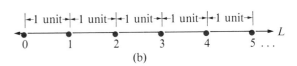

(b)

If the line L is drawn horizontally, the positive numbers are usually represented to the right of the origin O and the negative numbers to the left. An arbitrary unit length is chosen, and points corresponding to the positive and negative numbers are marked off from O (Figure 2).

The point associated with a number on a number line is called the **graph** of that number, and the number is called the **coordinate** of that point. For example, the coordinates of points A, B, C, and D on the number line in Figure 2 are $-3, 2, 4$, and 5, in that order.

Figure 2

```
            A                           B        C   D
  +────●────┼────┼────┼────┼────┼────●────┼────●───●──→
  -4   -3   -2   -1   O    1    2    3    4    5
```

The origin O is assigned the coordinate 0 (zero). If d is the distance between the origin and a point P, then the coordinate of P is d or $-d$, depending on whether P is to the right or to the left of the origin (Figure 3).

Figure 3

```
          |───── d ─────►|◄───── d ─────|
          P              O              P
  ◄───────●──────────────●──────────────●───────►
         -d            0 (zero)         d
```

In algebra, we encounter many different sets of numbers. Following is a summary of the common sets (and their symbols) used throughout this textbook. Note that we illustrate a set of numbers on a number line by *coloring* the points whose coordinates are members of the set.

1. The **natural numbers,** also called **counting numbers** or **positive integers,** are the numbers 1, 2, 3, 4, 5, . . . The set of natural numbers $\{1, 2, 3, 4, 5, \ldots\}$, represented by the symbol $\mathbb{N}$, is illustrated in Figure 4.

Figure 4

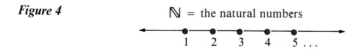

$\mathbb{N}$ = the natural numbers

2. The **integers** consist of all the natural numbers, the negatives of the natural numbers, and zero. The set of integers is represented by the symbol **I**; Thus $\mathbf{I} = \{\ldots, -4, -3, -2, -1, 0, 1, 2, 3, 4, \ldots\}$ (Figure 5).

Figure 5

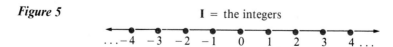

I = the integers

3. The **rational numbers** are the numbers that can be written in the form a/b, in which a and b are integers and $b \neq 0$. Because b may equal 1, every integer is a rational number. The set of all rational numbers in

which a and b are integers and $b \neq 0$ is often represented by $\mathbb{Q}$. Some numbers in the set $\mathbb{Q}$ are

$$\frac{1}{2}, \frac{4}{3}, \frac{30}{7}, \frac{20}{4}, -\frac{3}{4}, -\frac{11}{5}, \text{ and } -\frac{15}{3}$$

(Figure 6). In the expression a/b, a is called the **numerator** and b is called the **denominator.**

Figure 6

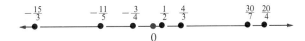

4. The **irrational numbers** are the numbers that cannot be expressed as a quotient of two integers. Some examples of these numbers are $\sqrt{2}$, $\sqrt{3}$, and π.

5. The **positive real numbers** correspond to points to the right of the origin (Figure 7a), and the **negative real numbers** correspond to points to the left of the origin (Figure 7b). The **real numbers** consist of the rational numbers and the irrational numbers. It can be shown that the graph of the set of real numbers is the entire number line and that every point on the number line corresponds to exactly one real number (Figure 7c). The set of all real numbers is represented by the symbol $\mathbb{R}$.

Figure 7

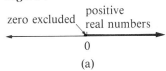
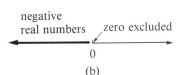

Decimals

We often find it useful to change rational numbers into decimal form. This makes it easier to compare two or more rational numbers. To express a rational number as a **decimal,** we divide the numerator by the denominator. The decimal form of a rational number may be **terminating,** as in the following examples:

$$\frac{2}{5} = 2 \div 5 = 0.4 \quad \text{and} \quad -\frac{3}{4} = -3 \div 4 = -0.75.$$

The decimal form of a rational number may also be **nonterminating** and **repeating,** as in the examples below:

$$\frac{2}{3} = 0.6666\ldots, \quad \frac{7}{9} = 0.7777\ldots, \quad \text{and} \quad \frac{1}{7} = 0.1428571428571\ldots.$$

A repeating decimal such as 0.6666 . . . is often written as $0.\overline{6}$, where the overbar indicates the block of digits that repeats:

$$\frac{2}{3} = 0.\overline{6}, \qquad 0.7777\ldots = 0.\overline{7}, \qquad \text{and}$$

$$0.142857142857\ldots = 0.\overline{142857}.$$

EXAMPLE **1**   Express each rational number as a decimal.

(a) $\frac{3}{8}$ (b) $\frac{10}{3}$ (c) $-\frac{5}{6}$ (d) $\frac{73}{99}$

SOLUTION   (a) The decimal form of $\frac{3}{8}$ is obtained by dividing 3 by 8; that is, $\frac{3}{8} = 0.375$.

(b) $\frac{10}{3} = 3.3333\ldots = 3.\overline{3}$

(c) $-\frac{5}{6} = -0.83333\ldots = -0.8\overline{3}$

(d) $\frac{73}{99} = 0.7373\ldots = 0.\overline{73}$

Example 1 shows that every rational number can be expressed as a decimal that is either terminating or repeating. It is also true that every terminating or repeating decimal represents a rational number.

EXAMPLE **2**   Express each terminating decimal as a quotient of integers.

(a) 0.7   (b) -0.53 (c) 1.025

SOLUTION   We change a decimal to a quotient of integers as follows: The numerator of the rational number will be the original number without the decimal point. The denominator will be some multiple of 10. It will contain as many zeros after the one as there are digits after the decimal point of the original form.

(a) $0.7 = \frac{7}{10}$ (b) $-0.53 = -\frac{53}{100}$ (c) $1.025 = \frac{1.025}{1.000} = \frac{41}{40}$

In Section 4.1 of Chapter 4, you will see how to rewrite a nonterminating repeating decimal as a quotient of two integers.

Irrational numbers cannot be written as decimals, although they can be *approximated* by decimals to any number of places. For instance, decimal representations of $\sqrt{2}$, $\sqrt{3}$, and π are:

$$\sqrt{2} = 1.4142135\ldots, \qquad \sqrt{3} = 1.7320508\ldots, \qquad \text{and}$$

$$\pi = 3.1415926\ldots.$$

EXAMPLE 3 Identify each number as being rational or irrational.

(a) $-\frac{4}{7}$ (b) $\sqrt{6}$ (c) $0.\overline{35}$

(d) $0.12112111211112\ldots$ (e) $\sqrt{36}$

SOLUTION (a) $-\frac{4}{7}$ is a quotient of two integers; thus it is rational.

(b) $\sqrt{6}$ is a square root of a positive integer that is not a perfect square; thus it is irrational.

(c) $0.\overline{35}$ is a repeating decimal; thus it is rational.

(d) $0.12112111211112\ldots$ is a nonterminating, nonrepeating decimal; thus it is irrational.

(e) $\sqrt{36} = 6$, which is an integer; therefore $\sqrt{36}$ is a rational number.

PROBLEM SET 1.1

In problems 1–10, represent each set on a number line.

1. $\{-4, -3, -2, -1, 2, 4\}$
2. $\{-5, -3, 0, 3, 5\}$
3. $\{-\frac{5}{3}, -\frac{2}{3}, -\frac{1}{3}, \frac{1}{3}, \frac{2}{3}, \frac{5}{3}\}$
4. $\{\frac{1}{5}, \frac{2}{5}, \frac{3}{5}, \frac{4}{5}\}$
5. $\{-2, -1, 0, \frac{1}{2}, \frac{3}{2}, \frac{7}{2}\}$
6. $\{-\frac{5}{9}, -\frac{4}{9}, -\frac{3}{9}, -\frac{2}{9}, -\frac{1}{9}\}$
7. $\{0, 2, 4, 6, \ldots\}$
8. $\{-\frac{3}{2}, -\frac{1}{2}, \frac{1}{2}, \frac{3}{2}, \frac{5}{2}, \ldots\}$
9. $\{\ldots, -5, -3, -1, 0\}$
10. $\{0, \frac{1}{7}, \frac{8}{7}, \frac{15}{7}, \frac{22}{7}, \frac{29}{7}, \ldots\}$

In problems 11–20, express each rational number as a decimal.

11. $\frac{3}{5}$
12. $-\frac{7}{4}$
13. $\frac{3}{2}$
14. $\frac{7}{2}$
15. $\frac{4}{5}$
16. $\frac{9}{100}$
17. $-\frac{5}{4}$
18. $\frac{5}{9}$
19. $-\frac{7}{3}$
20. $\frac{6}{7}$

In problems 21–30, express each decimal as a quotient of integers.

21. 0.27
22. 1.72
23. 2.64
24. 7.155
25. -0.125
26. -0.008
27. 0.0527
28. 0.0098
29. -0.00329
30. -0.00051

In problems 31–44, identify each number as being rational or irrational.

31. $-\frac{5}{9}$
32. $-\frac{3}{7}$
33. $\sqrt{14}$
34. $\sqrt{13}$
35. $0.\overline{27}$
36. $0.\overline{37}$
37. $3.464464446\ldots$
38. $4.575575557\ldots$
39. $\sqrt{16}$
40. $\sqrt{25}$
41. 0.374
42. 0.671
43. $-\sqrt{81}$
44. $-\sqrt{49}$

1.2 Properties of Real Numbers

In Section 1.1, we outlined the procedure for locating real numbers on a number line. In this section, we examine the basic properties of real numbers. These properties serve as a foundation for the algebraic steps we will use in later chapters. You may want to review how to add, subtract, multiply, and divide positive real numbers before you proceed. [The rules for signed (negative and positive) numbers are presented in Section 1.3.]

The basic properties of real numbers can be expressed in terms of the operations of addition and multiplication. If a and b are real numbers, there is a unique real number $a + b$, called their **sum.** The sum is formed by the process of **addition.** There is also a real number $a \cdot b$, called the **product** of a and b. The product is formed by the process of **multiplication.** The notation $a \times b$ for the product is not often used. The preferred notation is $a \cdot b$ or ab. Assume that the letters a, b, and c represent any real numbers.

1 The Commutative Properties

(i) **For addition:**	(ii) For multiplication:
$a + b = b + a$	$a \cdot b = b \cdot a$

For example,

$$13 + 20 = 20 + 13,$$

because

$$13 + 20 = 33 \quad \text{and} \quad 20 + 13 = 33.$$

Also,

$$3 \cdot 2 = 2 \cdot 3,$$

since

$$3 \cdot 2 = 6 \quad \text{and} \quad 2 \cdot 3 = 6.$$

2 The Associative Properties

(i) **For addition:**	(ii) For multiplication:
$a + (b + c) = (a + b) + c$	$(a \cdot b) \cdot c = a \cdot (b \cdot c)$

For example,

$$7 + (3 + 9) = (7 + 3) + 9,$$

because

$$7 + (3 + 9) = 7 + 12 = 19 \quad \text{and} \quad (7 + 3) + 9 = 10 + 9 = 19.$$

Also,

$$3 \cdot (5 \cdot 2) = (3 \cdot 5) \cdot 2,$$

since

$$3 \cdot (5 \cdot 2) = 3 \cdot 10 = 30 \quad \text{and} \quad (3 \cdot 5) \cdot 2 = 15 \cdot 2 = 30.$$

3 The Distributive Properties

(i) $a \cdot (b + c) = a \cdot b + a \cdot c$ (ii) $(b + c) \cdot a = b \cdot a + c \cdot a$

For example,

$$6 \cdot (5 + 7) = (6 \cdot 5) + (6 \cdot 7),$$

because

$$6 \cdot (5 + 7) = 6 \cdot 12 = 72 \quad \text{and} \quad (6 \cdot 5) + (6 \cdot 7) = 30 + 42 = 72.$$

Also,

$$(5 + 7) \cdot 2 = (5 \cdot 2) + (7 \cdot 2),$$

because

$$(5 + 7) \cdot 2 = 12 \cdot 2 = 24 \quad \text{and} \quad (5 \cdot 2) + (7 \cdot 2) = 10 + 14 = 24.$$

4 The Identity Properties

(i) For addition: (ii) For multiplication:

$$a + 0 = 0 + a = a \qquad\qquad a \cdot 1 = 1 \cdot a = a$$

Zero is called the **additive identity** and one is called the **multiplicative identity.** For example, $5 + 0 = 0 + 5 = 5$ and $6 \cdot 1 = 1 \cdot 6 = 6$.

5 The Inverse Properties

(i) For each real number a, there is a real number $-a$, called the **additive inverse** or **negative** of a, such that

$$a + (-a) = (-a) + a = 0.$$

(ii) For each real number $a \neq 0$, there is a real number $1/a$, called the **multiplicative inverse** or **reciprocal** of a, such that

$$a \cdot \frac{1}{a} = \frac{1}{a} \cdot a = 1.$$

For example,

$$3 + (-3) = (-3) + 3 = 0 \quad \text{and}$$

$$7 \cdot \frac{1}{7} = \frac{1}{7} \cdot 7 = 1.$$

EXAMPLE 1   State the properties that justify each of the following equalities.

(a) $3 \cdot (-4) = (-4) \cdot 3$ (b) $5 \cdot (6 \cdot 3) = (5 \cdot 6) \cdot 3$

(c) $14 \cdot (2 + \sqrt{3}) = 28 + 14 \cdot \sqrt{3}$ (d) $11 \cdot \frac{1}{11} = 1$

(e) $\frac{2}{3} \cdot (5 + \frac{3}{8}) = (5 + \frac{3}{8}) \cdot \frac{2}{3}$ (f) $7 + (-7) = 0$

SOLUTION

(a) Commutative property for multiplication

(b) Associative property for multiplication

(c) Distributive property

(d) Multiplicative inverse

(e) Commutative property for multiplication

(f) Additive inverse

These five properties will be used as reasons or justifications for much of what you do in algebra. Following is a list of other properties that can be *derived* from these properties.

6 The Cancellation Properties

(i) For addition:

if $a + c = b + c$, then $a = b$

(ii) For multiplication:

if $ac = bc$ and $c \neq 0$, then $a = b$

For example,

$$\text{if} \quad x + 3 = y + 3, \quad \text{then} \quad x = y.$$

Also,

$$\text{if} \quad 7x = 7y, \quad \text{then} \quad x = y.$$

7 The Negative Properties

(i) $-(-a) = a$	(ii) $(-a)b = a(-b) = -(ab)$
(iii) $(-a)(-b) = ab$	

For example,

$$-(-8) = 8 \qquad (-2)(4) = 2(-4) = -(2 \cdot 4) = -8,$$

and

$$(-3)(-7) = 3 \cdot 7 = 21.$$

8 The Zero-Factor Properties

(i) $a \cdot 0 = 0 \cdot a = 0$
(ii) If $a \cdot b = 0$, then $a = 0$ or $b = 0$ (or both)

It should be noted that *division by zero is not allowed*. For instance, if $a \neq 0$, $a \div 0 = a/0$ is undefined. Also $0 \div 0 = 0/0$ is undefined. However, if $b \neq 0$, then $0 \div b = 0/b = 0$.

PROBLEM SET 1.2

In problems 1–8, verify the commutative properties by performing actual computations.

1. $15 + 17$ and $17 + 15$ **2.** $21 + 9$ and $9 + 21$

3. $8 \cdot 9$ and $9 \cdot 8$ **4.** $14 \cdot 11$ and $11 \cdot 14$

5. $0.6 + 0.5$ and $0.5 + 0.6$ **6.** $2.25 + 3.50$ and $3.50 + 2.25$

7. $0.8 \cdot 0.41$ and $0.41 \cdot 0.8$ **8.** $0.76 \cdot 5.2$ and $5.2 \cdot 0.76$

In problems 9–16, verify the associative properties by performing actual computations.

9. $2 + (7 + 6)$ and $(2 + 7) + 6$ **10.** $10 + (9 + 3)$ and $(10 + 9) + 3$

11. $3 \cdot (8 \cdot 5)$ and $(3 \cdot 8) \cdot 5$ **12.** $6 \cdot (5 \cdot 2)$ and $(6 \cdot 5) \cdot 2$

13. $4.1 + (6.8 + 3.3)$ and $(4.1 + 6.8) + 3.3$ **14.** $2.25 + (3.5 + 1.75)$ and $(2.25 + 3.5) + 1.75$

15. $0.67 \cdot (0.5 \cdot 0.4)$ and $(0.67 \cdot 0.5) \cdot 0.4$ **16.** $4.2 \cdot (5.1 \cdot 3.6)$ and $(4.2 \cdot 5.1) \cdot 3.6$

In problems 17–22, verify the distributive properties by performing actual computations.

17. $3 \cdot (5 + 8)$ and $3 \cdot 5 + 3 \cdot 8$ **18.** $6.1 \cdot (7.6 + 4.8)$ and $6.1 \cdot 7.6 + 6.1 \cdot 4.8$

19. $(7 + 9) \cdot 3$ and $7 \cdot 3 + 9 \cdot 3$ **20.** $(15 + 12) \cdot 6$ and $15 \cdot 6 + 12 \cdot 6$

21. $2.1 \cdot (1.1 + 3.4)$ and $2.1 \cdot 1.1 + 2.1 \cdot 3.4$ **22.** $(3.25 + 2.33) \cdot 1.85$ and $3.25 \cdot 1.85 + 2.33 \cdot 1.85$

In problems 23–42, state the properties that justify each of the following statements.

23. $3 + (-8) = (-8) + 3$ **24.** $\frac{1}{2} \cdot 5 = 5 \cdot \frac{1}{2}$ **25.** $8 \cdot (\sqrt{5} \cdot 4) = (8 \cdot \sqrt{5}) \cdot 4$

26. $1 \cdot \sqrt{11} = \sqrt{11}$ **27.** $(7 + \sqrt{2}) \cdot 3 = 21 + 3\sqrt{2}$ **28.** $\sqrt{7} + (-\sqrt{7}) = 0$

29. $\left(\dfrac{1}{\sqrt{3}}\right) \cdot \sqrt{3} = 1$ **30.** $1 \cdot \frac{7}{9} = \frac{7}{9}$ **31.** $(-2) \cdot (-3) = 2 \cdot 3$

32. $-2 \cdot 5 = -10$ **33.** If $5a = 0$, then $a = 0$. **34.** If $-3x = -3y$, then $x = y$.

35. If $9x = 9y$, then $x = y$. **36.** $4 \cdot (u + v) = 4u + 4v$ **37.** $15 \cdot 0 = 0$

38. $\frac{3}{4} + (1 + \frac{1}{2}) = (\frac{3}{4} + 1) + \frac{1}{2}$ **39.** $3 + \sqrt{11} = \sqrt{11} + 3$ **40.** If $t + 3 = 5 + 3$, then $t = 5$.

41. $-8 + 0 = -8$ **42.** If $x \neq 0$, then $x \cdot \dfrac{1}{x} = 1$.

1.3 Operations with Signed Numbers

We will discuss the concepts and terminology commonly used with integers and then review the basic operations with integers.

Throughout this book, the negative integers are written with a negative sign affixed, and the positive integers are indicated by the absence of a sign. Hence, if a nonzero number has no sign in front of it, it is understood to be positive. For example, the coordinate of a point that corresponds to the number $+5$ is written as 5.*

The rules for adding, subtracting, multiplying, and dividing positive numbers were discussed in arithmetic. The concept of absolute value is helpful in extending these rules to include negative numbers.

Consider the distance between the origin and each of the points labeled 5 and -5 on a number line. Both of these points are located five units from the origin. The distance between the origin and the point with coordinate 5 is the same as the distance between the origin and the point with coordinate -5. This common distance is 5 (Figure 1). The distance between the origin and a point is always nonnegative and is independent of the direction from the origin to the point.

Figure 1

---

*Note: Some authors write the numbers preceded by the raised symbol $+$ for the points to the right of zero and the raised symbol $-$ for the points to the left of zero. For example, a point corresponding to the number "positive five" is denoted by $^+5$, and a point corresponding to the number "negative five" is denoted by $^-5$. However, in this book, we denote these numbers by 5 and -5.

The distance between the origin and a point whose coordinate is x is called the *absolute value* of x and is represented by $|x|$. Thus,

$$|-5| = 5 \quad \text{and} \quad |5| = 5.$$

More formally, we have the following definition:

DEFINITION   **Absolute Value**

> If x is a real number, then $|x|$, the **absolute value** of x, is defined by
> $$|x| = \begin{cases} x, \text{ if } x \text{ is positive or } x \text{ is zero} \\ -x, \text{ if } x \text{ is negative.} \end{cases}$$

For example,

$$|6| = 6, \text{ because 6 is positive,}$$
$$|-6| = -(-6) = 6, \text{ because } -6 \text{ is negative.}$$

and

$$|0| = 0.$$

Note that if x is negative, $-x$ means the *opposite* of x, so $-x$ is a positive number.

EXAMPLE 1   Find the value of each expression.

(a) $|8|$ (b) $\left|-\frac{2}{3}\right|$ (c) $\left|\frac{7}{5}\right|$ (d) $-|-17|$

SOLUTION   (a) $|8| = 8$, because 8 is positive.

(b) $\left|-\frac{2}{3}\right| = -\left(-\frac{2}{3}\right) = \frac{2}{3}$, because $-\frac{2}{3}$ is negative.

(c) $\left|\frac{7}{5}\right| = \frac{7}{5}$, since $\frac{7}{5}$ is positive.

(d) $|-17| = 17$, so $-|-17| = -17$.

Addition, Subtraction, Multiplication, and Division of Signed Numbers

The addition of real numbers can be illustrated on the number line. Using the properties of real numbers, we can represent addition by directed moves (changes of position) on the line. For example, to add 2 and 3 on the number line (Figure 2), we start at the origin and move two units to the right. Thus, the number 2 is represented by an arrow from 0 to 2. Next, start at the number 2 and move three units to the right, so that the number 3 is represented by an arrow between 2 and 5. Together, the sum of the two directed moves is $2 + 3 = 5$ (Figure 2a). It can also be seen that $3 + 2 = 5$ (Figure 2b). This illustrates the commutative property of addition: $3 + 2 = 2 + 3$.

Figure 2

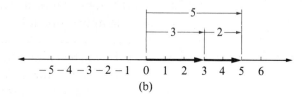

(a)   (b)

Now consider the sum $(-2) + (-3)$. Visualize the addition of a negative number as a movement to the left on the number line. The sum can be found by moving to the left of the origin: $(-2) + (-3) = -5$ (Figure 3).

Figure 3

Note that, in both cases illustrated above, the sums can be found by adding the absolute values of the numbers and retaining the common sign of the numbers. These results illustrate the following rule:

> To **add** two numbers with like signs, add their absolute values and keep their common sign.

For example,

$$5 + 7 = |5| + |7| = 12$$

and

$$(-5) + (-8) = -(|-5| + |-8|) = -(5 + 8) = -13.$$

The addition of numbers with different signs can easily be illustrated on a number line. For example, the sum $(-3) + 8$ can be interpreted as a movement of three units to the left and then eight units to the right (Figure 4a). The sum $(-3) + 8$ can also be interpreted as a movement of eight units to the right and then three units to the left (Figure 4b). In both cases, the same sum is obtained; that is, $(-3) + 8 = 5$.

Note that, in the example $-3 + 8 = 5$, the sum can be obtained by finding the difference of the absolute values of the numbers and retaining

Figure 4

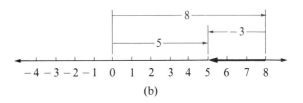

(a)   (b)

the sign of the number with the larger absolute value. This result is an example of the following rule:

> To **add** two numbers with unlike signs, find the difference of their absolute values by subtracting the smaller absolute value from the larger. Retain the sign of the number with the larger absolute value.

For example,

$$6 + (-2) = (|6| - |-2|) = (6 - 2) = 4$$

and

$$2 + (-5) = (-5) + 2 = -(|-5| - |2|) = -(5 - 2) = -3.$$

To subtract b from a, an operation denoted by $a - b$, we use the following definition:

DEFINITION   **Subtraction**

> If a and b are real numbers, the **difference** $a - b$ is defined to be $a + (-b)$, where $-b$ is the additive inverse of b. In other words, to subtract a number b, change its sign and add.

This definition, together with the rules for adding signed numbers, provides a method for subtracting signed numbers. For example, the difference $(-7) - 3$ can be found as follows:

$$(-7) - 3 = (-7) + (-3) = -10.$$

We can illustrate this subtraction on a number line by interpreting the subtraction of 3 from -7 as a movement of seven units to the left and then three units to the left; that is, $(-7) - 3 = -10$ (Figure 5).

Figure 5

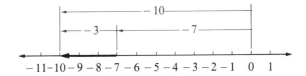

The rule for subtracting signed numbers can be stated as follows:

> To **subtract** one signed number from another, change the sign of the number to be subtracted, and then add by following the rules for adding signed numbers.

For example,

$$7 - 4 = 7 + (-4) = 3 \qquad \text{and} \qquad (-9) - 6 = (-9) + (-6) = -15.$$

EXAMPLE 2   Find the following sums.

(a) $9 + 7$ (b) $(-5) + (-7)$ (c) $7 + (-23)$

SOLUTION   (a) $9 + 7 = 16$ (b) $(-5) + (-7) = -(5 + 7) = -12$

(c) $7 + (-23) = -(23 - 7) = -16$

EXAMPLE 3   Find the following differences.

(a) $5 - 2$ (b) $(-15) - (-8)$

SOLUTION   (a) $5 - 2 = 5 + (-2) = 3$

(b) $(-15) - (-8) = (-15) + 8 = -7$

The multiplication of two positive integers is often described as repeated addition. For example, $3 \cdot 2$ can be interpreted as $3 + 3 = 6$ or $2 + 2 + 2 = 6$. The same approach can be used to describe the multiplication of any real number by a positive number. Thus, $(-6) \cdot 3$ can be interpreted as $(-6) + (-6) + (-6) = -18$. Also, we can write $(-5) \cdot (2)$ as $(-5) + (-5) = -10$.

Because multiplication is commutative, it must also be true that $3 \cdot (-6) = -18$ and $2 \cdot (-5) = -10$.

These examples suggest that:

> The product of a negative number and a positive number is a negative number. That is, if a and b are positive numbers, then
>
> $$a(-b) = -(ab)$$
>
> or
>
> $$(-a)b = -(ab).$$

In order to understand how to find the product of two negative numbers, it might be helpful to examine the following pattern:

$$(-4) \cdot 3 = -12$$
$$(-4) \cdot 2 = -8$$
$$(-4) \cdot 1 = -4$$
$$(-4) \cdot 0 = 0.$$

Notice that as the second number decreases by 1, the product increases by 4. Thus, you may expect the following:

$$(-4) \cdot (-1) = 4$$
$$(-4) \cdot (-2) = 8$$
$$(-4) \cdot (-3) = 12.$$

In fact, as these two examples suggest, it is true that:

> The product of two negative numbers is a positive number. That is, if both a and b are positive numbers, then
>
> $$(-a) \cdot (-b) = +(a \cdot b).$$

EXAMPLE 4   Find each product.

(a) $(-9) \cdot 6$ (b) $5 \cdot (-8)$ (c) $(-5) \cdot (-3)$

SOLUTION   (a) $(-9) \cdot 6 = -(9 \cdot 6) = -54$

(b) $5 \cdot (-8) = -(5 \cdot 8) = -40$

(c) $(-5) \cdot (-3) = 5 \cdot 3 = 15$

When multiplying more than two signed numbers, the numbers can be paired to determine the sign of the product. For example, the product $(-3) \cdot 2 \cdot (-5) \cdot 6$ can be written

$$[(-3) \cdot (-5)] \cdot [2 \cdot 6] = 15 \cdot 12 = 180.$$

In this example, there are two negative factors—that is, an even number of negative factors—and so the product is positive. In the case of the product $(-4) \cdot (-7) \cdot (-2) \cdot 3$, we have

$$[(-4) \cdot (-7)] \cdot [(-2) \cdot 3] = 28 \cdot (-6) = -168.$$

In this example, there are three negative factors—an odd number of negative factors—and so the product is negative. We can state a more general rule for finding the product of signed numbers:

> The product of signed numbers is positive if there is an even number of negative factors and negative if there is an odd number of negative factors.

EXAMPLE 5   Determine the following products.

(a) $3 \cdot (-2) \cdot 6 \cdot (-7) \cdot (-8) \cdot 2 \cdot (-4)$

(b) $(-0.5) \cdot 7 \cdot (-0.75) \cdot (-8) \cdot (-4) \cdot (-7)$

SOLUTION   (a) There is an even number of negative factors, so the product is positive:

$$3 \cdot (-2) \cdot 6 \cdot (-7) \cdot (-8) \cdot 2 \cdot (-4) = 16{,}128.$$

(b) There is an odd number of negative factors, so the product is negative:

$$(-0.5) \cdot 7 \cdot (-0.75) \cdot (-8) \cdot (-4) \cdot (-7) = -588.$$

To divide one signed number by another, we use the following definition: The **quotient** $a \div b$, in which $b \neq 0$, is defined to be $a \cdot (1/b)$ in which $1/b$ is the multiplicative inverse of b. That is,

$$a \div b = a \cdot \frac{1}{b}.$$

More formally, we have:

DEFINITION   **Division**

> $a \div b$ is the unique number c (if there is one) for which $c \cdot b = a$; that is, if $b \neq 0$,
> $$a \div b = c \quad \text{if and only if } c \cdot b = a.$$

For instance,

$12 \div 4 = 3$	because	$3 \cdot 4 = 12$
$12 \div (-4) = (-3)$	because	$(-3) \cdot (-4) = 12$
$(-12) \div 4 = (-3)$	because	$(-3) \cdot 4 = -12$
$(-12) \div (-4) = 3$	because	$3 \cdot (-4) = -12$

These examples suggest the following rules:

> (i) If two numbers have the same signs, their quotient is positive.
> (ii) If two numbers have different signs, their quotient is negative.

EXAMPLE 6   Perform each division.

(a) $45 \div (-9)$ (b) $(-42) \div 3$ (c) $(-6.5) \div (-1.3)$

SOLUTION   (a) $45 \div (-9) = -(45 \div 9) = -5$

(b) $(-42) \div 3 = -(42 \div 3) = -14$

(c) $(-6.5) \div (-1.3) = 6.5 \div 1.3 = 5$

The Distance Between Two Points on the Number Line

Consider the two points 3 and 7 on a number line (Figure 6). We see from Figure 6 that the distance d between 3 and 7 is four units:

$$d = 7 - 3 = 4 \quad \text{or} \quad d = |7 - 3| = |4| = 4.$$

Because $3 - 7 = -4$ and $|-4| = 4$, we may also state that the distance d between 3 and 7 is given by

$$d = |3 - 7| = |-4| = 4.$$

Figure 6

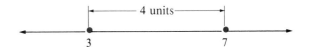

In this example you can see that the distance between these two points on the number line can be found by determining the absolute value of the difference of the coordinates of the two points.

In general, the distance d between the points whose coordinates are a and b is given by

$$\boxed{d = |a - b|.}$$

This formula holds no matter which point is to the left of the other (Figures 7a and b).

Figure 7

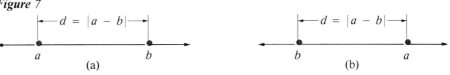

EXAMPLE 7   Find the distance d between the points whose coordinates are the given pairs of numbers.

(a) 7 and -1 (b) 5 and 0   (c) -4 and 0

SOLUTION   We substitute the given numbers for a and b in the formula $d = |a - b|$:

(a) $d = |7 - (-1)| = |7 + 1| = |8| = 8$

(b) $d = |5 - 0| = |5| = 5$

(c) $d = |-4 - 0| = |-4| = 4$

PROBLEM SET 1.3

In problems 1–10, find the value of each expression.

1. $|3|$ **2.** $|2.7|$ **3.** $|-11|$ **4.** $|-5\frac{1}{2}|$

5. $-|16|$ **6.** $-|-7.31|$ **7.** $-|-21|$ **8.** $-|0|$

9. $-|-\frac{5}{7}|$ **10.** $|-11.8|$

In problems 11–28, find the following sums.

11. $7 + (-3)$ **12.** $(-23) + 14$ **13.** $(-6) + (-8)$ **14.** $(-11) + (-7)$

15. $27 + (-39)$ **16.** $(-100) + 43$ **17.** $(-8) + 8$ **18.** $7 + (-7)$

19. $10 + (-17)$ **20.** $(-19) + 27$ **21.** $(-21) + 9$ **22.** $(-20) + (-17)$

23. $(-35) + (-18)$ **24.** $(-75) + 25$ **25.** $(15.2) + (-13.7)$ **26.** $(-21.3) + (-14.6)$

27. $(-9.7) + (-8.5)$ **28.** $7.28 + (-2.71)$

In problems 29–46, find the following differences.

29. $18 - (-19)$ **30.** $(-25) - 8$ **31.** $(-25) - (-45)$

32. $(-97) - (-39)$ **33.** $(-22) - (-7)$ **34.** $16 - (-30)$

35. $(-8) - 8$ **36.** $(-8) - (-5)$ **37.** $11.1 - (-0.9)$

38. $0 - (-2)$ **39.** $0.052 - (-0.007)$ **40.** $(-5.03) - (-4.83)$

41. $0 - (-8)$ **42.** $(-8.7) - 5.8$ **43.** $(-111) - (-17)$

44. $4.8 - (-3.71)$ **45.** $(-327) - (-481)$ **46.** $(-3,125) - (-1,241)$

In problems 47–62, find the following products.

47. $3 \cdot (-5)$ **48.** $(-6) \cdot 4$ **49.** $(-2) \cdot (-8)$

50. $(-12) \cdot 0.5$ **51.** $(-9) \cdot 0.3$ **52.** $0.4 \cdot 10$

53. $(-8) \cdot 0 \cdot (-4)$ **54.** $0.4 \cdot (-0.8)$ **55.** $(-3) \cdot (-2) \cdot (-4)$

56. $(-4) \cdot (-6) \cdot (-2)$ **57.** $(-6) \cdot (-6) \cdot 7$ **58.** $(-11) \cdot 0 \cdot (-7) \cdot (-6)$

59. $(-1) \cdot (-2) \cdot (-3) \cdot (-5)$ **60.** $(-2) \cdot (-3) \cdot 4 \cdot (-5)$ **61.** $5 \cdot (-7) \cdot 3 \cdot (-2)$

62. $3 \cdot (-2) \cdot (-10) \cdot (-4)$

In problems 63–78, find the following quotients.

63. $(-10) \div 5$

64. $27 \div (-9)$

65. $18 \div (-9)$

66. $(-10) \div (-2)$

67. $(-57) \div (-19)$

68. $(-32) \div 8$

69. $(-3.9) \div 0.3$

70. $(-\frac{5}{4}) \div (-2)$

71. $(-75) \div (-5)$

72. $12.5 \div (-1.25)$

73. $(22.5) \div (-0.015)$

74. $(-0.49) \div (-0.7)$

75. $(-350) \div (-70)$

76. $(-2,400) \div (-20)$

77. $(-4,400) \div (-1,100)$

78. $46.72 \div (-6.4)$

In problems 79–90, find the distance d between the points whose coordinates are the given pair of numbers.

79. 3 and 8

80. 15 and 3

81. -4 and 6

82. 19 and -8

83. -8 and -1

84. -11 and -39

85. 14 and 6

86. 3.6 and 2.7

87. 3.5 and -1.7

88. -7.3 and 5.7

89. 0 and -14

90. -6.8 and -9.2

91. The expression

$$\frac{(x + y) + |x - y|}{2}$$

always equals the *larger* of the numbers x and y. Verify this fact for $x = 9$ and $y = -2$.

92. The expression

$$\frac{(x + y) - |x - y|}{2}$$

always equals the *smaller* of the numbers x and y. Verify this fact for $x = 9$ and $y = -2$.

1.4 Calculators and Approximations

Today many students own or have access to an electronic calculator*. A scientific calculator with special keys will expedite many of the calculations required in this book. The symbol $\boxed{c}$ marks problems and examples, or groups of problems and examples, for which the use of a calculator is recommended. If you don't have access to a calculator, you can work most of these problems by using the tables provided in the appendixes in the back of the book.

---

*There are two types of calculators available, those using *algebraic notation* (AN) and those using *reverse Polish notation* (RPN). Advocates of AN claim that it is more "natural," while supporters of RPN say that RPN is just as "natural" but avoids the parentheses required when sequential calculations are made in AN. Before purchasing a scientific calculator, you should familiarize yourself with both AN and RPN so that you can make an intelligent decision based on your own preferences.

Learn to use your calculator properly by studying the instruction booklet furnished with it. In particular, practice performing chain calculations so you can do them as efficiently as possible, using whatever memory features your calculator may possess to store intermediate results. After you learn *how* to use a calculator, take care to learn *when* to use it and especially when *not* to use it.

EXAMPLE 1   ⓒ Change each rational number to a decimal.

(a) $\frac{7}{11}$ (b) $\frac{5}{17}$ (c) $6\frac{8}{21}$ (d) $\frac{11}{19}$

SOLUTION   If we use an eight-digit calculator, we find:

(a) $\frac{7}{11} = 0.63636364$ (b) $\frac{5}{17} = 0.29411765$

(c) $6\frac{8}{21} = 6.3809524$ (d) $\frac{11}{19} = 0.57894737$

EXAMPLE 2   ⓒ Change each square root to a decimal.

(a) $\sqrt{13}$ (b) $\sqrt{19}$ (c) $\sqrt{11.2}$ (d) $\sqrt{28.7}$

SOLUTION   If we use an eight-digit calculator, we find:

(a) $\sqrt{13} = 3.6055513$

(b) $\sqrt{19} = 4.3588989$

(c) $\sqrt{11.2} = 3.3466401$

(d) $\sqrt{28.7} = 5.3572381$

Approximations

Numbers produced by a calculator are often approximate, because the calculator can work only with a finite number of decimal places. For instance, a ten-digit calculator gives

$$\tfrac{2}{3} = 0.6666666667 \quad \text{and} \quad \sqrt{2} = 1.414213562,$$

both of which are **approximations** of the true values. Therefore, unless we explicitly ask for numerical approximations or indicate that a calculator is recommended, it's usually best to leave the answer in fractional form or in radical form.

Most numbers obtained from measurements of real-world quantities are subject to error and must also be regarded as approximations.

DEFINITION   **Significant Digits**

> The digits of a number beginning with the first nonzero digit on the left of the decimal point (or with the first nonzero digit after the decimal point if there is no nonzero digit to the left of the decimal point) and ending with the last digit to the right of the decimal point are known as **significant digits.**

Number	Significant Digits
0.0037	3, 7
3.8	3, 8
4.60	4, 6, 0

For instance, in the number 1.304, we understand that the digits 1, 3, 0, and 4 are significant, and we say that the number is accurate to four significant digits. In this case, the string of significant digits begins with the first nonzero digit and ends with the last digit specified.

Other examples of significant digits are illustrated in the table on the left:

To emphasize that a numerical value is only an approximation, we often use the symbol $\approx$. For instance,

$$\sqrt{2} \approx 1.414.$$

When it is impractical, undesirable, or impossible to write all of the available decimal places for a number, the number must be **rounded off** to, or approximated by, a number with fewer significant digits. Some scientific calculators can be set to round off all displayed numbers to a particular number of decimal places or significant digits. However, it's easy enough to round off numbers without a calculator: simply drop all unwanted digits to the right of the digits to be retained, and increase the last retained digit by 1 if the first dropped digit is 5 or greater. It may be necessary to replace dropped digits by zero in order to hold the decimal point. For instance, we round off 3,187.3 to the nearest hundred as 3,200.

Rounding off should be done in one step rather than digit by digit. Digit-by-digit rounding off may produce an incorrect result. For example, if 4.4248 is rounded off to four significant digits as 4.425, which in turn is rounded off to three significant digits, the result is 4.43. However, 4.4248 is correctly rounded off in one step to three significant digits as 4.42.

EXAMPLE **3**   Round off the given number as indicated.

(a) 6.438 to the nearest tenth

(b) 28.9825 to the nearest thousandth

(c) 3,782 to the nearest hundred.

SOLUTION   (a) To the nearest tenth, $6.438 \approx 6.4$.

(b) To the nearest thousandth, $28.9825 \approx 28.983$.

(c) To the nearest hundred, $3,782 \approx 3,800$.

EXAMPLE 4   Round off each number to three significant digits.

(a) 3.14159   (b) 2.831   (c) 3.796   (d) 7.075   (e) 0.027468

SOLUTION   (a) $3.14159 \approx 3.14$ (b) $2.831 \approx 2.83$ (c) $3.796 \approx 3.80$

(d) $7.075 \approx 7.08$ (e) $0.027468 \approx 0.0275$

EXAMPLE 5   © Use a calculator to change each number to a decimal, then round off the answer to three significant digits.

(a) $\frac{5}{6}$ (b) $3\frac{11}{13}$ (c) $\sqrt{17}$ (d) $\sqrt{31.7}$

SOLUTION   Using an eight-digit calculator, we find:

(a) $\frac{5}{6} = 0.83333333 \approx 0.833$ (b) $3\frac{11}{13} = 3.8461538 \approx 3.85$

(c) $\sqrt{17} = 4.1231056 \approx 4.12$ (d) $\sqrt{31.7} = 5.6302753 \approx 5.63$

PROBLEM SET 1.4

© In problems 1–8, use a calculator to change each number to a decimal.

1. $\frac{4}{29}$ 2. $\frac{8}{37}$ 3. $\frac{15}{17}$ 4. $6\frac{8}{23}$
5. $\sqrt{26}$ 6. $\sqrt{27.8}$ 7. $\sqrt{71.3}$ 8. $\sqrt{93.8}$

In problems 9–18, round off the given number to the nearest tenth.

9. 0.27   10. 0.553   11. 5.32   12. 6.07
13. 7.998   14. 11.349   15. 15.016   16. 22.991
17. 24.052   18. 4.96

In problems 19–28, round off the given number to the nearest hundredth.

19. 3.1872   20. 11.147   21. 14.3649   22. 223.5949
23. 21.0038   24. 42.703   25. 16.507   26. 156.155
27. 23.697   28. 17.006

In problems 29–38, round off each number to three significant digits.

29. 1.732   30. 4.809   31. 14.276   32. 121.5
33. 368.1   34. 97.04   35. 5,139   36. 5.1372
37. 27.98   38. 738,199

Ⓒ In problems 39–48, use a calculator to change the following rational numbers to decimals and then round off the results to three significant digits.

39. $\frac{4}{3}$ **40.** $\frac{6}{17}$ **41.** $\frac{1}{7}$ **42.** $\frac{10}{19}$

43. $2\frac{5}{13}$ **44.** $7\frac{4}{7}$ **45.** $\frac{11}{6}$ **46.** $\frac{19}{21}$

47. $5\frac{13}{27}$ **48.** $6\frac{3}{23}$

Ⓒ In problems 49–58, use a calculator to change the following square roots to decimals and then round off the results to four significant digits.

49. $\sqrt{3}$ **50.** $\sqrt{5}$ **51.** $\sqrt{7}$ **52.** $\sqrt{11}$

53. $\sqrt{23}$ **54.** $\sqrt{37}$ **55.** $\sqrt{47}$ **56.** $\sqrt{93}$

57. $\sqrt{111}$ **58.** $\sqrt{253}$

1.5 The Language of Algebra

Perhaps the most important characteristic of algebra is the use of symbols other than the usual numerals to stand for numbers. The symbols commonly employed are letters. By combining symbols, or symbols and numbers, in certain ways, we are able to make a few algebraic symbols take the place of many words. For instance, to represent the total cost of an unknown number of suits that cost $200 each, we may use some letter, say x, to designate the number of suits purchased. Then the total cost (in dollars) of the suits is given by the expression $200x$. Thus,

$$\text{if} \quad x = 3, \quad \text{the total cost is} \quad \$200(3) = \$600;$$

and

$$\text{if} \quad x = 10, \quad \text{the total cost is} \quad \$200(10) = \$2,000.$$

In the expression $200x$, the 200 is referred to as a constant, and x is called a variable. In general, a **variable** is a symbol that may represent different numbers. (These numbers may be chosen either at random or according to some law.) A **constant** represents one number, usually an explicit number.

Algebraic expressions are formed by using constants, variables, mathematical operations (such as addition, subtraction, multiplication, and division), other operations (such as raising to powers and taking roots) and grouping symbols.

For example,

$$8, \quad 3 + 2x, \quad 3(y + 5), \quad \frac{4x + 1}{3y + 2}, \quad \text{and} \quad 2\{t - 3[t + 2(t + 4)]\}$$

are all algebraic expressions.

The **value** of an algebraic expression is the number the expression represents when specific numbers are substituted for the letters in the expression. The process we use to find the value of the expression is called **evaluating** the expression.

For example, if x is replaced by -3 and y by 7, then the value of the algebraic expression $8x + 5y$ is given by

$$8x + 5y = 8(-3) + 5(7) = -24 + 35 = 11.$$

If an algebraic expression is formed by multiplying two or more expressions, each of these expressions is called a **factor** of the product. For instance, in the product

$$3uv$$

the factors are 3, u, and v. The factors of the product

$$5x(2y + 7)$$

are 5, x, and $2y + 7$.

If we wish to indicate a multiplication involving the same factor two times, we write $x \cdot x$; three times, we write $x \cdot x \cdot x$; and n times, we write

$$\overbrace{x \cdot x \cdot x \cdots x.}^{n \text{ times}}$$

We may also use exponents to provide an alternative notation for these products. For example, $x \cdot x$ can be written as x^2 and $x \cdot x \cdot x$ as x^3. In general, if n is a positive integer,

$$\overbrace{x \cdot x \cdot x \cdots x}^{n \text{ factors}} = x^n.$$

In using the **exponential notation** x^n, we refer to x as the **base** and n as the **exponent** or **power** to which the base is raised. The value of x^n is often referred to as the nth *power* of x.

EXAMPLE 1   Rewrite each expression using exponential notation.

(a) $3 \cdot 3 \cdot 3 \cdot 3 \cdot 3 \cdot 3$ 　　　　　　　　　(b) $b \cdot b \cdot b \cdot b \cdot c \cdot c \cdot c$

(c) $(-y) \cdot (-y) \cdot (-y) \cdot (-y) \cdot (-y)$

SOLUTION   (a) $3 \cdot 3 \cdot 3 \cdot 3 \cdot 3 \cdot 3 = 3^6$ 　　　(b) $b \cdot b \cdot b \cdot b \cdot c \cdot c \cdot c = b^4 c^3$

(c) $(-y) \cdot (-y) \cdot (-y) \cdot (-y) \cdot (-y) = (-y)^5$

EXAMPLE **2**   Rewrite each expression in expanded form.

(a) 2^4 (b) w^5 (c) $(-t)^7$ (d) $2^4 x^5 y^3$

SOLUTION   (a) $2^4 = 2 \cdot 2 \cdot 2 \cdot 2$

(b) $w^5 = w \cdot w \cdot w \cdot w \cdot w$

(c) $(-t)^7 = (-t) \cdot (-t) \cdot (-t) \cdot (-t) \cdot (-t) \cdot (-t) \cdot (-t)$

(d) $2^4 x^5 y^3 = 2 \cdot 2 \cdot 2 \cdot 2 \cdot x \cdot x \cdot x \cdot x \cdot x \cdot y \cdot y \cdot y$

Formulas

An *equation* is a mathematical statement which says that two expressions are equal. Thus,

$$3x + 2 = 5 \quad \text{and} \quad 7y - 1 = 4 + 3y$$

are equations. One important type of equation is a *formula*. A **formula** is an equation that expresses one variable in terms of one or more other variables. Formulas have many important applications in different fields. For example, if C represents the temperature on the Celsius scale, and F represents the temperature on the Fahrenheit scale, then a formula for C in terms of F is

$$C = \frac{5}{9}(F - 32).$$

Other examples follow:

EXAMPLE **3**   The formula

$$V = \frac{4}{3}\pi r^3$$

expresses the volume V of a sphere whose radius is r. Find the volume of a sphere of radius 3 centimeters. (Use 3.14 as an approximation for π.)

SOLUTION   Because $r = 3$ and $\pi \approx 3.14$, we have

$$V = \frac{4}{3}\pi r^3 \approx \frac{4}{3}(3.14)(3^3) = 113.04 \text{ cubic centimeters.}$$

EXAMPLE **4**   People who want to improve their level of cardiovascular fitness through exercise sometimes determine their maximum desirable heart rate. Medical researchers have found that the desirable maximum heart rate R (in beats per minute) of a person exercising is given by the formula

$$R = 220 - A$$

where A is the person's age. What is the desirable maximum heart rate for a 45-year-old while exercising?

SOLUTION   Here $A = 45$, so that

$$R = 220 - 45.$$

Therefore, the maximum desirable heart rate is 175 beats per minute.

EXAMPLE 5   The formula for the perimeter P of a rectangle of length l and width w is

$$P = 2l + 2w.$$

Find the perimeter P of a rectangle whose length is 5 centimeters and whose width is 3 centimeters.

SOLUTION   Here $l = 5$ and $w = 3$. Thus,

$$P = 2l + 2w = 2(5) + 2(3) = 10 + 6 = 16.$$

Therefore, the perimeter is 16 centimeters.

PROBLEM SET 1.5

In problems 1–10, rewrite each expression using exponential notation.

1. $5 \cdot 5 \cdot 5$
2. $7 \cdot 7 \cdot 7 \cdot 7 \cdot 7$
3. $x \cdot x \cdot x \cdot x$
4. $z \cdot z \cdot z \cdot z \cdot z \cdot z$
5. $(-t)(-t)(-t)(-t)(-t)$
6. $x \cdot x \cdot y \cdot y \cdot y$
7. $u \cdot u \cdot u \cdot v \cdot v \cdot v \cdot v \cdot v$
8. $(-a)(-a)(-a)(-b)(-b)$
9. $3 \cdot 3 \cdot x \cdot x \cdot x \cdot x \cdot y \cdot y$
10. $2 \cdot 2 \cdot 2 \cdot (-x)(-x)(-x)(-x)$

In problems 11–20, rewrite each expression in expanded form.

11. 8^4
12. $3 \cdot 9^4$
13. y^5
14. $5t^5$
15. $5^3 t^4$
16. $9^3 x^4 y$
17. $(-x)^4 y^3$
18. $(-t)^3(-s)^4$
19. $4^3 u^4 v$
20. $5^3 u^4 v^7$

In problems 21–26, use the formula $F = \frac{9}{5}C + 32$ to find the temperature F in degrees Fahrenheit when the temperature C in degrees Celsius is the given number.

21. $C = 0°$
22. $C = 5°$
23. $C = 10°$
24. $C = 15°$
25. $C = 20°$
26. $C = 25°$

In problems 27–32, use the formula $A = lw$ to find the area A of the rectangle with length l and width w.

27. $l = 7$ inches, $w = 5$ inches

28. $l = 7.3$ centimeters, $w = 6.1$ centimeters

29. $l = 8$ meters, $w = 3$ meters

30. $l = 14$ feet, $w = 10.3$ feet

31. $l = 15$ feet, $w = 10$ feet

32. $l = 9.3$ yards, $w = 8.6$ yards

In problems 33–38, use the formula $P = 2l + 2w$ to find the perimeter P of a rectangle with length l and width w.

33. $l = 6$ feet, $w = 5$ feet

34. $l = 19$ yards, $w = 11$ yards

35. $l = 14$ inches, $w = 12$ inches

36. $l = 10.3$ meters, $w = 6.5$ meters

37. $l = 7.4$ meters, $w = 4.3$ meters

38. $l = 23.8$ inches, $w = 15.2$ inches

In problems 39–44, use the formula $A = \pi r^2$ to find the area A of a circle with radius r. (Use $\pi \approx 3.14$.)

39. $r = 10$ inches

40. $r = 15$ feet

41. $r = 5$ feet

42. $r = 7$ yards

43. $r = 6.2$ meters

44. $r = 11.3$ centimeters

In problems 45–50, use the formula $d = rt$ to find the distance d an object travels when moving at a uniform rate r during time t.

45. $r = 50$ miles per hour, $t = 3$ hours

46. $r = 70$ miles per hour, $t = 2\frac{1}{2}$ hours

47. $r = 66$ feet per second, $t = 15$ seconds

48. $r = 44$ feet per second, $t = 21.5$ seconds

49. $r = 600$ centimeters per second, $t = 13$ seconds

50. $r = 0.7$ meter per second, $t = 25$ seconds

In problems 51–56, use the formula $A = s^2$ to find the area A of a square with side s.

51. $s = 9$ inches

52. $s = 3.4$ meters

53. $s = 4$ feet

54. $s = 0.7$ yard

55. $s = 15$ inches

56. $s = 5.3$ centimeters

In problems 57–61, use the formula $V = s^3$ to find the volume V of a cube with an edge length s.

57. $s = 3$ inches

58. $s = 2.3$ meters

59. $s = 6$ feet

60. $s = 5.2$ feet

61. $s = 10$ centimeters

62. In statistics, the following formula

$$v = \frac{(x_1 - \bar{x})^2 + (x_2 - \bar{x})^2 + (x_3 - \bar{x})^2}{3}$$

is used, where x_1, x_2, and x_3 represent real numbers, and

$$\bar{x} = \frac{x_1 + x_2 + x_3}{3}.$$

Compute v if $x_1 = 1.8$, $x_2 = 1.4$, and $x_3 = 3.3$.

REVIEW PROBLEM SET

In problems 1–4, represent each set on a number line.

1. $\{-5, -2, 3, 5\}$ **2.** $\{-\frac{3}{2}, \frac{1}{2}, 0, 2, 3\}$

3. $\{0, 3, 6, 9, \ldots\}$ **4.** $\{\ldots, -\frac{5}{2}, -\frac{3}{2}, -\frac{1}{2}\}$

In problems 5–8, express each rational number as a decimal.

5. $\frac{7}{5}$ **6.** $-\frac{5}{3}$

7. $3\frac{1}{8}$ **8.** $-\frac{11}{4}$

In problems 9–14, express each decimal as a quotient of integers.

9. 0.34 **10.** 5.22 **11.** 0.184

12. -4.7 **13.** -6.814 **14.** 0.226

In problems 15–26, identify each number as being rational or irrational.

15. $-\frac{2}{31}$ **16.** $\sqrt{11}$ **17.** $4.142242224\ldots$ **18.** $0.\overline{56}$

19. $-0.\overline{74}$ **20.** $\sqrt{81}$ **21.** $\sqrt{24.1}$ **22.** $5.374474447\ldots$

23. -5.7 **24.** 0.378 **25.** $8.6\overline{14}$ **26.** $\frac{11}{16}$

In problems 27–42, justify each statement by giving the appropriate property. Assume that all letters represent real numbers.

27. $x + 5 = 5 + x$ **28.** $a + (b + 3) = (a + b) + 3$

29. $3t = t3$ **30.** $u(v + w) = uv + uw$

31. $2(xy) = (2x)y$ **32.** $1 \cdot 4 = 4$

33. $5 + 0 = 5$ **34.** If $2x = 2y$, then $x = y$.

35. $0 \cdot k = 0$ **36.** $(-2) \cdot (-3) = 2 \cdot 3$

37. If $x + 9 = y + 9$, then $x = y$.   **38.** $uv = vu$

39. $3(x + 2) = 3x + 6$ **40.** $7(mn) = (7m)n$

41. $4t + 0 = 4t$ **42.** $1 \cdot y = y$

In problems 43–70, evaluate each expression.

43. $|-13|$ **44.** $-|14|$ **45.** $-|-21|$

46. $|-\frac{7}{3}|$ **47.** $(-5) + 2$ **48.** $(-13) + (-21)$

49. $(-6) + (-8)$ **50.** $(-15) + 9$ **51.** $5 - (-9)$

52. $(-5) + 13$ **53.** $17 - (-13)$ **54.** $0 - (-3)$

55. $(-7) + (-3) + (-2)$ **56.** $(-12) - 8$ **57.** $(-4) \cdot (-12)$

58. $(-11) \cdot (-10)$ **59.** $8 \cdot (-5)$ **60.** $(-6) \cdot 7$

61. $(-16) \cdot (-1) \cdot (-2)$ **62.** $(-7) \cdot 2 \cdot (-1) \cdot 6$ **63.** $(-3) \cdot (-7) \cdot 5$

64. $(-1) \cdot (-2) \cdot (-3) \cdot (-4)$ **65.** $(-63) \div 7$ **66.** $25 \div (-5)$

67. $(-40) \div (-5)$ **68.** $(-36) \div (-6)$ **69.** $21 \div (-7)$

70. $(-56) \div 7$

In problems 71–76, find the distance between the points whose coordinates are the given pair of numbers.

71. -5 and 11 **72.** -17 and -8 **73.** 4.7 and -2.3

74. 12 and 21 **75.** 8 and -6 **76.** -2.7 and -3.9

In problems 77–82, round off the given number to the nearest hundredth.

77. 1.078 **78.** 145.395 **79.** 3.816

80. 26.1632 **81.** 0.0371 **82.** 0.0482

In problems 83–88, round off each number to four significant digits.

83. 1.03468 **84.** 0.00137215 **85.** 1.0127415

86. 278.516 **87.** 29.3019 **88.** 3.001287

C In problems 89–100, use a calculator to change each number to a decimal, and then round off the result to three significant digits.

89. $\frac{5}{7}$ **90.** $\frac{11}{14}$ **91.** $\frac{16}{37}$ **92.** $\sqrt{32}$

93. $\sqrt{53}$ **94.** $4\frac{7}{13}$ **95.** $\sqrt{71.2}$ **96.** $\sqrt{63.7}$

97. $\frac{5}{113}$ **98.** $\frac{-3}{764}$ **99.** $\frac{8}{451}$ **100.** $\sqrt{118.4}$

In problems 101–106, rewrite each expression using exponential notation.

101. $2 \cdot 2 \cdot x \cdot x \cdot x$ **102.** $9 \cdot 9 \cdot w \cdot w \cdot w \cdot w$

103. $5 \cdot u \cdot u \cdot v \cdot v \cdot v$ **104.** $3 \cdot 3 \cdot m \cdot m \cdot m \cdot n \cdot n \cdot n \cdot n$

105. $(-2)(-2)(-t)(-t)(-s)(-s)$ **106.** $(-p)(-p)(-p)(-q)(-q)(-q)(-q)(-q)$

In problems 107–112, rewrite each expression in expanded form.

107. x^2y^4 **108.** $5t^3s^2$ **109.** $3^3u^2v^4$

110. $-3x^4y^5$ **111.** m^3n^4p **112.** p^3qt^6

In problems 113–120, evaluate the formula for the given values.

113. $F = \frac{9}{5}C + 32$, for $C = 30°$ **114.** $F = \frac{9}{5}C + 32$, for $C = 35°$

115. $A = lw$, for $l = 6$ inches, $w = 5$ inches   **116.** $A = lw$, for $l = 4.7$ meters, $w = 3.2$ meters

117. $P = 2l + 2w$, for $l = 4$ feet, $w = 3$ feet

118. $P = 2l + 2w$, for $l = 5.3$ centimeters, $w = 4.8$ centimeters

119. $A = \pi r^2$, for $r = 7$ inches (use $\pi \approx 3.14$)

120. $A = \pi r^2$, for $r = 9$ feet

2 The Algebra of Polynomials

We continue our study of algebra by applying the properties of real numbers to algebraic expressions known as *polynomials*. In this chapter, we learn to add, subtract, multiply, divide, and factor polynomials. Later in the book, we will see that many of the most practical applications of algebra require us to translate statements using words into statements using polynomials.

2.1 Polynomials

The following algebraic expressions are examples of polynomials:

$$4x, \qquad t - 5, \qquad y^2 + 4y + 7, \qquad \text{and} \qquad -z^5 + 3wz^4 + 2w^2 + 4.$$

More formally, a **polynomial** is an algebraic expression in which the variables appear only with nonnegative exponents. That is, all exponents of the variables of a polynomial are nonnegative integers, and there can be no division by a variable (no variable will appear in the denominator of a fraction).

The algebraic expressions

$$2x^4 + 3x - 5, \qquad -\frac{7}{2}t^3 + 5t^2 - 2t + 1, \qquad \frac{3}{2}y,$$

and

$$-w^3 + 5^{1/2}w^2 - w + 7$$

are polynomials.

The expressions

$$\frac{3}{x^2}, \qquad 4 - \frac{1}{y}, \qquad \text{and} \qquad \frac{2t - 1}{t + 7}$$

are *not* polynomials, because the variables x, y, and t appear in the denominators.

In the polynomial $2x^4 + 3x - 5$, the three expressions $2x^4$, $3x$, and -5 are called the **terms** of the polynomial. Similarly, $-\frac{7}{2}t^3$, $5t^2$, $-2t$, and 1 are the terms of $-\frac{7}{2}t^3 + 5t^2 - 2t + 1$. Any part of a term that is multiplied by the remaining part is called a **coefficient** of the remaining part. For instance, in the term $4x$, the coefficient of x is 4; in the term $-5xy$, the coefficient of x is $-5y$. A coefficient that contains no variables (the coefficient 4, for example) is called a **numerical coefficient.** The numerical coefficients of the terms of a polynomial are called the coefficients of the polynomial. For example, the coefficients of the following polynomial $-2x^5 + 3x^4 + 5x^3 - 4x^2 + 7x + 8$ are $-2, 3, 5, -4, 7$, and 8.

Three special types of polynomials are defined as follows:

(i) A polynomial containing only one term is called a **monomial.**
(ii) A polynomial containing exactly two terms is called a **binomial.**
(iii) A polynomial containing exactly three terms is called a **trinomial.**

For example,

$$4x \text{ is a monomial,}$$
$$3y - 5 \text{ is a binomial,} \qquad \text{and}$$
$$3z^2 + 4z + 7 \text{ is a trinomial.}$$

Each nonzero polynomial has a degree, which we define formally as follows:

DEFINITION   **Degree of a Polynomial**

The **degree of a term** in a polynomial is the sum of all the exponents of the variable factors in the term. The **degree of the polynomial** is the highest degree among all the terms in the polynomial that have nonzero coefficients.

Note that, when adding exponents to determine degree, a variable with no exponent is regarded as having exponent 1. A nonzero real number (such as 8) is a polynomial of degree zero. The real number zero is a polynomial with no degree assigned to it.

EXAMPLE 1   In each case, identify the polynomial as a monomial, binomial, or trinomial. Give the degree and the numerical coefficients of the polynomial.

(a) $-7y$ (b) $3w^4 - 2$ (c) $-2x^2 + 3x + 4$

SOLUTION   (a) Monomial of degree 1 with coefficient -7

                       (b) Binomial of degree 4 with coefficients 3 and -2

                       (c) Trinomial of degree 2 with coefficients $-2, 3$, and 4

EXAMPLE 2   Find the degree of each polynomial.

                       (a) $3x^3y - 2x^2y + 7xy^2 - y^3$

                       (b) $a^2b^3c + a^3b^2c^4 - 5$

SOLUTION   (a) The sums of the exponents in each term of $3x^3y - 2x^2y + 7xy^2 - y^3$ are

                       first term: $3 + 1 = 4$

                       second term: $2 + 1 = 3$

                       third term: $1 + 2 = 3$

                       fourth term: $3.$

Because the highest sum is 4, the degree of the polynomial is 4.

                       (b) The degree of the polynomial $a^2b^3c + a^3b^2c^4 - 5$ is 9, the sum of the exponents of the second term. (Why?)

Evaluation of a Polynomial

Polynomials such as $2x + 3$ can be evaluated for specific values of x. For example, if x is replaced by 4, we obtain

$$2x + 3 = 2(4) + 3.$$

Notice that the order in which we complete the above operation will determine the final results. For example, the expression

$$2(4) + 3$$

will be evaluated as $(2 \cdot 4) + 3 = 8 + 3 = 11$ if we multiply first and then add, or as $2 \cdot (4 + 3) = 2 \cdot 7 = 14$ if we add first and then multiply. To avoid this confusion, we agree that the correct order of operations is to multiply first and then perform the additions or subtractions. Thus, for $x = 4$ we have

$$
\begin{aligned}
2x + 3 &= 2(4) + 3 \\
&= 8 + 3 \\
&= 11.
\end{aligned}
$$

In general, when evaluating polynomials, we perform operations in the following order:

Order of Operations for Evaluating Polynomials

(i) Evaluate any expression involving exponents.

(ii) Perform all multiplications or divisions from left to right as they occur.

(iii) Perform all remaining additions or subtractions from left to right.

In Examples 3 and 4, evaluate the polynomials for the indicated values of the variables.

EXAMPLE **3**   (a) $40 - 9x^2$, for $x = 3$ (b) $96t - 16t^2$, for $t = -2$

SOLUTION   (a) We replace x by 3:

$$\begin{aligned} 40 - 9x^2 &= 40 - 9(3)^2 \\ &= 40 - 9(9) \quad \text{(We evaluated } (3)^2\text{)} \\ &= 40 - 81 \quad \text{(We multiplied)} \\ &= -41. \quad \text{(We subtracted)} \end{aligned}$$

(b) We replace t by -2:

$$\begin{aligned} 96t - 16t^2 &= 96(-2) - 16(-2)^2 \\ &= 96(-2) - 16 \cdot (4) \quad \text{(We evaluated } (-2)^2\text{)} \\ &= -192 - 64 \quad \text{(We multiplied)} \\ &= -256. \quad \text{(We subtracted)} \end{aligned}$$

EXAMPLE **4**   (a) $3x^2 - xy + y - 3$, for $x = 2$ and $y = 3$

(b) $4uv^3 - 2u^2v + 7$, for $u = -1$ and $v = -2$

SOLUTION   (a) We replace x by 2 and y by 3:

$$\begin{aligned} 3x^2 - xy + y - 3 &= 3(2)^2 - 2(3) + 3 - 3 \\ &= 3(4) - 2(3) + 3 - 3 \\ &= 12 - 6 + 3 - 3 = 6. \end{aligned}$$

(b) We replace u by -1 and v by -2:

$$\begin{aligned} 4uv^3 - 2u^2v + 7 &= 4(-1)(-2)^3 - 2(-1)^2(-2) + 7 \\ &= 4(-1)(-8) - 2(1)(-2) + 7 \\ &= 32 + 4 + 7 = 43. \end{aligned}$$

PROBLEM SET 2.1

In problems 1–10, identify the polynomial as a monomial, binomial, or trinomial. Give the degree and the coefficients.

1. $3x - 2$

2. -3

3. $4y^2$

4. $5x^3 - 5$

5. $t^2 - 5t + 6$

6. $w^3 - w - 1$

7. $2u^7 - 13$

8. $4 - 7y$

9. $-x^4 - x^2 + 13$

10. $\dfrac{1}{2}p^4 + \dfrac{3}{2}p^2 + 8$

In problems 11–20, find the degree of each polynomial.

11. $5x^2y + 16x$

12. $3m^2 + 5mnp - 1$

13. $t^2 - 5ts + 2s^2$

14. $13x^4 - 7xy + 5xz^4$

15. $7u^2 + 13u^2v^3 + 9v^4$

16. $4w^2 - 3wz^2 + 5wz^3$

17. $2x^3y - 8xy^5 + 5x^2y^3$

18. $-7u^5v^3 + 11u^7v - 4$

19. $11 - 4pq^6 - 6p^4q^5$

20. $14y^3z^4 - 16y^4z^3$

In problems 21–30, evaluate the polynomial for the given value of the variable.

21. $4x - 1$, for $x = 2$

22. $3z + 2$, for $z = -2$

23. $2u^2 - u + 4$, for $u = 3$

24. $3y^3 - 5y + 3$, for $y = 4$

25. $y^3 - 2y^2 - y + 1$, for $y = -1$

26. $-2x^4 + 3x^2 + 15$, for $x = -3$

27. $4 - t - t^2$, for $t = -2$

28. $3w^4 + 3w^2 + 15$, for $w = -1$

29. $x^5 - x^4 + x^3 - x^2 + x - 2$, for $x = 2$

30. $m^4 - 3m^2 + 17m + 23$, for $m = 3$

In problems 31–36, evaluate the polynomial for $x = 1$, $y = -1$, and $z = 2$.

31. $2xy^2 - yz$

32. $3x^2 + xy - z^2y$

33. $x^3y^2 - 2yz^2 + xy^2$

34. $4zy^2 + x^2z - 3yz^3$

35. $3xy^2 - 2y^2z - xy$

36. $2xyz + x^2z^3$

In problems 37–40, indicate whether or not the given algebraic expression is a polynomial.

37. $\dfrac{13}{x} + 75$

38. $7u^5 - 3u^5 - 3u^3 + 11$

39. $-\frac{1}{7}t^8 + \frac{1}{5}t^5 + 1$

40. $\dfrac{5}{y^2} + 9y + 2$

2.2 Addition and Subtraction of Polynomials

In Chapter 1, we solved many problems involving addition and subtraction of real numbers, and we used the commutative, associative, and distributive properties of real numbers. These properties can be extended to polynomials because polynomials are expressions that represent real numbers. In fact, if P, Q, and R represent polynomials, then:

(i) $P + Q = Q + P$

(ii) $P + (Q + R) = (P + Q) + R$

(iii) (a) $P(Q + R) = PQ + PR$

(b) $(P + Q)R = PR + QR$

Suppose that you want to add the monomials $3x^2$ and $4x^2$. By the distributive property, you have

$$3x^2 + 4x^2 = (3 + 4)x^2 = 7x^2.$$

The terms $3x^2$ and $4x^2$ are called *like* or *similar terms*. These can be defined more precisely as follows:

DEFINITION   **Like Terms**

In an algebraic expression, two or more terms that have the same variable part, although they may have different numerical coefficients, are called **like,** or **similar,** terms.

For example, $3x^2$ and $-5x^2/2$ are like terms, and so are $4y^7$ and $2y^7/3$. However, $4x^2$ and $3x$ are not like terms, since the exponents of the variables are not the same.

To add or subtract monomials, only like terms can be combined.

EXAMPLE 1   Perform each operation by combining like terms.

(a) $3x + 5x$ (b) $4t^2 - 2t^2$

(c) $2y^3 + 4y^3 - 3y^3$ (d) $-4w^4 - (-6w^4)$

(e) $3xy^2 + 7xy^2 + 2x^3y^2$ (f) $3a + 2b + 7c$

SOLUTION   (a) $3x + 5x = (3 + 5)x = 8x$

(b) Here, we use the fact $a - b = a + (-b)$, so that

$$4t^2 - 2t^2 = 4t^2 + (-2t^2) = [4 + (-2)]t^2 = 2t^2.$$

(c) $2y^3 + 4y^3 - 3y^3 = [2 + 4 + (-3)]y^3 = 3y^3$

(d) $-4w^4 - (-6w^4) = [-4 - (-6)]w^4 = 2w^4$

(e) The only like terms in the expression are the first two. Combining like terms, we have

$$3xy^2 + 7xy^2 + 2x^3y^2 = (3 + 7)xy^2 + 2x^3y^2 = 10xy^2 + 2x^3y^2.$$

(f) Because there are no like terms in the expression, no combining can be done.

Addition and subtraction of polynomials with more than one term can be performed by combining like terms. To do this, we use some of the basic properties of real numbers and the rules for addition and subtraction of real numbers. For example, to perform the addition $(3x^3 + 2x^2 + 4) + (7x^3 + 5x^2 + 8)$, we combine like terms, so that

$$\begin{aligned}(3x^3 + 2x^2 + 4) + (7x^3 + 5x^2 + 8) &= 3x^3 + 2x^2 + 4 + 7x^3 + 5x^2 + 8 \\ &= 3x^3 + 7x^3 + 2x^2 + 5x^2 + 4 + 8 \\ &= 10x^3 + 7x^2 + 12.\end{aligned}$$

With enough practice, writing the regrouping of like terms becomes unnecessary.

EXAMPLE 2   Perform the addition $(4y^3 + 7y - 3) + (y^3 - y + 2)$.

SOLUTION   We remove the parentheses and combine like terms:

$$\begin{aligned}(4y^3 + 7y - 3) + (y^3 - y + 2) &= 4y^3 + 7y - 3 + y^3 - y + 2 \\ &= (4 + 1)y^3 + (7 - 1)y + (-3 + 2) \\ &= 5y^3 + 6y - 1.\end{aligned}$$

To subtract $3x^2 + 7x - 11$ from $7x^2 - 9x + 13$ we extend the definition of subtraction to polynomials:

$$P - Q = P + (-Q).$$

Thus,

$$(7x^2 - 9x + 13) - (3x^2 + 7x - 11) = (7x^2 - 9x + 13) + [-(3x^2 + 7x - 11)].$$

Now, we form the opposite of $3x^2 + 7x - 11$ by taking the opposite of each term. That is,

$$-(3x^2 + 7x - 11) = -3x^2 - 7x + 11,$$

so that

$$\begin{aligned}(7x^2 - 9x + 13) - (3x^2 + 7x - 11) &= 7x^2 - 9x + 13 - 3x^2 - 7x + 11 \\ &= (7 - 3)x^2 + (-9 - 7)x + (13 + 11) \\ &= 4x^2 - 16x + 24.\end{aligned}$$

EXAMPLE 3   Perform the subtraction $(2w^3 - w^2 - 8w) - (-4w^3 + 5w^2 - 7w + 3)$.

SOLUTION

$$(2w^3 - w^2 - 8w) - (-4w^3 + 5w^2 - 7w + 3)$$
$$= (2w^3 - w^2 - 8w) + [-(-4w^3 + 5w^2 - 7w + 3)]$$
$$= 2w^3 - w^2 - 8w + 4w^3 - 5w^2 + 7w - 3$$
$$= (2 + 4)w^3 + (-1 - 5)w^2 + (7 - 8)w - 3$$
$$= 6w^3 - 6w^2 - w - 3.$$

When we add and subtract polynomials, it is often useful to rearrange like terms by using a "vertical scheme." To do this, we line up like terms beneath each other and then add or subtract numerical coefficients. The vertical scheme is most useful when we have several polynomials to add or subtract. For example, the polynomials $5x^2 + 3x + 2$ and $3x^2 + 2x + 7$ can be added as follows:

$$
\begin{array}{cccccc}
 & 5x^2 & + & 3x & + & 2 \\
(+) & 3x^2 & + & 2x & + & 7 \\
\hline
 & 8x^2 & + & 5x & + & 9.
\end{array}
$$

In Examples 4–7, perform each addition and subtraction by using the vertical scheme.

EXAMPLE 4 $(2x - 7x^2 + 8) + (-x^3 - x^2 - 1)$

SOLUTION   We line up like terms and add coefficients:

$$
\begin{array}{r}
-7x^2 + 2x + 8 \\
(+)\ \underline{-x^3 - x^2\qquad\ -1} \\
-x^3 - 8x^2 + 2x + 7.
\end{array}
$$

EXAMPLE 5 $(7 - 3t^2 + t) + (2t^2 + 8 + 13t) + (3t - 7 + 2t^2)$

SOLUTION   We arrange like terms in columns and combine:

$$
\begin{array}{r}
-3t^2 +\ \ t + 7 \\
2t^2 + 13t + 8 \\
(+)\ \underline{\ 2t^2 +\ 3t - 7} \\
t^2 + 17t + 8.
\end{array}
$$

EXAMPLE 6   Subtract $9t^2 - 13t + 21$ from $2t^2 + 5t - 43$.

SOLUTION   We arrange like terms in columns:

$$
\begin{array}{r}
2t^2 + 5t - 43 \\
(-)\ \underline{9t^2 - 13t + 21.}
\end{array}
$$

We form the opposite of the bottom polynomial (that is, $-(9t^2 - 13t + 21) = -9t^2 + 13t - 21$) and add:

$$
\begin{array}{r}
2t^2 + 5t - 43 \\
(+)\ \underline{-9t^2 + 13t - 21} \\
-7t^2 + 18t - 64.
\end{array}
$$

EXAMPLE 7 $(z^2 - 3z + 4) + (2z^2 - z + 2) - (z^2 - 5z + 8)$

SOLUTION   First we perform the addition of $z^2 - 3z + 4$ and $2z^2 - z + 2$:

$$
\begin{array}{r}
z^2 - 3z + 4 \\
(+)\ \underline{2z^2 - z + 2} \\
3z^2 - 4z + 6.
\end{array}
$$

Then we perform the subtraction by adding the opposite of the bottom polynomial.

$$
\begin{array}{r}
3z^2 - 4z + 6 \\
(-)\ \underline{z^2 - 5z + 8}
\end{array}
$$

$$
\text{or} \quad
\begin{array}{r}
3z^2 - 4z + 6 \\
(+)\ \underline{-z^2 + 5z - 8} \\
2z^2 + z - 2
\end{array}
$$

If polynomials contain more than one variable, addition and subtraction are still performed by combining like terms.

EXAMPLE 8   Subtract $3ab^2 - 5a^2b + 11$ from $-7a^2b + 8ab^2 - 18$.

SOLUTION   We arrange the like terms in columns. We have

$$
\begin{array}{r}
-7a^2b + 8ab^2 - 18 \\
(-)\ \underline{-5a^2b + 3ab^2 + 11.}
\end{array}
$$

Now we form the opposite of the bottom polynomial and add:

$$
\begin{array}{r}
-7a^2b + 8ab^2 - 18 \\
(+)\ \underline{5a^2b - 3ab^2 - 11} \\
-2a^2b + 5ab^2 - 29.
\end{array}
$$

PROBLEM SET 2.2

In problems 1–14, perform the indicated additions of the polynomials.

1. $5x^2 + 7x^2$

2. $3y^3 + 8y^3$

3. $2v^3 + 9v^3$

4. $4x^2 + 5x^2$

5. $-3t^2 + 7t^2$

6. $(-8z) + (-5z)$

7. $(3x + 4) + (5x + 3)$

8. $(-5u - 3) + (7u + 1)$

9. $(2z^2 + 3z + 1) + (5z^2 + 2z + 4)$

10. $(3x + 8x^3 + 5) + (2x^2 + 7x + 5)$

11. $(3 - 4x + 7x^2) + (2x - 3x^2 - 5)$

12. $(7w - 2w^2 - 3) + (5 + 3w^2 - 9w)$

13. $(5c^2 - 3c^3 - c + 2c^4) + (4c^3 + 3c^4 - c^2 + 2c)$

14. $(1 + 2x - 3x^2 + 4x^3) + (5x^3 - x^2 + 3x - 7)$

In problems 15–28, perform the indicated subtractions of the polynomials.

15. $7u - 3u$

16. $10y - 3y$

17. $3x^2 - x^2$

18. $4w^3 - w^3$

19. $4v^3 - (-2v^3)$

20. $5t^4 - (-2t^4)$

21. $(3t^3 + 2) - (-t^3 + 4)$

22. $(-11z^3 + 10z) - (9z - 12z^3)$

23. $(3x^2 + 5x + 8) - (x^2 + 3x + 4)$

24. $(4w^3 - 7w - 8) - (-2w^3 + 4w - 2)$

25. $(3s^4 - 4s^3 + 6s^2 + s - 1) - (4 - s + 2s^2 - 3s^3 - s^4)$

26. $(5y^3 - 3y^2 + 2y - 8) - (y - y^3 + 3y^2 - 1)$

27. $(-4t^3 + 8t^2 - 7) - (3t^2 - 4t + 11)$

28. $(x^6 - 2x^4 - 3x^2) - (x^5 - 2x^3 - 3x - 4)$

In problems 29–42, perform the indicated operations by combining like terms.

29. $xy + 3xy + 8xy$

30. $5xyz + xyz + 3xyz$

31. $10uv - 7uv$

32. $12t^2s - 9t^2s$

33. $5mn^2 + (-3mn^2) + 2m^2n$

34. $14x^3y - 4x^3y - 4xy^3 + 9xy^3$

35. $-5x^2y - (-3x^2y)$

36. $-11u^3v^2 - (-12u^3v^2)$

37. $(7ts - 4) + (-3ts - 2)$

38. $(-4w^2z + 6) + (5w^2z - 8)$

39. $(-8x^2y + 6xy - 7xy^2) + (3xy^2 - 4xy + 2x^2y)$

40. $(11mn - 9m^2n^2 + 13m^3n^3) + (3m^3n^3 + 5mn + 7m^2n^2 + 13)$

41. $(3w^2z + 4wz - 7wz^2) - (-2w^2z + 3wz^2 + wz)$

42. $(7 - 8xy - 5x^3y^3) - (2xy - 2x^3y^3)$

In problems 43–50, perform the indicated operations and combine like terms.

43. $(8x^2 + 3x - 7) + (-5x^2 + 2x + 1) + (2x^2 - 3x + 4)$

44. $(-2p^2 + 5p + 2) + (3p^2 - 2p + 3) + (p^2 - 3p - 7)$

45. $(t^3 - 2t^2 + 3t + 1) + (2t^3 + t^2 - 2t + 2) + (-t^3 + 3t^2 - 2t - 1)$

46. $(3y^4 - 4y^3 + y^2 - 2y + 3) + (7y^4 + 5y^3 + 2y^2 - y - 7) + (6y^3 - 6y + 5)$

47. $(2w^2 - 3w + 4) + (5w - 1 + w^2) - (w + 2w^2 - 6)$

48. $(-x^2 + 8x - 11) - (x^2 - 3x - 2) + (3x^2 - 10x + 3)$

49. $(7u^3v^2 - 3u^2v + 2w) - (4w + u^2v - u^3v^2) - (-2u^3v^2 + 5u^2v + 3w)$

50. $(2xy - 3xz + 4yz) - (5xz - 3yz + 4xy) - (-2yz + 3xy - xz)$

2.3 Properties of Positive Integral Exponents

Recall from Chapter 1, Section 1.5, that if n is a positive integer and x is a real number, then

$$x^n = \overbrace{x \cdot x \cdot x \cdots x,}^{n \text{ factors}}$$

where x is the **base** and n is the **power** or the **exponent.**
    If $n = 1$ in the above definition, we have

$$x^1 = x.$$

Thus, when no exponent is shown, we assume it to be the number 1. For example,

$$3 = 3^1,$$
$$a = a^1, \quad \text{and}$$
$$x^2y = x^2y^1.$$

Suppose that we wish to find the product of x^2 and x^3. We know that $x^2 = x \cdot x$ and $x^3 = x \cdot x \cdot x$. Therefore,

$$x^2 \cdot x^3 = (x \cdot x)(x \cdot x \cdot x)$$

and since this expression contains five factors of x, we have

$$x^2 \cdot x^3 = x^5.$$

Notice that the exponent of the product x^5 is the sum of the exponents of the factors x^2 and x^3, that is,

$$x^2 \cdot x^3 = x^{2+3}$$
$$= x^5.$$

Similarly, the expression $(x^2)^3$ can be written as a single power of x by applying the definition of positive exponents as follows:

$$
\begin{aligned}
(x^2)^3 &= x^2 \cdot x^2 \cdot x^2 && \text{(three factors of } x^2) \\
&= (x \cdot x)(x \cdot x)(x \cdot x) && \text{(replacing } x^2 \text{ by } x \cdot x) \\
&= x^6.
\end{aligned}
$$

The exponent in the result is the product of exponents 2 and 3, that is,

$$(x^2)^3 = x^{2 \cdot 3}$$
$$= x^6.$$

We may now generalize the following properties of exponents:

Properties of Exponents

If a and b are real numbers and m and n are positive integers, then:
  (i) **Multiplication property:** $a^m \cdot a^n = a^{m+n}$
 (ii) **Power-of-a-power property:** $(a^m)^n = a^{mn}$
(iii) **Power-of-a-product property:** $(ab)^n = a^n b^n$
 (iv) **Power-of-a-quotient property:** $\left(\dfrac{a}{b}\right)^n = \dfrac{a^n}{b^n}$ if $b \neq 0$
  (v) **Division property:** $\dfrac{a^m}{a^n} = a^{m-n}$ if $a \neq 0$ and m is greater than n

Notice that in Property (v), we assume that m is greater than n, so that $m - n$ represents a positive exponent. For instance,

$$\frac{u^8}{u^3} = u^{8-3} = u^5 \qquad \text{for} \qquad u \neq 0.$$

However, if $m = n$, we have

$$\frac{a^m}{a^n} = \frac{a^m}{a^m} = 1.$$

For example,

$$\frac{x^3}{x^3} = 1 \qquad \text{for} \qquad x \neq 0.$$

Also,

$$\frac{(-y)^5}{(-y)^5} = 1 \qquad \text{for} \qquad y \neq 0.$$

If m is less than n, we have

$$\frac{a^m}{a^n} = \frac{1}{a^{n-m}} \qquad \text{for} \qquad a \neq 0$$

where $n - m$ represents a positive exponent. For instance,

$$\frac{x^2}{x^5} = \frac{1}{x^{5-2}} = \frac{1}{x^3} \qquad \text{for} \qquad x \neq 0.$$

Also,

$$\frac{(-t)^7}{(-t)^{10}} = \frac{1}{(-t)^3} \qquad \text{for} \qquad t \neq 0.$$

In working with expressions like those in Properties (iv) and (v), always remember that the *denominator of a fraction cannot be zero*.

These five properties can be verified by applying the definition of positive integral exponents. We shall verify Properties (i), (iii), and (v) here and leave the verifications of Properties (ii) and (iv) to the reader as an exercise. (See Problems 62 and 64 of Problem Set 2.3.)

PROOF OF
PROPERTY (i)

$$a^m a^n = a^{m+n}$$

$$a^m a^n = \underbrace{(a \cdot a \cdots a)}_{m \text{ factors}} \underbrace{(a \cdot a \cdot a \cdots a)}_{n \text{ factors}} \qquad \text{(Definition)}$$

$$= \underbrace{a \cdot a \cdot a \cdot a \cdot a \cdots a \cdot a}_{m + n \text{ factors}}$$

$$= a^{m+n}. \qquad \text{(Definition)}$$

PROOF OF
PROPERTY (iii)

$$(ab)^n = a^n b^n$$

$$(ab)^n = \underbrace{(ab)(ab) \cdots (ab)}_{n \text{ factors}} \qquad \text{(Definition)}$$

$$= \underbrace{(a \cdot a \cdots a)}_{n \text{ factors}} \underbrace{(b \cdot b \cdots b)}_{n \text{ factors}} \qquad \text{(Why?)}$$

$$= a^n b^n \qquad \text{(Definition)}$$

PROOF OF
PROPERTY (v)

$$\frac{a^m}{a^n} = a^{m-n} \qquad \text{if } m \text{ is greater than } n \text{ and } a \neq 0$$

$$\frac{a^m}{a^n} = \frac{\overbrace{a \cdot a \cdot a \cdot a \cdot a \cdots a}^{m \text{ factors}}}{\underbrace{a \cdot a \cdots a}_{n \text{ factors}}} = \frac{\overbrace{(a \cdot a \cdots a)}^{n \text{ factors}} \overbrace{(a \cdot a \cdot a \cdots a)}^{m - n \text{ factors}}}{\underbrace{a \cdot a \cdots a}_{n \text{ factors}}}$$

$$= \underbrace{a \cdot a \cdots a}_{m - n \text{ factors}}$$

$$= a^{m-n}.$$

We can use the five properties to rewrite algebraic expressions containing exponents as compactly as possible. When we do this we say that the expression has been **simplified.** Although the word "simplify" has no precise mathematical definition, the meaning it conveys is usually clear from the context.

EXAMPLE 1   Use Property (i) to simplify the following expressions.

(a) $2^3 \cdot 2^4$ (b) $x^4 \cdot x^6$ (c) $(-a)^2 \cdot (-a)^4$

SOLUTION   To apply Property (i), we keep the common base and add the exponents. Thus:

(a) $2^3 \cdot 2^4 = 2^{3+4} = 2^7 = 128$

(b) $x^4 \cdot x^6 = x^{4+6} = x^{10}$

(c) $(-a)^2 \cdot (-a)^4 = (-a)^{2+4}$
$$= (-a)^6$$

Because the negative sign within the parentheses is part of the base, we have

$$(-a)^6 = (-a)(-a)(-a)(-a)(-a)(-a) = a^6.$$

Warning: It is a common error to try to find a product like $4^3 \cdot 2^5$ by multiplying the bases ($4 \cdot 2 = 8$) and then adding the exponents ($3 + 5 = 8$). You would then arrive at the solution $4^3 \cdot 2^5 \overset{?}{=} 8^8$. This is wrong, of course, since $4^3 \cdot 2^5 = 64 \cdot 32 \neq 8^8$.

EXAMPLE 2   Use Property (ii) to remove the parentheses in the following expressions.

(a) $(3^2)^3$

(b) $(x^4)^3$

(c) $(y^7)^n$ n any positive integer

SOLUTION   Applying Property (ii), we have:

(a) $(3^2)^3 = 3^{2 \cdot 3} = 3^6 = 729$

(b) $(x^4)^3 = x^{4 \cdot 3} = x^{12}$

(c) $(y^7)^n = y^{7 \cdot n} = y^{7n}$

EXAMPLE 3   Use Property (iii) to remove the parentheses in the following expressions.

(a) $(3x)^3$ (b) $(-x)^5$ (c) $(xyz)^4$

SOLUTION   Applying Property (iii), we have:

(a) $(3x)^3 = 3^3 \cdot x^3 = 27x^3$

(b) $(-x)^5 = [(-1)x]^5 = (-1)^5 x^5 = (-1)x^5 = -x^5$

(c) $(xyz)^4 = [x(yz)]^4$ (associative property)
$$= x^4(yz)^4 = x^4 y^4 z^4$$

EXAMPLE 4   Use Property (iv) to remove the parentheses in the following expressions.

(a) $\left(\dfrac{2}{3}\right)^3$ (b) $\left(\dfrac{x}{y}\right)^5$ (c) $\left(\dfrac{-a}{3}\right)^2$

SOLUTION   Applying Property (iv), we raise both the numerator and the denominator of the quotient to the indicated power. Thus:

(a) $\left(\dfrac{2}{3}\right)^3 = \dfrac{2^3}{3^3} = \dfrac{8}{27}$ (b) $\left(\dfrac{x}{y}\right)^5 = \dfrac{x^5}{y^5}$ (c) $\left(\dfrac{-a}{3}\right)^2 = \dfrac{(-a)^2}{3^2} = \dfrac{a^2}{9}$

EXAMPLE 5   Use Property (v) to simplify the following expressions.

(a) $\dfrac{2^5}{2^2}$ (b) $\dfrac{x^7}{x^2}$ (c) $\dfrac{(-a)^5}{(-a)^2}$

SOLUTION   Applying Property (v), that is, $a^m/a^n = a^{m-n}$, where m is larger than n, we keep the common base and subtract exponents:

(a) $\dfrac{2^5}{2^2} = 2^{5-2} = 2^3 = 8$

(b) $\dfrac{x^7}{x^2} = x^{7-2} = x^5$

(c) $\dfrac{(-a)^5}{(-a)^2} = (-a)^{5-2} = (-a)^3 = -a^3$

EXAMPLE 6   Use Properties (i)–(v) to simplify the given expressions.

(a) $(3x^2)^3$ (b) $\left(\dfrac{x^3}{y^2}\right)^4$ (c) $\dfrac{x^7}{x^3 \cdot x^2}$ (d) $\left(\dfrac{x^3 y^5}{2xy^2}\right)^7$

SOLUTION   (a) $(3x^2)^3 = 3^3(x^2)^3$ [Property (iii)]
$\qquad\qquad = 27x^{2\cdot3}$ [Property (ii)]
$\qquad\qquad = 27x^6$

(b) $\left(\dfrac{x^3}{y^2}\right)^4 = \dfrac{(x^3)^4}{(y^2)^4}$ [Property (iv)]

$\qquad\qquad = \dfrac{x^{3\cdot4}}{y^{2\cdot4}}$ [Property (ii)]

$\qquad\qquad = \dfrac{x^{12}}{y^8}$

(c) $\dfrac{x^7}{x^3 \cdot x^2} = \dfrac{x^7}{x^{3+2}} = \dfrac{x^7}{x^5}$ [Property (i)]

$= x^{7-5}$ [Property (v)]

$= x^2$

(d) $\left(\dfrac{x^3 y^5}{2xy^2}\right)^7 = \left(\dfrac{x^{3-1} y^{5-2}}{2}\right)^7$ [Property (v) applied to x and y within the parentheses]

$= \left(\dfrac{x^2 y^3}{2}\right)^7$

$= \dfrac{(x^2)^7 (y^3)^7}{2^7}$ [Properties (iii) and (iv)]

$= \dfrac{x^{2 \cdot 7} y^{3 \cdot 7}}{128}$ [Property (ii)]

$= \dfrac{x^{14} y^{21}}{128}$

PROBLEM SET 2.3

In problems 1–10, use Property (i) to simplify each expression.

1. $3^2 \cdot 3^3$ **2.** $2^4 \cdot 2^2$ **3.** $(-2)^3 \cdot (-2)^2$

4. $(-3) \cdot (-3)^2 (-3)^3$ **5.** $x^5 \cdot x^7$ **6.** $y^6 \cdot y^4$

7. $t^3 \cdot t^4 \cdot t^5$ **8.** $(-u)^9 \cdot (-u)^3 \cdot (-u)$ **9.** $(-v)^3 \cdot (-v)^5$

10. $(-x)^7 \cdot (-x)^3 \cdot x^2$

In problems 11–20, use Property (ii) to remove the parentheses and brackets in each expression.

11. $(2^2)^3$ **12.** $(3^2)^4$ **13.** $[(-2)^3]^2$

14. $[(-3)^2]^2$ **15.** $(x^7)^5$ **16.** $(y^3)^{12}$

17. $(t^2)^{11}$ **18.** $(z^4)^5$ **19.** $[(-w)^3]^4$

20. $[(c^2)^3]^5$

In problems 21–30, use Property (iii) to remove the parentheses and brackets in each expression.

21. $(2x)^4$ **22.** $(5y)^2$ **23.** $(uv)^5$

24. $(ab)^4$ **25.** $(xyz)^7$ **26.** $(3mn)^5$

27. $(-2w)^3$ **28.** $(-2xyz)^4$ **29.** $[-3(-x)y]^3$

30. $[-3(-x)(-y)]^2$

In problems 31–40, use Property (iv) to remove the parentheses in each expression.

31. $\left(\dfrac{3}{4}\right)^2$ **32.** $\left(\dfrac{1}{2}\right)^5$ **33.** $\left(\dfrac{-2}{3}\right)^3$ **34.** $\left(-\dfrac{3}{2}\right)^3$

35. $\left(\dfrac{x}{y}\right)^4$ **36.** $\left(\dfrac{y}{z}\right)^7$ **37.** $\left(\dfrac{a}{-b}\right)^5$ **38.** $\left(-\dfrac{m}{n}\right)^8$

39. $\left(\dfrac{-t}{s}\right)^6$ **40.** $\left(-\dfrac{w}{c}\right)^9$

In problems 41–50, use Property (v) to simplify each expression.

41. $\dfrac{3^5}{3^2}$ **42.** $\dfrac{(-2)^6}{(-2)^3}$ **43.** $\dfrac{4^9}{4^6}$ **44.** $\dfrac{(-3)^4}{(-3)^6}$

45. $\dfrac{x^8}{x^3}$ **46.** $\dfrac{m^{25}}{m^{11}}$ **47.** $\dfrac{y^{25}}{y^{20}}$ **48.** $\dfrac{(-t)^5}{(-t)^5}$

49. $\dfrac{w^{13}}{w^{10}}$ **50.** $\dfrac{x^n}{x^2}$ (*n* is a positive integer greater than 2)

In problems 51–60, use Properties (i)–(v) to simplify each expression.

51. $(3x^2y^3)^4$ **52.** $(-2z^4)^3$ **53.** $\dfrac{u^4v^7}{uv^3}$ **54.** $\dfrac{x^{16}y^5}{x^5y^9}$

55. $\left(\dfrac{w^2}{2z}\right)^3$ **56.** $\left(\dfrac{3m^2}{4n}\right)^2$ **57.** $\left(\dfrac{3a^5b^3}{2a^2b^6}\right)^3$ **58.** $\left(\dfrac{4t^2s^3}{8t^3s}\right)^5$

59. $\dfrac{(-4xy^4z^5)^3}{(2x^2yz^3)^2}$ **60.** $\dfrac{(8u^4v^7)^3}{(4u^2v^3)^2}$

61. Find the value of $(-1)^n$ if *n* is an even positive integer.
62. Verify Property (ii), page 41.
63. Find the value of $(-1)^n$ if *n* is an odd positive integer.
64. Verify Property (iv), page 41.

2.4 Multiplication of Polynomials

To multiply polynomials, we extend the commutative and associative properties for the multiplication of real numbers to polynomials. That is, if *P*, *Q*, and *R* are polynomial expressions, then:

| (i) $PQ = QP$ | (ii) $P(QR) = (PQ)R$ |

For example, to multiply the monomials $8x^3$ and $2x^2$, we apply the associative and commutative properties:

$$(8x^3)(2x^2) = (8 \cdot 2)(x^3 \cdot x^2)$$
$$= 16x^5.$$

Note that we found the product of the two monomials essentially by regrouping the factors, multiplying the coefficients, and applying the rule for multiplying like bases by adding exponents.

EXAMPLE 1   Perform each multiplication and simplify the results.

(a) $(4x^2)(7x)$

(b) $(-3y)(2y^3)$

(c) $(-\frac{2}{3}a^2b^3)(-6a^2b)$

SOLUTION   We regroup the coefficients together and the variables together, and then multiply:

(a) $(4x^2)(7x) = (4 \cdot 7)(x^2 \cdot x) = 28x^3$

(b) $(-3y)(2y^3) = (-3 \cdot 2)(y \cdot y^3) = -6y^4$

(c) $(-\frac{2}{3}a^2b^3)(-6a^2b) = [(-\frac{2}{3}) \cdot (-6)](a^2 \cdot a^2 \cdot b^3 \cdot b) = 4a^4b^4$

To multiply a monomial by a polynomial, we apply the distributive property to polynomials, that is, if P, Q, and R are polynomials, then:

(i) $P(Q + R) = PQ + PR$	(ii) $(P + Q)R = PR + QR$

For example, to find the product of $3x^2$ and $2x^3 + 4x$, we apply the above distributive property, so that

$$3x^2(2x^3 + 4x) = (3x^2)(2x^3) + (3x^2)(4x)$$
$$= 6x^5 + 12x^3.$$

EXAMPLE 2   Find the following products.

(a) $4x(3x - 2)$ (b) $2y^3(y^2 - 3y + 4)$ (c) $-2t^4(-3t^2 - 7t + 4)$

SOLUTION   Using the distributive property, we have:

(a) $4x(3x - 2) = (4x)(3x) + (4x)(-2)$
$$= 12x^2 - 8x$$

(b) Multiplying the expression $2y^3$ (outside the parentheses) by each term inside the parentheses, we have

$$2y^3(y^2 - 3y + 4) = (2y^3)(y^2) + (2y^3)(-3y) + (2y^3)(4)$$
$$= 2y^5 - 6y^4 + 8y^3.$$

(c) $-2t^4(-3t^2 - 7t + 4) = (-2t^4)(-3t^2) + (-2t^4)(-7t) + (-2t^4)(4)$
$$= 6t^6 + 14t^5 - 8t^4$$

To multiply one polynomial by another, use the distributive property to reduce the given multiplication to a multiplication of monomials.

For example, to multiply $x + 2$ by $x + 4$, we can think of $x + 4$ as one *term*—call it u for the time being—and then apply the distributive property, so that

$$(x + 2)(x + 4) = (x + 2)u$$

$= xu + 2u$	(We used the distributive property)
$= x(x + 4) + 2(x + 4)$	(We replaced u by $x + 4$)
$= x^2 + 4x + 2x + 8$	(We used the distributive property)
$= x^2 + 6x + 8.$	

Although it is important to remember that the distributive property is what allows us to multiply polynomials, at times it takes too long to write out all the steps involved. A shorter procedure follows:

> To **multiply** two polynomials, multiply each term of one polynomial by each term of the other and then simplify the result by combining like terms.

EXAMPLE 3   Find each product.

(a) $(x + 2)(2x - 3)$

(b) $(3t - 1)(2t^2 + 7t + 4)$

SOLUTION   (a) Multiply each term of $x + 2$ by each term of $2x - 3$:

$$(x + 2)(2x - 3) = (x)(2x) + (x)(-3) + (2)(2x) + (2)(-3)$$
$$= 2x^2 - 3x + 4x - 6$$
$$= 2x^2 + x - 6.$$

(b) $(3t - 1)(2t^2 + 7t + 4) = (3t)(2t^2) + (3t)(7t) + (3t)(4) + (-1)(2t^2)$
$$+ (-1)(7t) + (-1)(4)$$
$$= 6t^3 + 21t^2 + 12t - 2t^2 - 7t - 4$$
$$= 6t^3 + 19t^2 + 5t - 4$$

The actual computations can become quite tedious in many multiplications involving polynomials (as we have seen in the previous examples). Therefore, any device to help perform the computations and to simplify the work is desirable. One such device, the vertical scheme, is illustrated by the following example. To find the product of $2x - 1$ and $x^3 + x^2 - 2x - 1$, we can arrange the polynomials in a vertical scheme:

$$
\begin{array}{l}
x^3 + x^2 - 2x - 1 \\
\underline{2x - 1} \qquad \text{multiply} \\
2x^4 + 2x^3 - 4x^2 - 2x \qquad\qquad [(2x)(x^3 + x^2 - 2x - 1)]\text{(partial product)} \\
\underline{ - x^3 - x^2 + 2x + 1} \quad \text{add} \quad [(-1)(x^3 + x^2 - 2x - 1)]\text{(partial product)} \\
2x^4 + x^3 - 5x^2 + 1. \qquad\qquad \text{(product)}
\end{array}
$$

This shortcut involves arranging the partial products so that like terms are in the same column, ready for the final step of addition. Note that the "mechanics" of this method are based on properties you have already learned. This was also true for the vertical scheme for addition.

In Examples 4 and 5, use the vertical scheme to multiply each of the following.

EXAMPLE 4 $(x + 4)(2x - 3)$

SOLUTION   Using the vertical scheme, we have

$$
\begin{array}{l}
x + 4 \\
\underline{2x - 3} \quad \text{multiply} \\
2x^2 + 8x \\
\underline{ - 3x - 12} \quad \text{add} \\
2x^2 + 5x - 12.
\end{array}
$$

EXAMPLE 5 $(x^2 - 2x + 1)(x^2 + x + 2)$

SOLUTION   Using the vertical scheme, we have

$$
\begin{array}{l}
x^2 - 2x + 1 \\
\underline{x^2 + x + 2} \quad \text{multiply} \\
x^4 - 2x^3 + x^2 \\
 \quad x^3 - 2x^2 + x \\
\underline{ 2x^2 - 4x + 2} \quad \text{add} \\
x^4 - x^3 + x^2 - 3x + 2.
\end{array}
$$

We can also multiply polynomials with more than one variable. In doing so, it is good practice to express the mixed-letter parts of terms in alphabetical order. This makes it easier to spot like terms.

In Examples 6 and 7 find each product.

EXAMPLE 6 $(x - y)(3x^3 - 5xy + 7y^2)$

SOLUTION   Using the vertical scheme, we have

$$3x^3 - 5xy + 7y^2$$
$$\underline{ x - y}$$
$$3x^4 - 5x^2y + 7xy^2$$
$$\underline{ 5xy^2 - 3x^3y - 7y^3 \quad \text{add}}$$
$$3x^4 - 5x^2y + 12xy^2 - 3x^3y - 7y^3.$$

EXAMPLE 7 $(u^2 + 2uv - v^2)(u + uv + v)$

SOLUTION   Applying the vertical scheme, we have

$$u^2 + 2uv - v^2$$
$$\underline{u + uv + v}$$
$$u^3 + 2u^2v - uv^2$$
$$ u^3v + 2u^2v^2 - uv^3$$
$$\underline{ u^2v + 2uv^2 - v^3 \quad \text{add}}$$
$$u^3 + 3u^2v + uv^2 + u^3v + 2u^2v^2 - uv^3 - v^3.$$

We can also apply a device for simplifying the computations required when multiplying two binomials in which the first and second terms are like terms. For example, consider the product $(x + 2)(x + 3)$. If we apply distributive properties, we have

$$(x + 2)(x + 3) = x(x + 3) + 2(x + 3)$$
$$= x^2 + 3x + 2x + 6$$
$$= x^2 + 5x + 6.$$

The result of the above multiplication is a trinomial whose terms are determined as follows:

First term: $(x + 2)(x + 3) \qquad = \qquad x^2 + 5x + 6$

$(x)(x)$

Middle term: $(x + 2)(x + 3) \qquad = \qquad x^2 + 5x + 6$

$\oplus \qquad (3)(x) + (2)(x)$

Last term: $(x + 2)(x + 3) \qquad = \qquad x^2 + 5x + 6$

$(2)(3)$

This method enables us to multiply two binomials and write the product directly, without having to show any intermediate steps.

In Examples 8–10, find the following products by using the method just illustrated.

EXAMPLE **8** $(x + 4)(2x + 1)$

SOLUTION

First term: $(x + 4)(2x + 1) \quad = \quad 2x^2 + \underline{\hspace{1cm}} + \underline{\hspace{1cm}}$

$(x)(2x)$

Middle term: $(x + 4)(2x + 1) \quad = \quad 2x^2 + 9x + \underline{\hspace{1cm}}$

$(1)(x) + (4)(2x)$

Last term: $(x + 4)(2x + 1) \quad = \quad 2x^2 + 9x + 4$

$(4)(1)$

Therefore,

$$(x + 4)(2x + 1) = 2x^2 + 9x + 4.$$

EXAMPLE **9** $(3x - 4)(2x + 3)$

SOLUTION

First term: $(3x - 4)(2x + 3) \quad = \quad 6x^2 + \underline{\hspace{1cm}} + \underline{\hspace{1cm}}$

$(3x)(2x)$

Middle term: $(3x - 4)(2x + 3) \quad = \quad 6x^2 + \quad x \quad + \underline{\hspace{1cm}}$

$(3x)(3) + (-4)(2x)$

Last term: $(3x - 4)(2x + 3) \quad = \quad 6x^2 + \quad x \quad + (-12)$

$(-4)(3)$

Therefore,

$$(3x - 4)(2x + 3) = 6x^2 + x - 12.$$

EXAMPLE **10** $(4x - 3y)(x - y)$

SOLUTION   We can obtain this product, using the following scheme:

$$(4x - 3y)(x - y) \quad = \quad 4x^2 - 7xy + 3y^2$$

First term: $(4x)(x) = 4x^2$

Middle term: $(4x)(-y) + (-3y)(x) = -7xy$

Last term: $(-3y)(-y) = 3y^2$

Special Products

Certain products occur often enough in algebra and in applications of algebra to be worthy of special consideration. When these special situations arise, you can use time-saving shortcuts to multiply polynomials.

Consider the problem of squaring the binomial $A + B$. We have

$$(A + B)^2 = (A + B)(A + B) = A^2 + AB + BA + B^2 = A^2 + 2AB + B^2.$$

We state this result as follows:

Special Product (i)

Let A and B be real numbers, then
$$(A + B)^2 = A^2 + 2AB + B^2.$$

A similar result occurs in $(A - B)^2$:

$$(A - B)^2 = (A - B)(A - B) = A^2 - AB - BA + B^2$$
$$= A^2 - 2AB + B^2.$$

Therefore, we have

Special Product (ii)

Let A and B be real numbers, then
$$(A - B)^2 = A^2 - 2AB + B^2.$$

EXAMPLE 11   Use Special Products (i) or (ii) to carry out the multiplications.

(a) $(x + 3)^2$

(b) $(c + 2d)^2$

(c) $(3u - 2)^2$

(d) $(2m^2n^3 - \frac{1}{2})^2$

SOLUTION   (a) Substituting x for A and 3 for B in Special Product (i), we have

$$(x + 3)^2 = (x)^2 + 2(x)(3) + (3)^2$$
$$= x^2 + 6x + 9.$$

(b) Substituting c for A and $2d$ for B in Special Product (i), we have

$$(c + 2d)^2 = (c)^2 + 2(c)(2d) + (2d)^2$$
$$= c^2 + 4cd + 4d^2.$$

(c) Substituting $3u$ for A and 2 for B in Special Product (ii), we have

$$(3u - 2)^2 = (3u)^2 - 2(3u)(2) + (2)^2$$
$$= 9u^2 - 12u + 4.$$

(d) Substituting $2m^2n^3$ for A and $\frac{1}{2}$ for B in Special Product (ii), we have

$$\left(2m^2n^3 - \frac{1}{2}\right)^2 = (2m^2n^3)^2 - 2(2m^2n^3)\left(\frac{1}{2}\right) + \left(\frac{1}{2}\right)^2$$

$$= 4m^4n^6 - 2m^2n^3 + \frac{1}{4}.$$

Another special product that occurs frequently in mathematics and the sciences has the form $(A - B)(A + B)$. You will notice that the middle term of this product is zero.

Thus

$$(A - B)(A + B) = A^2 + AB - BA - B^2$$
$$= A^2 - B^2.$$

Hence, we have:

Special Product (iii)

Let A and B be real numbers, then
$$(A - B)(A + B) = A^2 - B^2.$$

Notice that the binomial on the right side of the equation in Special Product (iii) is the **difference of two squares.**

EXAMPLE 12   Use Special Product (iii) to carry out the multiplications.

(a) $(c + 5)(c - 5)$ (b) $(2u + 3v)(2u - 3v)$

SOLUTION   (a) We substitute c for A and 5 for B in Special Product (iii):

$$(c + 5)(c - 5) = c^2 - 5^2$$
$$= c^2 - 25.$$

(b) $(2u + 3v)(2u - 3v) = (2u)^2 - (3v)^2$
$$= 4u^2 - 9v^2.$$

Four more special products occur frequently enough to be mentioned here. They are as follows:

Special Products (iv)–(vii)

Let A and B be real numbers, then

$$\text{(iv) } (A + B)^3 = A^3 + 3A^2B + 3AB^2 + B^3.$$

$$\text{(v) } (A - B)^3 = A^3 - 3A^2B + 3AB^2 - B^3.$$

$$\text{(vi) } (A + B)(A^2 - AB + B^2) = A^3 + B^3.$$

$$\text{(vii) } (A - B)(A^2 + AB + B^2) = A^3 - B^3.$$

These special products can be verified by direct multiplication and are left as exercises (see Problems 71–74 in Problem Set 2.4).

EXAMPLE 13   Use Special Products (iv)–(vii) to carry out the multiplications.

(a) $(x + 2)^3$

(b) $(2x - y)^3$

(c) $(2u + v)(4u^2 - 2uv + v^2)$

(d) $(3y - 2z)(9y^2 + 6yz + 4z^2)$

SOLUTION   (a) We substitute x for A and 2 for B in Special Product (iv):

$$(x + 2)^3 = (x)^3 + 3(x)^2(2) + 3(x)(2)^2 + (2)^3$$
$$= x^3 + 6x^2 + 12x + 8.$$

(b) We substitute $2x$ for A and y for B in Special Product (v):

$$(2x - y)^3 = (2x)^3 - 3(2x)^2(y) + 3(2x)(y)^2 - (y)^3$$
$$= 8x^3 - 12x^2y + 6xy^2 - y^3.$$

(c) We substitute $2u$ for A and v for B in Special Product (vi). Note that $4u^2 = (2u)^2$.

$$(2u + v)(4u^2 - 2uv + v^2) = (2u)^3 + (v)^3$$
$$= 8u^3 + v^3.$$

(d) Substitute $3y$ for A and $2z$ for B in Special Product (vii). Note that $9y^2 = (3y)^2$ and $4z^2 = (2z)^2$.

$$(3y - 2z)(9y^2 + 6yz + 4z^2) = (3y)^3 - (2z)^3$$
$$= 27y^3 - 8z^3.$$

PROBLEM SET 2.4

In problems 1–10, find the products of the given monomials.

1. $(2x^2)(3x^4)$ **2.** $(3y^3)(5y^2)$ **3.** $(-5t^3)(6t^4)$

4. $(7m^2)(-8m^5)$ **5.** $(7u^2v^3)(-4u^3v^4)$ **6.** $(-3xy^7)(6x^5y)$

7. $(-3x^2yz^3)(-4xy^2z)$ **8.** $(-10p^3q^4r^5)(-2pq^2r^3)$ **9.** $(2ab)(3a^2c)(-4b^2c^3)$

10. $(3x^2yz^3)(5xyz^2)(-2yz)$

In problems 11–20, use the distributive property to find each product.

11. $x(x + 1)$ **12.** $-y(2y - 3)$ **13.** $t^2(t + 2)$

14. $x^3(2x - 4)$ **15.** $3w(2w^2 - 4)$ **16.** $5u(3u^4 + 7u^2)$

17. $-2xy^2(2x^2 - 3xy + 5y^2)$ **18.** $m^2n^3(4m^2n - 4mn + 5mn^2)$

19. $4c^2d(3c^3d - 2c^2d^2 + cd^3)$ **20.** $-3x^4z^3(-2xz^2 - 4x^2y + 5x^3 - 3)$

In problems 21–30, use the vertical scheme to find each product.

21. $(x + y)(x + y - 1)$ **22.** $(u - v)(u - v + 1)$

23. $(2t + s)(t^2 + 2ts + s^2)$ **24.** $(x - 2y)(x^2 - 3xy - y^2)$

25. $(m^2 + 3)(m^3 + 2m^2 - 3m + 4)$ **26.** $(4 - x^2)(x^4 - 2x^2 - 3x + 4)$

27. $(y^2 - 5y + 6)(y^2 + 4y + 4)$ **28.** $(w^2 - 2w + 1)(w^2 + 2w + 1)$

29. $(x^2 + 2xy + y^2)(x^3 - 3x^2y + 3xy^2 - y^3)$ **30.** $(2u^2 - uv + v^2)(u^3 - u^2v + uv^2 - v^3)$

In problems 31–50, use the method illustrated in Examples 8, 9, and 10, page 51, to find each product.

31. $(x + 1)(x + 2)$ **32.** $(w + 3)(w + 4)$ **33.** $(u - 4)(u - 5)$

34. $(z - 6)(z - 8)$ **35.** $(t + 5)(t - 2)$ **36.** $(x + 3y)(3x + y)$

37. $(y + 3)(y - 6)$ **38.** $(4t - 7)(t + 4)$ **39.** $(3x - 1)(2x + 1)$

40. $(9c + 2)(c - 3)$ **41.** $(5w + 4)(w - 1)$ **42.** $(11x - 3y)(2x + y)$

43. $(2x + y)(x + 3y)$ **44.** $(3m - 2p)(2m + 3p)$ **45.** $(6m - 5n)(4m + 3n)$

46. $(2y - 1)(y - 3)$ **47.** $(7x + 3y)(4x - 5y)$ **48.** $(12x - 5y)(3x - 2y)$

49. $(10v - 7)(5v - 8)$ **50.** $(9u - 7v)(5u - 4v)$

In problems 51–60, use Special Product (i) or (ii) to find each product.

51. $(x + 1)^2$ **52.** $(w + 5)^2$ **53.** $(2s + t)^2$ **54.** $(x + 3y)^2$

55. $(u - 3v)^2$ **56.** $(m - 2n)^2$ **57.** $(3x - 5)^2$ **58.** $(7z - 4)^2$

59. $(4y + 5z)^2$ **60.** $(9y + 8z)^2$

In problems 61–70, use Special Product (iii) to find each product.

61. $(x + y)(x - y)$ **62.** $(t - s)(t + s)$ **63.** $(w - 7)(w + 7)$

64. $(x + 11)(x - 11)$ **65.** $(2m + 9)(2m - 9)$ **66.** $(10 - 3y)(10 + 3y)$

67. $(8x - y)(8x + y)$ **68.** $(6w - 7z)(6w + 7z)$ **69.** $(5u + 6v)(5u - 6v)$

70. $(9t + 2s)(9t - 2s)$ **71.** Verify Special Product (iv).   **72.** Verify Special Product (v).

73. Verify Special Product (vi).   **74.** Verify Special Product (vii).

In problems 75–90, use Special Products (iv)–(vii) to find each product.

75. $(x + 1)^3$

76. $(u + v)^3$

77. $(x + 1)(x^2 - x + 1)$

78. $(m + n)(m^2 - mn + n^2)$

79. $(c - 2d)^3$

80. $(2x - 5y)^3$

81. $(u - 3v)(u^2 + 3uv + 9v^2)$

82. $(3x - 4y)(9x^2 + 12xy + 16y^2)$

83. $(5t + 2s)^3$

84. $(w + 7)(w^2 - 7w + 49)$

85. $(2x + 3y)(4x^2 - 6xy + 9y^2)$

86. $(4t - 3s)^3$

87. $(x^3 - 2)^3$

88. $(u^2 + v^2)^3$

89. $(u^2 - v^2)(u^4 + u^2v^2 + v^4)$

90. $(2x^2 + y^2)(4x^4 - 2x^2y^2 + y^4)$

2.5 Factoring Polynomials

In mathematics, it is often useful to represent a polynomial as a product of two or more polynomials. Each polynomial that is multiplied to form the product is called a **factor** of the product. For example, consider the product

$$a(2a - b)(a + b).$$

The factors are a, $2a - b$, and $a + b$.

We often start with a product (in its expanded form) and wish to find its factors. This is done by a process called **factoring;** and we say that the polynomial is **factorable.** A polynomial that has no factors other than itself and 1, or its negative and -1, is said to be **prime.** When a polynomial is written as a product of prime factors, we say that the polynomial is **factored completely.**

Common Factors

The distributive property,

$$P(Q + R) = PQ + PR,$$

provides a bridge between factors and products. If you write the distributive property in the "reverse" order,

$$PQ + PR = P(Q + R),$$

you obtain a general principle of common factoring. Using the example above, P, on the left side of the equation, is a common factor of PQ and PR. The factors on the right side of the equation are P and $Q + R$.

The polynomial expression $x^2 + x$ can be written as $x \cdot x + x \cdot 1$. Thus, by applying the distributive property, we can factor out the common factor x. Therefore, the polynomial is factored completely as follows:

$$x^2 + x = x(x + 1).$$

This procedure for factoring is further illustrated by the following examples.

In Examples 1 and 2, factor each polynomial by factoring out the common factor.

EXAMPLE 1   (a) $2y^3 + 4y^2$

(b) $4u^5 + 8u^3$

(c) $3x^3y^4 - 9x^2y^3 - 6xy^2$

(d) $-5x^3 - 25x$

SOLUTION   (a) The factored forms of each term are

$$2y^3 = 2 \cdot y \cdot y \cdot y \quad \text{and} \quad 4y^2 = 2 \cdot 2 \cdot y \cdot y.$$

We see that $2y^2$ is a common factor of the two terms. Therefore,

$$2y^3 + 4y^2 = (2y^2)(y) + (2y^2)(2)$$
$$= 2y^2(y + 2).$$

(b) Here $4u^3$ is a common factor of the two terms, since

$$4u^5 = (4u^3)(u^2) \quad \text{and} \quad 8u^3 = (4u^3)(2).$$

Therefore,

$$4u^5 + 8u^3 = (4u^3)(u^2) + (4u^3)(2)$$
$$= 4u^3(u^2 + 2).$$

(c) Here $3xy^2$ is a common factor of the three terms, since

$$3x^3y^4 = (3xy^2)(x^2y^2), \quad 9x^2y^3 = (3xy^2)(3xy), \quad \text{and} \quad 6xy^2 = (3xy^2)(2).$$

Therefore,

$$3x^3y^4 - 9x^2y^3 - 6xy^2 = (3xy^2)(x^2y^2) - (3xy^2)(3xy) - (3xy^2)(2)$$
$$= 3xy^2(x^2y^2 - 3xy - 2).$$

(d) $-5x^3 - 25x = -5x(x^2 + 5)$

Some algebraic expressions are written so that all the terms have a common binomial factor that can be factored out by the methods shown above. The following example will illustrate this.

EXAMPLE 2   (a) $5x(y + z) + 2(y + z)$

(b) $a(c - d) + 2(d - c)$

SOLUTION   (a) Because the binomial $y + z$ is a common factor of each term, we have
$$5x(y + z) + 2(y + z) = (5x + 2)(y + z).$$

(b) Because $d - c = -(c - d)$, we have
$$a(c - d) + 2(d - c) = a(c - d) - 2(c - d) = (a - 2)(c - d).$$

Factoring by Grouping

It is sometimes possible to factor polynomials that do not contain factors common to every term by using the method of common factors. To do this, we first group the terms of the polynomials, and then we look for common polynomial factors in each group. For example, consider the polynomial $3xm + 3ym - 2x - 2y$. We note that two terms of the expression contain a factor of x and that the other two terms contain a factor of y. Grouping these terms accordingly, we obtain
$$3xm + 3ym - 2x - 2y = (3xm - 2x) + (3ym - 2y).$$

Factoring out the common monomial x from the first group and the common monomial y from the second group, we have
$$(3xm - 2x) + (3ym - 2y) = (3m - 2)x + (3m - 2)y.$$

We factor out the common binomial $3m - 2$:
$$(3m - 2)x + (3m - 2)y = (3m - 2)(x + y).$$

Therefore,
$$3xm + 3ym - 2x - 2y = (3m - 2)(x + y).$$

EXAMPLE 3   Factor each expression by grouping the terms in a suitable way.

(a) $ac - d - c + ad$

(b) $3u + uv + 3u^2 + v$

(c) $2ax^2 + 2ay^2 - bx^2 - by^2$

(d) $3ax + 3ay + 3az - 2bx - 2by - 2bz$

(e) $pq - pr - sr + sq$

SOLUTION   (a) $ac - d - c + ad = (ac - c) + (ad - d)$
$$= c(a - 1) + d(a - 1) = (c + d)(a - 1)$$

(b) $3u + uv + 3u^2 + v = (3u + 3u^2) + (uv + v)$
$$= 3u(1 + u) + v(u + 1)$$
$$= (3u + v)(1 + u)$$

(c) $2ax^2 + 2ay^2 - bx^2 - by^2 = (2ax^2 + 2ay^2) + (-bx^2 - by^2)$
$$= 2a(x^2 + y^2) - b(x^2 + y^2)$$
$$= (2a - b)(x^2 + y^2)$$

(d) $3ax + 3ay + 3az - 2bx - 2by - 2bz$
$$= (3ax + 3ay + 3az) + (-2bx - 2by - 2bz)$$
$$= 3a(x + y + z) - 2b(x + y + z)$$
$$= (3a - 2b)(x + y + z)$$

(e) $pq - pr - sr + sq = (pq - pr) + (-sr + sq)$
$$= p(q - r) + s(-r + q)$$
$$= p(q - r) + s(q - r)$$
$$= (p + s)(q - r)$$

PROBLEM SET 2.5

In problems 1–20, factor each polynomial by factoring out the common factor.

1. $x^2 - x$ **2.** $4x^2 + 2x$ **3.** $9x^2 + 3x$

4. $10x^2 - 5x$ **5.** $4x^2 + 7xy$ **6.** $9x^3 + 3x^4$

7. $a^2b - ab^2$ **8.** $17x^3y^2 - 34x^2y$ **9.** $6p^2q + 24pq^2$

10. $12a^3b^2 + 36a^2b^3$ **11.** $6ab^2 + 30a^2b$ **12.** $5abc + 20abc^2$

13. $12x^3y - 48x^2y^2$ **14.** $4x^3 - 2x^2 + x$ **15.** $2a^3b - 8a^2b^2 - 6ab^3$

16. $4xy^2z + x^2y^2z^2 - x^3y^3$ **17.** $x^3y^2 + x^2y^3 + 2xy^4$ **18.** $3x^2y^2 + 6x^2z^2 - 9x^2$

19. $9m^2n + 18mn^2 - 27mn$ **20.** $8xy^2 + 24x^2y^3 + 4xy^3$

In problems 21–30, factor out the common binomial factor in each expression.

21. $3x(2a + b) + 5y(2a + b)$ **22.** $(2m + 3)x - (2m + 3)y$ **23.** $5x(a + b) + 9y(a + b)^2$

24. $x(y - z) - (z - y)$ **25.** $m(x - y) + (y - x)$ **26.** $(x - y) + 5(x - y)^2$

27. $7x(2a + 7b) + 14(2a + 7b)^2 + (2a + 7b)^3$ **28.** $(x + y)^3 + x(x + y)^2 + 5y(x + y)$

29. $y(xy + 2)^3 - 5x(xy + 2)^2 + 7(xy + 2)$ **30.** $x^2(a + b)^3 - 4xy(a + b)^2 + 3(a + b)$

In problems 31–45, factor each expression by grouping.

31. $ax + ay + bx + by$ **32.** $x^2a + x^2b + a + b$ **33.** $x^5 + 3x^4 + x + 3$

34. $ax^5 + b - bx^5 - a$ **35.** $yz + 2y - z - 2$ **36.** $a^2x - 1 - a^2 + x$

37. $ab^2 - b^2c - ad + cd$ **38.** $2x^2 - yz^2 - x^2y + 2z^2$ **39.** $2ax + by - 2ay - bx$

40. $x^3 + x^2 - 5x - 5$ **41.** $x^2 - ax + bx - ab$ **42.** $x^2 + ax + bx + ab$

43. $ax + bx + ay + by + a + b$ **44.** $2ax - b + 2bx - c + 2cx - a$

45. $2x^3 + x^2y - x^2 + 2xy + y^2 - y$

2.6 Factoring by Recognizing Special Products

When Special Products (i)–(vii) in Section 2.4 are read from right to left, they reveal patterns useful for factoring. We restate some of these products below for reference.

(i) **Difference between two squares**

$$A^2 - B^2 = (A - B)(A + B)$$

EXAMPLE 1   Factor each expression completely.

(a) $a^2 - 25$

(b) $16s^2 - 25t^2$

(c) $16u^2 - (v - 2w)^2$

(d) $16x^4 - y^4$

SOLUTION   (a) We write $a^2 - 25 = a^2 - 5^2$, which is the difference between two squares. Then we use the special product $A^2 - B^2 = (A - B)(A + B)$ with $A = a$ and $B = 5$. We have

$$a^2 - 25 = (a - 5)(a + 5).$$

(b) Notice that $16s^2 = (4s)^2$ and $25t^2 = (5t)^2$. Thus, the given expression is the difference between two squares. Use the special product $A^2 - B^2 = (A - B)(A + B)$ with $A = 4s$ and $B = 5t$. We have

$$16s^2 - 25t^2 = (4s)^2 - (5t)^2 = (4s - 5t)(4s + 5t).$$

(c) The expression is the difference between two squares. We have

$$\begin{aligned} 16u^2 - (v - 2w)^2 &= (4u)^2 - (v - 2w)^2 \\ &= [4u - (v - 2w)][4u + (v - 2w)] \\ &= (4u - v + 2w)(4u + v - 2w). \end{aligned}$$

(d) The expression is the difference between two squares, so that

$$\begin{aligned} 16x^4 - y^4 &= (4x^2)^2 - (y^2)^2 \\ &= (4x^2 - y^2)(4x^2 + y^2). \end{aligned}$$

However, $4x^2 - y^2$ is also the difference between two squares, so that

$$4x^2 - y^2 = (2x - y)(2x + y).$$

Thus, the original expression is completely factored as follows:

$$16x^4 - y^4 = (2x - y)(2x + y)(4x^2 + y^2).$$

Although it is usually difficult to factor third-degree polynomials, polynomials representing the *sum* or the *difference* of two cubes can be factored directly. Rewriting Special Products (vi) and (vii) from Section 2.4, we have:

(ii) **Sum of two cubes**

$$A^3 + B^3 = (A + B)(A^2 - AB + B^2)$$

(iii) **Difference of two cubes**

$$A^3 - B^3 = (A - B)(A^2 + AB + B^2)$$

EXAMPLE 2   Factor each of the following expressions.

(a) $x^3 + 27$ (b) $8u^3 - v^3$ (c) $(x + y)^3 - (z - w)^3$

SOLUTION   (a) Because $27 = 3^3$, the expression $x^3 + 27$ is a sum of two cubes. We use the special product $A^3 + B^3 = (A + B)(A^2 - AB + B^2)$ with $A = x$ and $B = 3$:

$$x^3 + 27 = x^3 + 3^3 = (x + 3)(x^2 - 3x + 9).$$

(b) Because $8u^3 = (2u)^3$, the expression $8u^3 - v^3$ is a difference of two cubes. We use the special product $A^3 - B^3 = (A - B)(A^2 + AB + B^2)$:

$$8u^3 - v^3 = (2u)^3 - v^3 = (2u - v)[(2u)^2 + (2u)v + v^2]$$
$$= (2u - v)(4u^2 + 2uv + v^2).$$

(c) $(x + y)^3 - (z - w)^3$
$$= [(x + y) - (z - w)][(x + y)^2 + (x + y)(z - w) + (z - w)^2]$$
$$= (x + y - z + w)[(x + y)^2 + (x + y)(z - w) + (z - w)^2]$$

Factoring by Combined Methods

Until now we have discussed factoring by recognizing common factors, factoring by grouping, and factoring by special products. Factoring some polynomials may require the application of more than one of these methods. To accomplish this, we suggest the following steps:

Step 1. Factor out common factors (if there are any).

Step 2. Examine the remaining polynomial factors to see if each is prime or if a special product can be applied directly.

Step 3. If step 2 fails, determine if factoring by grouping can be applied.

In Examples 3–8, factor each of the following expressions.

EXAMPLE 3 $75b^2 - 243$

SOLUTION   First we factor out 3 so that

$$75b^2 - 243 = 3(25b^2 - 81).$$

The expression $25b^2 - 81$ is a difference between two squares:

$$25b^2 - 81 = (5b)^2 - 9^2$$
$$= (5b - 9)(5b + 9).$$

The original expression is factored completely as follows:

$$75b^2 - 243 = 3(25b^2 - 81)$$
$$= 3(5b - 9)(5b + 9).$$

EXAMPLE 4 $5t^6 + 40$

SOLUTION   First we factor out 5 so that

$$5t^6 + 40 = 5(t^6 + 8).$$

We see that $t^6 + 8 = (t^2)^3 + 2^3$. Thus, the expression is the sum of two cubes. Therefore,

$$t^6 + 8 = (t^2 + 2)[(t^2)^2 - 2t^2 + 2^2]$$
$$= (t^2 + 2)(t^4 - 2t^2 + 4).$$

The original expression is factored completely as follows:

$$5t^6 + 40 = 5(t^6 + 8)$$
$$= 5(t^2 + 2)(t^4 - 2t^2 + 4).$$

EXAMPLE 5 $64u^6 - 1$

SOLUTION   The expression $64u^6 - 1$ can be factored as the difference of two squares:

$$64u^6 - 1 = (8u^3)^2 - 1^2$$
$$= (8u^3 - 1)(8u^3 + 1).$$

However,

$$8u^3 - 1 = (2u)^3 - 1^3$$
$$= (2u - 1)(4u^2 + 2u + 1)$$

and

$$8u^3 + 1 = (2u)^3 + 1^3$$
$$= (2u + 1)(4u^2 - 2u + 1).$$

The complete factorization is

$$64u^6 - 1 = (2u - 1)(2u + 1)(4u^2 + 2u + 1)(4u^2 - 2u + 1).$$

EXAMPLE 6 $x^2 + 4x + 4$

SOLUTION   We can write Special Product (i), Section 2.4, page 52, in reverse:

$$A^2 + 2AB + B^2 = (A + B)^2.$$

If we set $A = x$ and $B = 2$, we have

$$x^2 + 4x + 4 = x^2 + 2(x)(2) + 2^2 = (x + 2)^2.$$

EXAMPLE 7 $x^2 - 2xy + y^2 - a^2 - 2ab - b^2$

SOLUTION   Steps 1 and 2 fail. Using step 3, we have

$$x^2 - 2xy + y^2 - a^2 - 2ab - b^2 = (x^2 - 2xy + y^2) + (-a^2 - 2ab - b^2)$$
$$= (x^2 - 2xy + y^2) - (a^2 + 2ab + b^2).$$

We apply Special Products (i) and (ii) of Section 2.4:

$$(x^2 - 2xy + y^2) - (a^2 + 2ab + b^2) = (x - y)^2 - (a + b)^2.$$

Because this latter form is the difference of two squares, the complete factorization is

$$x^2 - 2xy + y^2 - a^2 - 2ab - b^2 = (x - y)^2 - (a + b)^2$$
$$= [(x - y) - (a + b)][(x - y) + (a + b)]$$
$$= (x - y - a - b)(x - y + a + b).$$

EXAMPLE 8 $x^4 + 2x^2y^2 + 9y^4$

SOLUTION   If the middle term were $6x^2y^2$ rather than $2x^2y^2$, we could factor the expression as

$$x^4 + 6x^2y^2 + 9y^4 = (x^2 + 3y^2)^2.   \text{(Special Product (i), Section 2.4)}$$

But the middle term can be changed to $6x^2y^2$ by adding $4x^2y^2$ to $2x^2y^2$ (already there) and then subtracting $4x^2y^2$ at the end of the expression. (This will *not* change the value of the expression.) Thus,

$$x^4 + 2x^2y^2 + 9y^4 = (x^4 + 6x^2y^2 + 9y^4) - 4x^2y^2$$
$$= (x^2 + 3y^2)^2 - 4x^2y^2$$
$$= (x^2 + 3y^2)^2 - (2xy)^2$$
$$= [(x^2 + 3y^2) - 2xy][(x^2 + 3y^2) + 2xy]$$
$$= (x^2 - 2xy + 3y^2)(x^2 + 2xy + 3y^2).$$

PROBLEM SET 2.6

In problems 1–20, use the difference between two squares to factor each expression completely.

1. $x^2 - 4$

2. $100 - x^2$

3. $1 - 9y^2$

4. $25 - 4a^2$

5. $36 - 25t^2$

6. $16x^2y^2 - 9$

7. $16u^2 - 25v^2$

8. $25m^2 - 49n^2$

9. $a^2b^2 - c^2$

10. $49w^2 - 81z^2$

11. $(a - b)^2 - 100c^2$

12. $144p^2 - (q - 3)^2$

13. $81x^4 - 1$

14. $256x^4 - y^4$

15. $u^8 - v^8$

16. $625w^4 - 81z^4$

17. $(x + y)^2 - (a - b)^2$

18. $(3x + 2y)^2 - 25z^2$

19. $t^4 - 81(r + s)^4$

20. $16(x + y)^4 - 81(w - z)^4$

In problems 21–38, use the sum or difference of two cubes to factor each expression completely.

21. $x^3 + 1$

22. $y^3 + 125$

23. $64 - t^3$

24. $27m^3 - n^3$

25. $27w^3 + z^3$

26. $x^3y^3 - 64$

27. $8x^3 - 27y^3$

28. $125t^3 - 216$

29. $w^3 - 8y^3z^3$

30. $64a^3 + 27b^3$

31. $(x + 2)^3 - y^3$

32. $(c + 3)^3 + d^3$

33. $(y + 1)^3 + (w + 2)^3$

34. $64 - (t - 1)^3$

35. $w^6 + 8z^6$

36. $u^6 - 27$

37. $x^9 - 1$

38. $v^9 + 512$

In problems 39–62, use common factors, special products, or grouping to factor each expression.

39. $8x^3 - 2xy^2$

40. $3u^3v - 27uv^3$

41. $64y - 4y^3$

42. $36s^2t^2 - 4s^4$

43. $3u^4v - 24uv$

44. $216x^4y^3 + 27xy^3$

45. $7x^7y + 7xy^7$

46. $3t^8 + 81t^2$

47. $t^6 - 1$

48. $64 - x^6$

49. $2u^7 - 128u$

50. $64w^6 - 729$

51. $y^2 + 8y + 16$

52. $t^2 - 12t + 36$

53. $9u^2 - 42uv + 49v^2$

54. $25x^2 - 40xy + 16y^2$

55. $x^2 + y^2 - z^2 - 9 + 2xy - 6z$

56. $u^2 + 4uv + 4v^2 - 4a^2 + 4ab - b^2$

57. $w^2 - y^2 + 2w - 2yz + 1 - z^2$

58. $t^2 - 10t - 16s^2 + 24sr + 25 - 9r^2$

59. $x^4 + x^2y^2 + y^4$

60. $9w^4 + 2w^2z^2 + z^4$

61. $4m^4 + n^4$

62. $25x^4 + 4x^2y^2 + 4y^4$

2.7 Factoring Trinomials of the Form $ax^2 + bx + c$

Some trinomial expressions of the form

$$ax^2 + bx + c,$$

in which a, b, and c are integers and $a \neq 0$, can be factored into a product of two binomials.

Recall from Section 2.4 that the product of two binomials can be a trinomial. Review the process of multiplying two binomials. This should suggest an idea for factoring. For instance,

$$(3x + 4)(2x + 5) = 6x^2 + 15x + 8x + 20$$
$$= 6x^2 + (15 + 8)x + 20$$
$$= 6x^2 + 23x + 20.$$

Notice that the first term in the trinomial is the product of the first terms in each binomial. The last term in the trinomial is the product of the last terms in each binomial. The middle term in the trinomial is found by adding the product of the outside terms to the product of the inside terms.

The relationship between the coefficients of the trinomial and the coefficients of the factors is illustrated in the diagrams below. The coefficients 6 and 20 are obtained as follows:

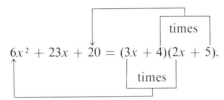

$$6x^2 + 23x + 20 = (3x + 4)(2x + 5).$$

The coefficient of the middle term in the trinomial, 23, is obtained as follows:

$$6x^2 + (15 + 8)x + 20 = (3x + 4)\quad(2x + 5).$$

These diagrams illustrate how to factor a trinomial of the form $ax^2 + bx + c$.

We begin by writing

$$ax^2 + bx + c = (\underline{\quad}x + \underline{\quad})(\underline{\quad}x + \underline{\quad}).$$

Next, we fill in the blanks with the numbers so that:

1. The product of the first term of each binomial is ax^2.

2. The sum of the product of the outside terms and the product of the inside terms is bx.

3. The product of the last terms is c.

We try all possible choices of factors of a and c until we find a combination that gives us the desired middle term. If all the coefficients of the trinominal are positive, it is only necessary to try combinations of positive integers. If some of the coefficients of the trinomial are negative, we will have to try combinations that include negative integers. For example, to factor the trinomial $x^2 + 7x + 10$, we only try combinations of positive integers. Since the product of the first terms of the binomials is x^2, we have 1 as the coefficient of each first term, so that

$$x^2 + 7x + 10 = (x + \underline{\quad})(x + \underline{\quad}).$$

Using only positive integer factors of 10, we have

$$10 = 10 \cdot 1 \quad \text{or} \quad 10 = 2 \cdot 5.$$

That is, there are only two possible ways to fill in the remaining blanks:

$$(x + 10)(x + 1) \quad \text{or} \quad (x + 5)(x + 2).$$

We test the two factorizations to see whether either one produces the correct middle term, $7x$:

Possible Combinations	Products
$(x + 10)(x + 1)$	$x^2 + 11x + 10$
$(x + 5)(x + 2)$	$x^2 + 7x + 10$

Note that the combination $(x + 5)(x + 2)$ gives $7x$ as the middle term. Therefore, the correct factoring is

$$x^2 + 7x + 10 = (x + 5)(x + 2).$$

To factor $3x^2 + x - 2$, first factor the first term, $3x^2$, as $3x \cdot x$. Because the sign of the last term is negative, we know that the signs of the second terms in the binomials must be opposites of one another. (If both were positive or both were negative, the sign of the last term in the trinomial would be positive.) Therefore, we have

$$3x^2 + x - 2 = (3x + \underline{\quad})(x - \underline{\quad}) \quad \text{or} \quad (3x - \underline{\quad})(x + \underline{\quad}).$$

Next, factor the last term, 2. (Ignore the negative sign now because we've already accounted for it by setting up the signs in the binomials.) Thus,

$$2 = 1 \cdot 2 \quad \text{or} \quad 2 = 2 \cdot 1.$$

Note that we might have to try both $1 \cdot 2$ and $2 \cdot 1$, since the order of the factors makes a difference when we substitute the factors in the blanks. For example, $(3x + 1)(x - 2)$ has a different product than $(3x + 2)(x - 1)$. Try the two different factorizations of 2 in the blanks. We will see that the possible combinations of binomial factors are:

Possible Combinations	Products
$(3x + 1)(x - 2)$	$3x^2 - 5x - 2$
$(3x - 1)(x + 2)$	$3x^2 + 5x - 2$
$(3x + 2)(x - 1)$	$3x^2 - x - 2$
$(3x - 2)(x + 1)$	$3x^2 + x - 2$

The combination that gives the correct middle term, x, is $(3x - 2)(x + 1)$. Therefore, the correct factorization is

$$3x^2 + x - 2 = (3x - 2)(x + 1).$$

In Examples 1–6, factor each trinomial.

EXAMPLE 1 $x^2 + 6x + 8$

SOLUTION   Since all the coefficients are positive, we only consider combinations of positive integers. The first term, x^2, is factored as $x \cdot x$, so the trinomial can be expressed in the form

$$x^2 + 6x + 8 = (x + \underline{\quad})(x + \underline{\quad}).$$

The last term, 8, is factored as

$$8 = 8 \cdot 1 \qquad \text{or} \qquad 8 = 2 \cdot 4.$$

Trying different possible combinations of the factors of 8 in the blanks, we see that $(x + 4)(x + 2)$ produces the correct middle term, $6x$:

$$(x + 4)(x + 2) \qquad\qquad 6x \qquad \text{(middle term)}$$
$$x \cdot 2 + 4 \cdot x = 2x + 4x$$

Therefore, the factors of the trinomial are $x + 4$ and $x + 2$, and

$$x^2 + 6x + 8 = (x + 4)(x + 2).$$

EXAMPLE 2 $y^2 - 9y + 20$

SOLUTION   Factoring the first term, we have $y^2 = y \cdot y$. Since the sign of the last term is positive, the signs of the second terms in the binomials must either both be positive or both be negative. The sign of the middle term of the polynomial is negative, which tells us that the signs in the binomials must be negative. So we have

$$y^2 - 9y + 20 = (y - \underline{\quad})(y - \underline{\quad}).$$

Next we factor the last term, 20, in the following ways:

$$20 = 20 \cdot 1, \qquad 20 = 10 \cdot 2, \qquad \text{or} \qquad 20 = 5 \cdot 4.$$

Substituting different possible combinations of the factors of 20, we see that $(y - 5)(y - 4)$ produces the correct middle term, $-9y$:

$$(y - 5)(y - 4) \qquad\qquad\qquad -9y \qquad \text{(middle term)}$$

$$y \cdot (-4) + (-5) \cdot y = -4y - 5y$$

Therefore, $y - 5$ and $y - 4$ are the factors of the trinomial, and

$$y^2 - 9y + 20 = (y - 5)(y - 4).$$

EXAMPLE 3 $2r^2 + 5r - 3$

SOLUTION   We begin by factoring the first term, $2r^2$, as $2r \cdot r$. Since the sign of the last term is negative, the signs of the second terms in the binomials are opposites of one another, so we have

$$2r^2 + 5r - 3 = (2r + \underline{\hspace{0.5cm}})(r - \underline{\hspace{0.5cm}}) \qquad \text{or} \qquad (2r - \underline{\hspace{0.5cm}})(r + \underline{\hspace{0.5cm}}).$$

Next we factor the last term, 3. (Again, ignore the negative sign because we've already accounted for it by setting up the signs in the binomial.) We have

$$3 = 3 \cdot 1.$$

Substituting the possible combinations of the factors of 3 in the blanks, we see that $(2r - 1)(r + 3)$ gives the correct middle term, $5r$:

$$(2r - 1)(r + 3) \qquad\qquad\qquad 5r \qquad \text{(middle term)}$$

$$2r \cdot 3 + (-1) \cdot r = 6r - r$$

Therefore,

$$2r^2 + 5r - 3 = (2r - 1)(r + 3).$$

EXAMPLE 4 $6m^2 - 11m - 10$

SOLUTION   Factoring the first term, $6m^2$, we have

$$6m^2 = 6m \cdot m \qquad \text{or} \qquad 6m^2 = 3m \cdot 2m.$$

Because the sign of the last term is negative, the signs of the second terms of the binomials are opposites of one another, so that

$$6m^2 - 11m - 10 = (6m + \underline{\hspace{0.5cm}})(m - \underline{\hspace{0.5cm}}) \qquad \text{or} \qquad (6m - \underline{\hspace{0.5cm}})(m + \underline{\hspace{0.5cm}}),$$

or

$$6m^2 - 11m - 10 = (3m + \underline{\quad})(2m - \underline{\quad}) \quad \text{or} \quad (3m - \underline{\quad})(2m + \underline{\quad}).$$

Next we factor the last term, 10, as

$$10 = 10 \cdot 1 \quad \text{or} \quad 10 = 2 \cdot 5.$$

We try different combinations of the factors of 10, and we find that $(3m + 2)$ $(2m - 5)$ gives the desired middle term, $-11m$:

$(3m + 2)(2m - 5)$ $-11m$ (middle term)

$$3m \cdot (-5) + 2 \cdot 2m = -15m + 4m$$

Therefore,

$$6m^2 - 11m - 10 = (3m + 2)(2m - 5).$$

EXAMPLE **5** $7u^2 - 37uv + 10v^2$

SOLUTION   We start by factoring the first term, $7u^2$, as $7u \cdot u$. Because the sign of the last term is positive, the signs of the second terms in the binomials must either both be positive or both be negative. The sign of the middle term in the polynomial is negative. Therefore, the signs in the binomials must be negative. Thus, we have

$$7u^2 - 37uv + 10v^2 = (7u - \underline{\quad})(u - \underline{\quad}).$$

Note that the middle term of the trinomial includes the factors uv and that the last term includes v^2. If these variables appear in the trinomial, the binomials must both have a v in their second terms. Thus, we can factor the last term, $10v^2$, in the following ways:

$$10v^2 = 10v \cdot v \quad \text{or} \quad 10v^2 = 2v \cdot 5v.$$

We try different possible combinations of the factors of $10v^2$ in the blanks, and we find that $(7u - 2v)(u - 5v)$ gives the desired middle term, $-37uv$:

$(7u - 2v)(u - 5v)$ $-37uv$ (middle term)

$$7u \cdot (-5v) + (-2v) \cdot u = -35uv - 2uv$$

Therefore,

$$7u^2 - 37uv + 10v^2 = (7u - 2v)(u - 5v).$$

  The trinomials we are now factoring are more complicated than the trinomials we started with. However, it is still important that we begin by looking for common factors in the terms. The following example makes this clear.

EXAMPLE **6**   Factor the trinomial $60a^3b + 25a^2b^2 - 15ab^3$.

SOLUTION   First we take out the common factor $5ab$:

$$60a^3b + 25a^2b^2 - 15ab^3 = 5ab(12a^2 + 5ab - 3b^2).$$

Next, we try to factor the trinomial $12a^2 + 5ab - 3b^2$. We start by factoring the first term, $12a^2$, as $12a \cdot a$, $6a \cdot 2a$, or $4a \cdot 3a$. Because the sign of the last term is negative, the signs of the second terms in the binomials are opposites of one another, so that

$$12a^2 + 5ab - 3b^2 = (12a + \underline{\quad})(a - \underline{\quad}) \quad \text{or} \quad (12a - \underline{\quad})(a + \underline{\quad}),$$

or

$$12a^2 + 5ab - 3b^2 = (6a + \underline{\quad})(2a - \underline{\quad}) \quad \text{or} \quad (6a - \underline{\quad})(2a + \underline{\quad}),$$

or

$$12a^2 + 5ab - 3b^2 = (4a + \underline{\quad})(3a - \underline{\quad}) \quad \text{or} \quad (4a - \underline{\quad})(3a + \underline{\quad}).$$

Factoring the last term, $3b^2$, we have

$$3b^2 = 3b \cdot b.$$

Trying the different possible combinations of the factors of $3b^2$, we find that $(4a + 3b)(3a - b)$ gives the correct middle term, $5ab$:

$$
(4a + 3b)(3a - b) \qquad\qquad 5ab \qquad \text{(middle term)}
$$
$$
4a \cdot (-b) + 3b \cdot 3a
$$

Therefore,

$$12a^2 + 5ab - 3b^2 = (4a + 3b)(3a - b),$$

and the complete factorization of the original trinomial is

$$60a^3b + 25a^2b^2 - 15ab^3 = 5ab(4a + 3b)(3a - b).$$

PROBLEM SET 2.7

In problems 1–40, factor each trinomial completely.

1. $x^2 + 4x + 3$ **2.** $y^2 + 5y + 6$
3. $t^2 - 3t + 2$ **4.** $x^2 - 3x - 4$
5. $y^2 + 15y + 36$ **6.** $z^2 + 3z - 10$
7. $x^2 - 2x - 15$ **8.** $a^2 - 3ab - 28b^2$
9. $u^2 - 16u + 63$ **10.** $x^2 - 3x - 40$
11. $z^2 + 11zw + 30w^2$ **12.** $w^2 - 9wz - 10z^2$

13. $x^2 - 7x - 18$

14. $x^2 - 17x + 30$

15. $m^2 + 2mn - 120n^2$

16. $m^2 - 13mn - 30n^2$

17. $12 - x^2 - 4x$

18. $40 - 3x - x^2$

19. $-5t + 36 - t^2$

20. $16 - y^2 - 6y$

21. $2w^2 + 7w + 3$

22. $2w^2 + w - 6$

23. $3x^2 + 5x - 2$

24. $2y^2 + 9y - 5$

25. $5y^2 - 11y + 2$

26. $4x^2 - 35xy - 9y^2$

27. $3c^2 + 7cd + 2d^2$

28. $10t^2 - 19t + 6$

29. $6x^2 + 13x + 6$

30. $6y^2 - y - 7$

31. $6z^2 + 5zy - 6y^2$

32. $6x^2 - 7xy - 5y^2$

33. $12v^2 + 17v - 5$

34. $6c^2 - 7cd - 3d^2$

35. $56x^2 - 83x + 30$

36. $42z^2 + z - 30$

37. $12 - 2w^2 - 5w$

38. $18x^2 + 101x + 90$

39. $6rs + 5r^2 - 8s^2$

40. $24x^2 - 67xy + 8y^2$

In problems 41–50, use common factors and the factoring of trinomials to factor the polynomials.

41. $5x^3 - 55x^2 + 140x$

42. $x^2yz^2 - xyz^2 - 12yz^2$

43. $128st^3 - 32s^2t^2 + 2s^3t$

44. $bx^2c + 7bcx + 12bc$

45. $x^2y^2 + 10xy^2 + 21y^2$

46. $a^2x^2z^2 + 5a^2xz^2 - 14a^2z^2$

47. $4m^2n^2 + 24m^2n - 28m^2$

48. $7hkx^2 + 21hkx + 14hk$

49. $wx^2y - 9wxy + 14wy$

50. $pq^2x^2y - 2pq^2xy - 15pq^2y$

2.8 Division of Polynomials

In Section 2.3, we introduced the property of exponents,

$$\frac{a^m}{a^n} = a^{m-n}.$$

This property is used in the division of monomials. For example, to divide $4x^5y^3$ by $2x^2y$, we have

$$\frac{4x^5y^3}{2x^2y} = 2x^{5-2}y^{3-1} = 2x^3y^2.$$

Similarly, we divide x^3yz^2 by xyz^5 as follows:

$$\frac{x^3yz^2}{xyz^5} = (x^{3-1})(1)\left(\frac{1}{z^{5-2}}\right) = \frac{x^2}{z^3}.$$

To divide a polynomial by a monomial, we divide each term of the polynomial by the monomial. For instance, to divide $x^3 + 2x^2 + x$ by x, we write

$$\frac{x^3 + 2x^2 + x}{x} = \frac{x^3}{x} + \frac{2x^2}{x} + \frac{x}{x}$$

$$= x^2 + 2x + 1.$$

The following examples illustrate this procedure.

EXAMPLE 1   Divide $5x^3 + 3x^2 + 7x$ by x and simplify.

SOLUTION   We divide each term of the polynomial by the monomial:

$$\frac{5x^3 + 3x^2 + 7x}{x} = \frac{5x^3}{x} + \frac{3x^2}{x} + \frac{7x}{x}$$

$$= 5x^2 + 3x + 7.$$

EXAMPLE 2   Divide $x^3 + 2x^2 - 3x + 7$ by x^2 and simplify.

SOLUTION   We divide each term of the polynomial by the monomial:

$$\frac{x^3 + 2x^2 - 3x + 7}{x^2} = \frac{x^3}{x^2} + \frac{2x^2}{x^2} - \frac{3x}{x^2} + \frac{7}{x^2}$$

$$= x + 2 - \frac{3}{x} + \frac{7}{x^2}.$$

EXAMPLE 3   Divide $-4u^5v^6 - 12u^4v^5 + 8u^3v^4$ by $2uv^3$ and simplify.

SOLUTION
$$\frac{-4u^5v^6 - 12u^4v^5 + 8u^3v^4}{2uv^3} = \frac{-4u^5v^6}{2uv^3} - \frac{12u^4v^5}{2uv^3} + \frac{8u^3v^4}{2uv^3}$$

$$= -2u^4v^3 - 6u^3v^2 + 4u^2v.$$

Warning: An *error* often made when dividing a polynomial by a monomial is to divide only one term of the polynomial by the monomial. Consider the following example:

$$\frac{8x^3 + 2x}{x} \overset{?}{=} 8x^2 + 2x.$$

This is *not* correct. The correct method is:

$$\frac{8x^3 + 2x}{x} = \frac{8x^3}{x} + \frac{2x}{x}$$

$$= 8x^2 + 2.$$

To divide one polynomial by another polynomial, we use a method similar to the "long-division" method used in arithmetic; for example, $6{,}741 \div 21$. We usually do this long division in the following way:

$$
\begin{array}{r}
321 \longleftarrow \text{quotient} \\
\text{divisor} \longrightarrow 21 \overline{)6741} \longleftarrow \text{dividend} \\
\underline{63} \qquad (= 21 \cdot 3) \\
44 \\
\underline{42} \qquad (= 21 \cdot 2) \\
21 \\
\underline{21} \qquad (= 21 \cdot 1) \\
0 \longleftarrow \text{remainder}
\end{array}
$$

The result of this calculation can be expressed as $6{,}741 = (21)(321) + 0$, that is,

---

dividend = (divisor)(quotient) + remainder.

We can also perform this division by changing the dividend 6,741 and the divisor 21 to their expanded forms first:

$$21 = 2 \cdot 10 + 1$$

and

$$6{,}741 = 6 \cdot 10^3 + 7 \cdot 10^2 + 4 \cdot 10 + 1$$

and then dividing:

$$
\begin{array}{r}
3 \cdot 10^2 + 2 \cdot 10 + 1 \longleftarrow \text{quotient} \\
\text{divisor} \longrightarrow 2 \cdot 10 + 1 \overline{)6 \cdot 10^3 + 7 \cdot 10^2 + 4 \cdot 10 + 1} \longleftarrow \text{dividend} \\
\underline{6 \cdot 10^3 + 3 \cdot 10^2} \qquad [= (2 \cdot 10 + 1)(3 \cdot 10^2)] \\
4 \cdot 10^2 + 4 \cdot 10 \\
\underline{4 \cdot 10^2 + 2 \cdot 10} \qquad [= (2 \cdot 10 + 1)(2 \cdot 10)] \\
2 \cdot 10 + 1 \\
\underline{2 \cdot 10 + 1} \qquad [= (2 \cdot 10 + 1)(1)] \\
0 \longleftarrow \text{remainder}
\end{array}
$$

Note that the expanded forms of the numbers in this example are polynomials. Each term of the polynomials has a known base of 10.

Let us change this problem slightly by changing base 10 to base x. We now have the long division of $6x^3 + 7x^2 + 4x + 1$ by $2x + 1$:

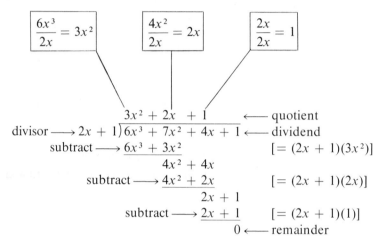

$$\frac{6x^3}{2x} = 3x^2 \qquad \frac{4x^2}{2x} = 2x \qquad \frac{2x}{2x} = 1$$

$$
\begin{array}{r}
3x^2 + 2x + 1 \quad\longleftarrow \text{quotient} \\
\text{divisor} \longrightarrow 2x + 1\overline{)6x^3 + 7x^2 + 4x + 1} \quad\longleftarrow \text{dividend} \\
\text{subtract} \longrightarrow \underline{6x^3 + 3x^2} \qquad\qquad\qquad [= (2x + 1)(3x^2)] \\
4x^2 + 4x \\
\text{subtract} \longrightarrow \underline{4x^2 + 2x} \qquad\qquad [= (2x + 1)(2x)] \\
2x + 1 \\
\text{subtract} \longrightarrow \underline{2x + 1} \qquad [= (2x + 1)(1)] \\
0 \quad\longleftarrow \text{remainder}
\end{array}
$$

This example illustrates the following systematic step-by-step procedure for dividing one polynomial by another:

Procedure for Long Division

Step 1. Arrange both polynomials in descending powers of one variable, and write the missing terms of the dividend with zero coefficients.

Step 2. Find the first term of the quotient by dividing the first term of the dividend by the first term of the divisor.

Step 3. Multiply the quotient term obtained in step 2 by the entire divisor.

Step 4. Subtract the product obtained in step 3 from the dividend; and bring down the next term of the original dividend to form the new dividend.

Step 5. Repeat the procedure in steps 2, 3, and 4 for the new dividend; keep repeating the steps until the degree of the remainder is less than the degree of the divisor.

Step 6. Check the calculation: does

$$\text{(divisor)(quotient)} + \text{remainder} = \text{dividend?}$$

EXAMPLE 4   Divide $3x^2 + 3x^3 + x^4 - 2x - 3$ by $x + 1$.

SOLUTION   Follow the procedure given:

Step 1 Arrange both polynomials in descending powers of x, so that

$$\text{divisor} \longrightarrow x + 1 \overline{\smash{)}x^4 + 3x^3 + 3x^2 - 2x - 3} \longleftarrow \text{dividend}$$

Step 2 Divide x^4, the first term of the dividend, by x, the first term of the divisor, to obtain

$$\frac{x^4}{x} = x^3.$$

Step 3 Multiply x^3, the first quotient term, by $x + 1$, the divisor, to get

$$x^3(x + 1) = x^4 + x^3.$$

Step 4 Put the product $x^4 + x^3$ under the dividend. Subtract and bring down the next term to obtain a new dividend, so that

$$
\begin{array}{r}
x^3 \\
x + 1 \overline{\smash{)}x^4 + 3x^3 + 3x^2 - 2x - 3} \\
\text{subtract} \longrightarrow \underline{x^4 + x^3} \\
2x^3 + 3x^2 \longleftarrow \text{new dividend}
\end{array}
$$

Step 5 Repeat the procedure in steps 2, 3, and 4 for the new dividend to obtain

$$\boxed{\frac{x^4}{x} = x^3} \qquad \boxed{\frac{2x^3}{x} = 2x^2} \qquad \boxed{\frac{x^2}{x} = x} \qquad \boxed{\frac{-3x}{x} = -3}$$

$$
\begin{array}{r}
x^3 + 2x^2 + x - 3 \longleftarrow \text{quotient} \\
x + 1 \overline{\smash{)}x^4 + 3x^3 + 3x^2 - 2x - 3} \\
\text{subtract} \longrightarrow \underline{x^4 + x^3} \\
2x^3 + 3x^2 \\
\text{subtract} \longrightarrow \underline{2x^3 + 2x^2} \qquad [= 2x^2(x + 1)] \\
x^2 - 2x \\
\text{subtract} \longrightarrow \underline{x^2 + x} \qquad [= x(x + 1)] \\
-3x - 3 \\
\text{subtract} \longrightarrow \underline{-3x - 3} \qquad [= -3(x + 1)] \\
0 \longleftarrow \text{remainder}
\end{array}
$$

Therefore, the quotient is $x^3 + 2x^2 + x - 3$, and the remainder is 0.

Step 6 To check, we calculate

$$
\begin{aligned}
\text{(divisor)(quotient)} + \text{remainder} &= (x + 1)(x^3 + 2x^2 + x - 3) + 0 \\
&= x^4 + 3x^3 + 3x^2 - 2x - 3 \\
&= \text{dividend.}
\end{aligned}
$$

EXAMPLE 5   Divide $2t^4 + 5 - 3t^3 - 3t$ by $2t - 1$.

SOLUTION   First arrange both polynomials in descending powers of t, and write the missing term of the dividend with a zero coefficient to obtain:

$$\text{divisor} \longrightarrow 2t - 1 \overline{)2t^4 - 3t^3 + 0t^2 - 3t + 5} \longleftarrow \text{dividend}$$

The remaining steps are shown as follows:

$$\boxed{\frac{2t^4}{2t} = t^3} \qquad \boxed{\frac{4t^3}{2t} = 2t^2} \qquad \boxed{\frac{2t^2}{2t} = t} \qquad \boxed{\frac{-2t}{2t} = -1}$$

$$
\begin{array}{r}
t^3 + 2t^2 + t - 1 \quad \longleftarrow \text{quotient} \\
2t - 1 \overline{)2t^4 + 3t^3 + 0t^2 - 3t + 5} \\
\text{subtract} \longrightarrow \underline{2t^4 - t^3} \\
4t^3 + 0t^2 \\
\text{subtract} \longrightarrow \underline{4t^3 - 2t^2} \\
2t^2 - 3t \\
\text{subtract} \longrightarrow \underline{2t^2 - t} \\
-2t + 5 \\
\text{subtract} \longrightarrow \underline{-2t + 1} \\
4 \longleftarrow \text{remainder}
\end{array}
$$

Therefore, the quotient is $t^3 + 2t^2 + t - 1$, and the remainder is 4. To check, we calculate

$$
\begin{aligned}
(\text{divisor})(\text{quotient}) + \text{remainder} &= (2t - 1)(t^3 + 2t^2 + t - 1) + 4 \\
&= (2t^4 + 3t^3 - 3t + 1) + 4 \\
&= 2t^4 + 3t^3 - 3t + 5 \\
&= \text{dividend.}
\end{aligned}
$$

EXAMPLE 6   Divide $-3w^3 + 2w^4 + 5w^2 + 2w + 7$ by $-w + w^2 + 1$.

SOLUTION   First we arrange both polynomials in descending powers of w. Then we follow the procedures in the examples above to obtain

$$\boxed{\frac{2w^4}{w^2} = 2w^2} \qquad \boxed{\frac{-w^3}{w^2} = -w} \qquad \boxed{\frac{2w^2}{w^2} = 2}$$

$$
\begin{array}{r}
2w^2 - w + 2 \quad \longleftarrow \text{quotient} \\
w^2 - w + 1 \overline{)2w^4 - 3w^3 + 5w^2 + 2w + 7} \\
\text{subtract} \longrightarrow \underline{2w^4 - 2w^3 + 2w^2} \\
-w^3 + 3w^2 + 2w \\
\text{subtract} \longrightarrow \underline{-w^3 + w^2 - w} \\
2w^2 + 3w + 7 \\
\text{subtract} \longrightarrow \underline{2w^2 - 2w + 2} \\
5w + 5 \longleftarrow \text{remainder}
\end{array}
$$

Therefore, the quotient is $2w^2 - w + 2$, and the remainder is $5w + 5$. To check, we calculate

$$\begin{aligned}
\text{(divisor)(quotient)} + \text{remainder} &= (w^2 - w + 1)(2w^2 - w + 2) + 5w + 5 \\
&= (2w^4 - 3w^3 + 5w^2 - 3w + 2) + 5w + 5 \\
&= 2w^4 - 3w^3 + 5w^2 + 2w + 7 \\
&= \text{dividend.}
\end{aligned}$$

We can apply this method to the division of polynomials involving more than one variable. First we arrange the dividend and the divisor in descending powers of one of the variables, and then we divide as illustrated above.

EXAMPLE 7   Divide $x^4 - y^4 + 3xy^3 - 3x^3y$ by $x + y$.

SOLUTION   After arranging the dividend and the divisor in descending powers of x, we have

$$(x^4 - 3x^3y + 0x^2y^2 + 3xy^3 - y^4) \div (x + y).$$

Again, we follow the outlined procedure:

$$\boxed{\dfrac{x^4}{x} = x^3} \qquad \boxed{\dfrac{-4x^3y}{x} = -4x^2y} \qquad \boxed{\dfrac{4x^2y^2}{x} = 4xy^2} \qquad \boxed{\dfrac{-xy^3}{x} = -y^3}$$

$$
\begin{array}{r}
x^3 - 4x^2y + 4xy^2 - y^3 \\
x + y \overline{)\,x^4 - 3x^3y + 0x^2y^2 + 3xy^3 - y^4} \\
\text{subtract} \longrightarrow \underline{x^4 + x^3y} \\
-4x^3y + 0x^2y^2 \\
\text{subtract} \longrightarrow \underline{-4x^3y - 4x^2y^2} \\
4x^2y^2 + 3xy^3 \\
\text{subtract} \longrightarrow \underline{4x^2y^2 + 4xy^3} \\
-xy^3 - y^4 \\
\text{subtract} \longrightarrow \underline{-xy^3 - y^4} \\
0
\end{array}
$$

Therefore, the quotient is

$$x^3 - 4x^2y + 4xy^2 - y^3,$$

and the remainder is

$$0.$$

To check, we calculate

$$\begin{aligned}
\text{(divisor)(quotient)} + \text{remainder} &= (x + y)(x^3 - 4x^2y + 4xy^2 - y^3) + 0 \\
&= x^4 - 3x^3y + 3xy^3 - y^4 \\
&= \text{dividend.}
\end{aligned}$$

PROBLEM SET 2.8

In problems 1–12, divide as indicated and simplify.

1. $6x^5$ by $2x^2$

2. $10y^7$ by $4y^3$

3. $12x^6y^7$ by $-4x^4y^9$

4. $-24t^5s^3$ by $-8t^2s$

5. $14u^3vw^4$ by $7u^5vw^2$

6. $30xy^3z^5$ by $-15x^3y^3z$

7. $9mn^2 - 6m^2$ by $3m$

8. $15u^2v^5 - 25u^4v^3$ by $-5uv^2$

9. $4x^2y^3 - 16xy^3 + 4xy$ by $2xy$

10. $4(a + b)^4 + 12(a + b)^3 - 8(a + b)$ by $2(a + b)$

11. $6a^3b^5 + 3a^2b^4 - 12ab + 9$ by $3a^2b^3$

12. $35(u - v)^3 - 15(u - v)^2 - 5(u - v)$ by $5(u - v)^2$

In problems 13–34, divide as indicated and check the result.

13. $x^2 - 7x + 10$ by $x - 5$

14. $y^2 - 5y + 8$ by $y - 2$

15. $2v^2 - 5v - 6$ by $2v - 1$

16. $1 - 4x + 4x^2$ by $2x - 1$

17. $w^3 + 3w^2 - 2w - 5$ by $w + 2$

18. $2x^4 + 3x^3 + 2x - 5x^2 - 1$ by $1 + x$

19. $3t^3 - 5t^2 + 2t + 2t^4 - 1$ by $t + 1$

20. $5y^5 - 2y^3 + 1$ by $y - 1$

21. $x^4 + 5x^3 + 9x^2 + 5x - 4$ by $x^2 + 2x - 1$

22. $5w^3 - 2w^2 + 3w - 4$ by $w^2 - 2w + 1$

23. $y^3 + 16y + 52 - 3y^2$ by $y^2 + 26 - 5y$

24. $m^3 + 2m - 4m^2 + 1$ by $m^2 - 1 - 3m$

25. $x^5 + 2x^2 - 24 + 3x^4$ by $x^3 - 2x^2 + x^4 + 6x - 12$

26. $3t^4 + 2t^2 - 28$ by $3t^3 + 6t^2 + 14t + 28$

27. $u^3 + 2v^3 - u^2v - uv^2$ by $u + v$

28. $3x^2y - 2xy^2 - 8y^3 + x^3$ by $x + 2y$

29. $2m^3 + mn^2 + m^2n + 4n^3$ by $m + n$

30. $w^2z - 6w^3 - 6z^3 - 12wz^2$ by $2w - 3z$

31. $2x^4 + 3x^3y - 6x^2y^2 + 4xy^3 - 7y^4$ by $2x - y$

32. $y^5 + z^5$ by $y + z$

33. $x^3 - y^3$ by $x - y$

34. $32x^5 - y^5$ by $2x - y$

REVIEW PROBLEM SET

In problems 1–6, identify the polynomial as a monomial, binomial, or trinomial. Give the degree and the coefficients.

1. $4y^2 - 3y + 2$

2. $8a - 5$

3. $-7x^2 + 3x$

4. $16m^3$

5. $10u^3v^2$

6. $\frac{1}{3}y^3 - \frac{2}{3}yz + 5z^2$

In problems 7–12, evaluate the polynomial for the given values of the variables.

7. $3x^2 + 5, x = -2$

8. $5t^2 - 3t + 2, t = 3$

9. $4w^2 - 3w - 6, w = 4$

10. $-2y^3 + 3y^2 - 4y + 2, y = 2$

11. $2u^2 - 3uv - v^2, u = 2, v = -1$

12. $5x^2 + 2xy + 3y^2, x = -2, y = 1$

In problems 13–20, perform the additions and subtractions.

13. $5w^2 + 7w^2 - 3w^2$

14. $-3xy^2 + 8xy^2 + xy^2$

15. $(2x^2 - 4) + (-5x^2 + 3)$

16. $(4x^3 + 3x^2) + (-5x^2 + 2x)$

17. $(3v^2 + 7v + 8) - (2v^2 + 3v + 2)$

18. $(-y^3 + 2y^2 + 3y - 7) - (-3y^3 + y^2 - 5y - 2)$

19. $(5x^2 - 3x + 2) + (2x^2 + 5x - 7) - (3x^2 - 4x - 1)$

20. $(7t^3 - 3t^2s + s^3) - (5t^3 - 4ts^2 + s^3) - (t^3 - s^3)$

In problems 21–36, use the properties of exponents to simplify each expression.

21. x^3x^8

22. $(-t)^5(-t)^4$

23. $w^2w^7w^3$

24. $(-y)^2(-y)^3(-y)^7$

25. $(m^3)^9$

26. $(-x^4)^6$

27. $(-vu^2)^3$

28. $(3p^2q^3)^4$

29. $(2y^3z)^2$

30. $(-4x^2y^3z)^3$

31. $\left(\dfrac{3t}{s^3}\right)^2$

32. $\left(\dfrac{2x^3}{y^2}\right)^4$

33. $\dfrac{x^{12}}{x^9}$

34. $\dfrac{3y^2}{9y^5}$

35. $\dfrac{(3x^2z^4)^3}{(3x^3z)^2}$

36. $\dfrac{(-2ab^2c^4)^2}{(2a^2bc^3)^3}$

In problems 37–48, find each product.

37. $3t^2(2t^3 - 4t)$

38. $-4y^3(-5y^2 - 2y + 1)$

39. $-2xy^3(-3x^2y + 5xy^2 - y)$

40. $3mn^2p^3(7mn - 2m^2p + 5n)$

41. $(a - b)(a^2 - 2ab + b^2)$

42. $(u^2 - u + 1)(2u^2 - 3u + 2)$

43. $(w + 7)(w + 3)$

44. $(3t + 7)(2t + 5)$

45. $(2x - 3)(x + 5)$

46. $(4y + 9x)(3y - 9x)$

47. $(4u - 5v)(u - 3v)$

48. $(7 - 5pq)(4 - 3pq)$

In problems 49–62, use special products to find each product.

49. $(t + 8)(t - 8)$

50. $(2s - 3)^2$

51. $(3x + 7)^2$

52. $(2v - 1)^3$

53. $(t + 2)(t^2 - 2t + 4)$

54. $(m^2 + 5)(m^2 - 5)$

55. $(y + 3)^3$

56. $(3s - 4)(9s^2 + 12s + 16)$

57. $(4 - 3s)^2$

58. $(2p^2 + q^2)^2$

59. $(3w - 2)^3$

60. $(x^2 - y^2)^3$

61. $(2x - z)(4x^2 + 2xz + z^2)$

62. $(m^2 - 3)(m^4 + 3m^2 + 9)$

In problems 63–68, factor each expression by finding common factors.

63. $7x^2y - 21xy^3$

64. $8uv^3 + 16u^2v$

65. $26a^3b^2 + 39a^5b^4 - 52a^2b^3$

66. $25mnp^2 - 50mn^2 - 50mn^2p^2 + 75m^2n^2p^2$

67. $2y(y + z) - 4x(y + z)$

68. $7x(y - z) + 14x^2(y - z)$

In problems 69–74, use the method of grouping to factor each expression.

69. $3x - 3y + xz - yz$

70. $a^4 - b^4 - c^4 - 2b^2c^2$

71. $2ux + xv - 6uy - 3vy$

72. $2x^3 - 6x + x^2z - 3z$

73. $5am^2 - bn + 5an - bm^2$

74. $t^3 + t^2s + t^2 + ts + 2t + 2s$

In problems 75–82, use special products to factor each expression.

75. $25m^2 - 9n^2$

76. $16(u + v)^2 - 81$

77. $x^3 + 64$

78. $y^3 - 216$

79. $16t^4 - 81$

80. $x^8 - 256$

81. $(x + y)^2 - (z - 1)^2$

82. $121x^2y^4 - z^6$

In problems 83–92, use common factors, special products, and/or grouping to factor each expression.

83. $9u^3 - 81uv^2$

84. $100y^3w - 25yw^3$

85. $5st^3 + 320s$

86. $-32w^2 + 4w^5$

87. $64x^6 - y^6$

88. $m^6p^6 - q^6$

89. $x^2 + y^2 - w^2 - z^2 + 2xy - 2wz$

90. $9t^2 + s^2 - r^4 - 6ts$

91. $t^4 + t^2 + 1$

92. $x^4 - 3x^2 + 9$

In problems 93–106, factor each trinomial.

93. $x^2 + 2xy - 3y^2$

94. $t^2 + 7t - 18$

95. $m^2 - 5m - 36$

96. $z^2 - zy - 12y^2$

97. $3u^2 + 17u + 10$

98. $6x^2 - 29x + 35$

99. $2y^2 - y - 6$

100. $33w^2 + 14wz - 40z^2$

101. $20x^2 - 31xy + 12y^2$

102. $68 - 31y - 15y^2$

103. $w^3 + 9w^2 - 22w$

104. $2a^3b - 10a^2b^2 - 28ab^3$

105. $x^2yz - 6xy^2z - 16y^3z$

106. $2m^3np^2 + 7m^2n^2p^2 - 15mn^3p^2$

In problems 107–118, divide as indicated.

107. $18u^4$ by $3u^2$

108. $-12x^3y^4$ by $-6xy^3$

109. $32w^5z^3 - 16w^4z^2$ by $4w^3z$

110. $100t^3s^2 - 90t^4s^3 + 70t^5s^4$ by $10t^2s^3$

111. $5x^2 - 16x + 3$ by $x - 3$

112. $2y^2 + 17y + 21$ by $3 + 2y$

113. $x^3 - 6x^2 + 12x - 8$ by $x^2 - 4x + 4$

114. $x^2 + y^2 + 2xy - 6x - 6y + 9$ by $x + y - 3$

115. $t^4 + 3t^3s + 2t^2s^2 + ts^3 - s^4$ by $t^2 + ts + s^2$

116. $x^4 - 3xy^3 - 2y^4$ by $x^2 - xy - y^2$

117. $w^{15} + 1$ by $w^5 + 1$

118. $m^6 - n^6$ by $m - n$

3 The Algebra of Fractions

In this chapter, you will learn to use the rules that govern numerical fractions to reduce, multiply, divide, add, and subtract **algebraic fractions**—fractions that contain variables.

3.1 Rational Expressions

In Chapter 1, Section 1.1, we defined a rational number as a number that can be written in the form

$$\frac{a}{b}, \qquad \text{where } a \text{ and } b \text{ are integers} \qquad (b \neq 0).$$

We now define a **rational expression** as an algebraic expression that can be written in the form

$$\frac{P}{Q}, \qquad \text{where } P \text{ and } Q \text{ are polynomials} \qquad (Q \neq 0).$$

This expression is also called an **algebraic fraction** with **numerator** P and **denominator** Q. Throughout this book, we use the term "fraction" to mean either a common arithmetic fraction or an algebraic fraction.

Examples of rational expressions are:

$$\frac{1}{x}, \qquad \frac{y + 1}{5y - 2}, \qquad \frac{5}{t^2 - 7}, \qquad \frac{3w^2 + 4w - 1}{1}, \qquad \text{and} \qquad \frac{3c^2 + c}{5c + 1}.$$

Rational expressions have specific values when numbers are substituted for the variables (letters).

For example, the value of the rational expression

$$\frac{x}{x - 2},$$

when 3 is substituted for x, is

$$\frac{3}{3-2} = \frac{3}{1} = 3.$$

The value of the expression, when x is replaced by -1, is

$$\frac{-1}{-1-2} = \frac{-1}{-3} = \frac{1}{3}.$$

However,

$$\frac{x}{x-2}$$

does not have a value when $x = 2$, because

$$\frac{2}{2-2} = \frac{2}{0}$$

is not defined. (The symbol $2/0$ does not represent a number.)

We can make the following general statement:

A rational expression represents a real number when any values are assigned to the variable except those values that make the denominator zero. If the denominator of a rational expression is equal to zero, we say that the rational expression is *undefined* for that value of the variable.

For example, if we substitute $x = 1$ in the fraction

$$\frac{(x-1)^2}{(x-1)(x-2)}$$

we obtain

$$\frac{(1-1)^2}{(1-1)(1-2)} = \frac{0}{0(-1)} = \frac{0}{0}.$$

This is undefined. If we substitute $x = 2$ in the fraction, we have

$$\frac{(2-1)^2}{(2-1)(2-2)} = \frac{1}{1(0)} = \frac{1}{0}.$$

This is also undefined.

Thus,

$$\frac{(x-1)^2}{(x-1)(x-2)}$$

does not represent a real number when $x = 1$ or when $x = 2$.

In this text we assume that the variables in any fraction may not be assigned values that will result in a value of zero for the denominator (a division by zero).

EXAMPLE 1 For what value of x is

$$\frac{x-2}{x+3}$$

undefined?

SOLUTION The expression

$$\frac{x-2}{x+3}$$

is undefined when the denominator is zero. That is, when

$$x + 3 = 0 \quad \text{or} \quad x = -3.$$

In earlier mathematics courses you learned how to determine the equality of rational numbers. For example, you know that

$$\frac{2}{3} = \frac{10}{15},$$

and that therefore $\frac{2}{3}$ can be used interchangeably with $\frac{10}{15}$ (or with $\frac{4}{6}$, or with $\frac{6}{9}$, etc.) in any expression in which $\frac{2}{3}$ appears. In order to extend this concept to rational expressions we must define the equality (or equivalence) of algebraic fractions.

We say that two algebraic fractions are **equivalent** if they give the same real numbers for every value assigned to their variables for which both fractions are defined. When two fractions are equivalent, we can represent this by writing an *equals* sign between them.

For example,

$$\frac{3y}{xy} = \frac{3}{x},$$

because both fractions have the same values for all nonzero values of x and y. However,

$$\frac{3y}{xy} \neq \frac{3y}{x}$$

because these fractions have different values for some nonzero values of x and y (for example, if $x = 4$ and $y = 2$).

Notice that in the two equalities,

$$\frac{2}{3} = \frac{10}{15} \quad \text{and} \quad \frac{3y}{xy} = \frac{3}{x},$$

the "cross products" are equal:

$$(2)(15) = (3)(10) \quad \text{and} \quad (3y)(x) = (3)(xy).$$

These examples illustrate the following property:

Property 1

Two fractions

$$\frac{P}{Q} \quad \text{and} \quad \frac{R}{S}$$

are **equivalent** if and only if $PS = QR$, provided that $Q \neq 0$ and $S \neq 0$.
That is,

$$\frac{P}{Q} = \frac{R}{S} \text{ if and only if } PS = QR.$$

If two algebraic fractions are equal (or equivalent) they can be used interchangeably in any expression in which either fraction appears.

EXAMPLE **2**   Indicate whether the given pair of fractions are equivalent.

(a) $\dfrac{4}{6}$ and $\dfrac{2}{3}$

(b) $\dfrac{14a}{56a^3}$ and $\dfrac{1}{4a^2}$

(c) $\dfrac{14y}{7y + 21}$ and $\dfrac{2y}{y + 3}$

(d) $\dfrac{15x}{10x + 4}$ and $\dfrac{3x}{2x + 1}$

SOLUTION   (a) Because $4 \cdot 3 = 6 \cdot 2 = 12$, it follows that $\dfrac{4}{6} = \dfrac{2}{3}$.

(b) Because $(14a)(4a^2) = (56a^3)(1) = 56a^3$, it follows that $\dfrac{14a}{56a^3} = \dfrac{1}{4a^2}$.

(c) Because $(14y)(y + 3) = (7y + 21)(2y) = 14y^2 + 42y$, it follows that

$$\frac{14y}{7y + 21} = \frac{2y}{y + 3}.$$

(d) $(15x)(2x + 1) = 30x^2 + 15x$ and $(10x + 4)(3x) = 30x^2 + 12x$. Thus, $(15x)(2x + 1) \neq (10x + 4)(3x)$, and

$$\frac{15x}{10x + 4} \neq \frac{3x}{2x + 1}.$$

Property 1 verifies the following principle:

Fundamental Principle of Fractions

If $Q \neq 0$ and $K \neq 0$, then $\dfrac{PK}{QK} = \dfrac{P}{Q}.$

The fundamental principle of fractions states that when the numerator and the denominator of a given fraction are divided by the same *nonzero* expression, an equivalent fraction is obtained. In other words, we can "divide out" common factors from the numerator *and* the denominator of a fraction without changing the value of the fraction.

A fraction is said to be **reduced to lowest terms,** or **simplified,** if the numerator and the denominator have no common factors (other than 1 and -1). Thus, to **simplify** a fraction, first we factor both the numerator and the denominator. Then we divide both the numerator and the denominator by any factors they have in common.

For example, to reduce

$$\frac{9x + 15}{3x^2 + 5x}$$

to lowest terms, we begin by factoring both the numerator and the denominator:

$$\frac{9x + 15}{3x^2 + 5x} = \frac{3(3x + 5)}{x(3x + 5)}.$$

Next, we divide both the numerator and the denominator by $3x + 5$:

$$\frac{3(3x + 5)}{x(3x + 5)} = \frac{3\cancel{(3x + 5)}}{x\cancel{(3x + 5)}} = \frac{3}{x}.$$

The slanted lines drawn through $3x + 5$ in the numerator and the denominator indicate that we have divided through by $3x + 5$.

In Examples 3–9, reduce each fraction to lowest terms.

EXAMPLE 3 $\dfrac{36}{44}$

SOLUTION $\dfrac{36}{44} = \dfrac{(9)\cancel{(4)}}{(11)\cancel{(4)}} = \dfrac{9}{11}$

EXAMPLE 4 $\dfrac{28x^3y}{21xy^2}$

SOLUTION $\dfrac{28x^3y}{21xy^2} = \dfrac{4x^2\cancel{(7xy)}}{3y\cancel{(7xy)}} = \dfrac{4x^2}{3y}$

EXAMPLE 5 $\dfrac{y^2 - 3y}{y^2 - 9}$

SOLUTION   Factoring the numerator and the denominator, we obtain

$$\dfrac{y^2 - 3y}{y^2 - 9} = \dfrac{y\cancel{(y - 3)}}{(y + 3)\cancel{(y - 3)}} = \dfrac{y}{y + 3}.$$

EXAMPLE 6 $\dfrac{c^2 + 4c - 21}{c^2 - c - 6}$

SOLUTION   Factoring, we obtain

$$\dfrac{c^2 + 4c - 21}{c^2 - c - 6} = \dfrac{(c + 7)\cancel{(c - 3)}}{(c + 2)\cancel{(c - 3)}} = \dfrac{c + 7}{c + 2}.$$

EXAMPLE 7 $\dfrac{25 - w^2}{w^2 - 3w - 10}$

SOLUTION   Factoring, we obtain

$$\dfrac{25 - w^2}{w^2 - 3w - 10} = \dfrac{(5 - w)(5 + w)}{(w - 5)(w + 2)}.$$

Notice that $5 - w = -(w - 5)$. Thus, we can write

$$\dfrac{(5 - w)(5 + w)}{(w - 5)(w + 2)} = \dfrac{-(w - 5)(5 + w)}{(w - 5)(w + 2)}.$$

Therefore,

$$\frac{25 - w^2}{w^2 - 3w - 10} = \frac{(5 - w)(5 + w)}{(w - 5)(w + 2)} = \frac{-(w - 5)(5 + w)}{(w - 5)(w + 2)} = \frac{-(5 + w)}{w + 2} = \frac{-5 - w}{w + 2}.$$

EXAMPLE 8 $\dfrac{6p^2 - 7p - 3}{4p^2 - 8p + 3}$

SOLUTION   Factoring, we obtain

$$\frac{6p^2 - 7p - 3}{4p^2 - 8p + 3} = \frac{(3p + 1)(2p - 3)}{(2p - 1)(2p - 3)} = \frac{3p + 1}{2p - 1}.$$

EXAMPLE 9 $\dfrac{x^2 y^4 - x^4 y^2}{x^2 y^4 + 2x^3 y^3 + x^4 y^2}$

SOLUTION   Factoring, we obtain

$$\frac{x^2 y^4 - x^4 y^2}{x^2 y^4 + 2x^3 y^3 + x^4 y^2} = \frac{x^2 y^2 (y^2 - x^2)}{x^2 y^2 (y^2 + 2xy + x^2)} = \frac{x^2 y^2 (y - x)(y + x)}{x^2 y^2 (y + x)(y + x)}$$

$$= \frac{y - x}{y + x}.$$

If we are given a fraction P/Q, and we multiply the numerator and the denominator of the fraction by the same nonzero expression K, we will always obtain a fraction that is equivalent to the original one, that is:

> If $Q \neq 0$ and $K \neq 0$, then $\dfrac{P}{Q} = \dfrac{PK}{QK}.$

The following example illustrates this rule.

EXAMPLE 10   Find the missing expression for each of the equivalent fractions.

(a) $\dfrac{3}{4} = \dfrac{?}{16}$

(b) $\dfrac{3a}{b} = \dfrac{?}{ba}$

(c) $\dfrac{x + 3}{x - 2} = \dfrac{?}{x^2 - 4}$

SOLUTION

(a) First we must determine by what factor the original denominator, 4, was multiplied to give the new denominator, 16. Then we multiply the original numerator, 3, by this same factor to obtain the missing numerator. Because $16 = 4 \cdot 4$, we have

$$\frac{3}{4} = \frac{3 \cdot 4}{4 \cdot 4} = \frac{12}{16},$$

and the unknown number is 12.

(b) The original denominator, b, must have been multiplied by a to get the new denominator, ba. Therefore, we multiply the original numerator, $3a$, by a:

$$\frac{3a}{b} = \frac{3a(a)}{b(a)} = \frac{3a^2}{ba},$$

and the unknown expression is $3a^2$.

(c) We can factor $x^2 - 4$ as $(x - 2)(x + 2)$. Therefore, we multiply the numerator and the denominator of the original fraction by $x + 2$:

$$\frac{x + 3}{x - 2} = \frac{(x + 3)(x + 2)}{(x - 2)(x + 2)} = \frac{x^2 + 5x + 6}{x^2 - 4},$$

and the unknown expression is $x^2 + 5x + 6$.

Every fraction has three signs associated with it: the sign of the numerator, the sign of the denominator, and the sign of the fraction. Two of these signs may be changed without changing the value of the fraction. In general, we have the following rules:

$$\text{(i)} \quad \frac{-P}{Q} = \frac{P}{-Q} = -\frac{P}{Q}$$

$$\text{(ii)} \quad \frac{P}{Q} = -\frac{-P}{Q} = -\frac{P}{-Q}$$

$$\text{(iii)} \quad \frac{P}{Q} = \frac{-P}{-Q}$$

EXAMPLE 11

Write each of the following fractions as an equivalent fraction with denominator $x - y$.

(a) $\dfrac{1}{y - x}$ (b) $\dfrac{-xy}{y - x}$ (c) $\dfrac{a - b}{y - x}$ (d) $-\dfrac{1}{y - x}$ (e) $-\dfrac{-1}{y - x}$

SOLUTION

(a) $\dfrac{1}{y-x} = \dfrac{-1}{-(y-x)} = \dfrac{-1}{x-y}$ (By rule iii)

(b) $\dfrac{-xy}{y-x} = \dfrac{xy}{-(y-x)} = \dfrac{xy}{x-y}$ (By rule i)

(c) $\dfrac{a-b}{y-x} = \dfrac{-(a-b)}{-(y-x)} = \dfrac{b-a}{x-y}$ (By rule iii)

(d) $-\dfrac{1}{y-x} = \dfrac{1}{-(y-x)} = \dfrac{1}{x-y}$ (By rule i)

(e) $-\dfrac{-1}{y-x} = -\dfrac{1}{-(y-x)} = -\dfrac{1}{x-y}$ (By rule ii)

PROBLEM SET 3.1

In problems 1–10, determine all values of the variable for which the fraction is not defined.

1. $\dfrac{7x}{x-3}$ **2.** $\dfrac{y+2}{y+6}$ **3.** $\dfrac{2}{y+10}$

4. $\dfrac{-11}{12-6x}$ **5.** $\dfrac{2x-4}{(x-6)(x+7)}$ **6.** $\dfrac{x^2-1}{(x+3)(x-2)}$

7. $\dfrac{t+2}{(t+2)(t-8)}$ **8.** $\dfrac{1-u}{(2-u)(6+u)}$ **9.** $\dfrac{v^2+2v-3}{(v+8)(v+9)(v-4)}$

10. $\dfrac{t^3+1}{(t+2)(t-5)(t+7)}$

In problems 11–20, use the property that $P/Q = R/S$ if and only if $PS = QR$, to determine whether the given pairs of fractions are equivalent.

11. $\dfrac{5}{4}$ and $\dfrac{10}{8}$ **12.** $\dfrac{-7}{6}$ and $\dfrac{21}{-18}$ **13.** $\dfrac{7}{9}$ and $\dfrac{8}{10}$

14. $\dfrac{14}{-9}$ and $\dfrac{-12}{7}$ **15.** $\dfrac{5}{x}$ and $\dfrac{15x}{3x^2}$ **16.** $\dfrac{a+b}{x}$ and $\dfrac{7a+b}{7x}$

17. $\dfrac{v+2}{v^2-4}$ and $\dfrac{3}{3v-12}$ **18.** $\dfrac{m+2n}{5}$ and $\dfrac{2m^2-8n^2}{10m-20n}$ **19.** $\dfrac{x^2-9}{x-3}$ and $\dfrac{2x+6}{2}$

20. $\dfrac{x+y}{1}$ and $\dfrac{x^3+y^3}{x^2-xy+y^2}$

In problems 21–40, reduce each fraction to lowest terms.

21. $\dfrac{15}{18}$ **22.** $\dfrac{65}{26}$ **23.** $\dfrac{25x^2y^5}{45x^3y^2}$

24. $\dfrac{14a^5b^2c^6}{7a^2b^6c^3}$

25. $\dfrac{m^2 + m}{m^2 - m}$

26. $\dfrac{x^2 - 9}{x^2 - 6x + 9}$

27. $\dfrac{t^2 + 2t - 3}{t^2 + 5t + 6}$

28. $\dfrac{y^2 + 4y}{y^2 + 6y + 8}$

29. $\dfrac{4x^2 - 9}{6x^2 - 9x}$

30. $\dfrac{4v^2 - 1}{8v^3 - 1}$

31. $\dfrac{v^2 + v - 12}{v^2 + 4v - 21}$

32. $\dfrac{x + x^2 - y - xy}{x^2 - 2xy + y^2}$

33. $\dfrac{3x^2 + 7x + 4}{3x^2 - 5x - 12}$

34. $\dfrac{10t^2 + 29t - 21}{4t^2 + 12t - 7}$

35. $\dfrac{4u^2 - 9}{8u^3 - 27}$

36. $\dfrac{7x^2 - 5xy}{49x^3 - 25xy^2}$

37. $\dfrac{3x^3 - 3xy^2}{3xy^2 + 3x^2y - 6x^3}$

38. $\dfrac{3 + 13m - 10m^2}{2m^2 + 5m - 12}$

39. $\dfrac{xz + xw - yz - yw}{xy + xz - y^2 - yz}$

40. $\dfrac{x^3 - 2x^2 + 5x - 10}{3x^5 + 15x^3 - x^2 - 5}$

In problems 41–56, find the missing numerator that will make the two fractions equivalent.

41. $\dfrac{9}{16} = \dfrac{?}{64}$

42. $\dfrac{13}{25} = \dfrac{?}{75}$

43. $\dfrac{5m^3y}{7my^3} = \dfrac{?}{28m^5y^6}$

44. $\dfrac{6}{11st} = \dfrac{?}{33s^4t}$

45. $\dfrac{3}{u + v} = \dfrac{?}{5(u + v)^2}$

46. $\dfrac{9(a - b)}{7a} = \dfrac{?}{7a^4(a - b)}$

47. $\dfrac{6}{x + 4} = \dfrac{?}{x^2 + 3x - 4}$

48. $\dfrac{mn}{m - n} = \dfrac{?}{m^3 - mn^2}$

49. $\dfrac{2t}{2t - 1} = \dfrac{?}{4t^2 - 1}$

50. $\dfrac{2x - 3}{x^2 + 3x + 9} = \dfrac{?}{x^3 - 27}$

51. $\dfrac{y - 3}{y^2 - y + 1} = \dfrac{?}{y^3 + 1}$

52. $\dfrac{2v}{u + 5v} = \dfrac{?}{2u^3 - 50uv^2}$

53. $\dfrac{x + 2}{x - 2} = \dfrac{?}{2x^2 - 7x + 6}$

54. $\dfrac{ts}{5t + s} = \dfrac{?}{10t^2 + 17ts + 3s^2}$

55. $\dfrac{2y + 3x}{3y + 5x} = \dfrac{?}{6y^2 + yx - 15x^2}$

56. $\dfrac{1}{x - y} = \dfrac{?}{x^4y - xy^4}$

In problems 57–62, write an equivalent fraction with the denominator $x - y$.

57. $\dfrac{-7}{y - x}$

58. $\dfrac{8}{-(x - y)}$

59. $-\dfrac{-xy}{y - x}$

60. $\dfrac{c - d}{y - x}$

61. $\dfrac{3 - x}{y - x}$

62. $\dfrac{x + 2}{y - x}$

3.2 Multiplication and Division of Fractions

The rules for multiplying and dividing rational expressions are the same as the rules for multiplying and dividing numerical fractions.

Multiplication of Fractions

The product of two fractions P/Q and R/S is a fraction whose numerator is the product of the two given numerators, $P \cdot R$, and whose denominator is the product of the two given denominators, $Q \cdot S$. That is,

$$\frac{P}{Q} \cdot \frac{R}{S} = \frac{P \cdot R}{Q \cdot S}$$

For example,

$$\frac{3}{14} \cdot \frac{35}{12} = \frac{(3)(35)}{(14)(12)} = \frac{105}{168} = \frac{\cancel{3}\cancel{7}(5)}{(2)\cancel{7}\cancel{3}(4)} = \frac{5}{8}$$

and

$$\frac{2x}{5y^3} \cdot \frac{25y^4}{4x^2} = \frac{(2x)(25y^4)}{(5y^3)(4x^2)} = \frac{50xy^4}{20x^2y^3} = \frac{(5y)\cancel{(10xy^3)}}{(2x)\cancel{(10xy^3)}} = \frac{5y}{2x}.$$

In practice, it is easier to divide out factors common to both the numerator and the denominator before performing the multiplication. In the last example, $2x$ and $5y^3$ are the common factors. Thus,

$$\frac{2x}{5y^3} \cdot \frac{25y^4}{4x^2} = \frac{\overset{}{\cancel{2x}}}{\underset{}{\cancel{5y^3}}} \cdot \frac{\overset{5y}{\cancel{25y^4}}}{\underset{2x}{\cancel{4x^2}}} = \frac{5y}{2x}.$$

In Examples 1–5, perform each multiplication and simplify the result.

EXAMPLE 1 $\dfrac{7}{8} \cdot \dfrac{5}{6}$

SOLUTION $\dfrac{7}{8} \cdot \dfrac{5}{6} = \dfrac{7 \cdot 5}{8 \cdot 6} = \dfrac{35}{48}$

EXAMPLE 2 $\dfrac{a^3b^2}{a^2b} \cdot \dfrac{ab^3}{a^3b^2}$

SOLUTION   We begin by finding the factors that are common to both the numerators and the denominators of the fractions, and we divide out these factors:

$$\frac{a^3b^2}{a^2b} \cdot \frac{ab^3}{a^3b^2} = \frac{\overset{}{\cancel{a^3b^2}}}{\underset{a}{\cancel{a^2b}}} \cdot \frac{\overset{b^2}{\cancel{ab^3}}}{\underset{}{\cancel{a^3b^2}}} = \frac{b^2}{a}.$$

EXAMPLE 3 $\dfrac{c}{c-1} \cdot \dfrac{c^2-1}{c^3}$

SOLUTION   First we factor the numerators and the denominators of the fractions. Then we divide out the common factors:

$$\frac{c}{c-1} \cdot \frac{c^2-1}{c^3} = \frac{\cancel{c}}{\cancel{c-1}} \cdot \frac{\cancel{(c-1)}(c+1)}{\underset{c^2}{\cancel{c^3}}} = \frac{c+1}{c^2}.$$

EXAMPLE 4 $\dfrac{6t-6}{t^2+2t} \cdot \dfrac{t^2+4t+4}{2t^2+2t-4}$

SOLUTION   First we factor the numerators and the denominators of the fractions. Then we divide out the common factors:

$$\frac{6t-6}{t^2+2t} \cdot \frac{t^2+4t+4}{2t^2+2t-4} = \frac{\overset{3}{\cancel{6}}\cancel{(t-1)}}{t\cancel{(t+2)}} \cdot \frac{\cancel{(t+2)}\cancel{(t+2)}}{\cancel{2}\cancel{(t+2)}\cancel{(t-1)}} = \frac{3}{t}.$$

EXAMPLE 5 $\dfrac{x^2+xy-2y^2}{xy} \cdot \dfrac{5xy^2}{x+2y} \cdot \dfrac{1}{5x-5y}$

SOLUTION   We factor the numerators and the denominators of the fractions, and then we divide out the common factors:

$$\frac{x^2+xy-2y^2}{xy} \cdot \frac{5xy^2}{x+2y} \cdot \frac{1}{5x-5y} = \frac{\cancel{(x+2y)}\cancel{(x-y)}}{\cancel{xy}} \cdot \frac{\overset{y}{\cancel{5xy^2}}}{\cancel{x+2y}} \cdot \frac{1}{\cancel{5(x-y)}} = y.$$

Division of Fractions

To divide two numerical fractions, such as

$$\frac{4}{5} \div \frac{7}{13},$$

we invert the divisor (the fraction after the division sign) and multiply:

$$\frac{4}{5} \div \frac{7}{13} = \frac{4}{5} \cdot \frac{13}{7} = \frac{52}{35}.$$

This rule can be generalized to cover algebraic fractions:

If P/Q and R/S are fractions, with $R/S \neq 0$, then

$$\frac{P}{Q} \div \frac{R}{S} = \frac{P}{Q} \cdot \frac{S}{R} = \frac{PS}{QR}.$$

We must be careful *not* to divide out any factor common to the numerators and the denominators of the fractions we are going to divide. This can only be done if the operation is multiplication. Therefore, when you divide one fraction by another, first change the division to multiplication (invert the divisor), and then divide out the common factors.

In Examples 6–11, find the following quotients and simplify.

EXAMPLE 6 $\dfrac{34}{57} \div \dfrac{51}{95}$

SOLUTION   First invert the divisor, changing the division to multiplication:

$$\frac{34}{57} \div \frac{51}{95} = \frac{34}{57} \cdot \frac{95}{51}.$$

Next, divide out all the factors that are common to both the numerators and the denominators of the fractions, and then multiply:

$$\frac{34}{57} \cdot \frac{95}{51} = \frac{(2)(\cancel{17})}{(3)(\cancel{19})} \cdot \frac{(5)(\cancel{19})}{(3)(\cancel{17})} = \frac{10}{9}.$$

EXAMPLE 7 $\dfrac{3a^2}{5b} \div \dfrac{2a^3}{6b^3}$

SOLUTION   Use the division rule and divide out the common factors:

$$\frac{3a^2}{5b} \div \frac{2a^3}{6b^3} = \frac{3a^2}{5b} \cdot \frac{\overset{3b^2}{\cancel{6b^3}}}{\cancel{2a^3}}$$

$$= \frac{9b^2}{5a}.$$

EXAMPLE 8 $\dfrac{a+b}{2} \div \dfrac{(a+b)^2}{6}$

SOLUTION   We perform the division operation—invert the divisor, changing the division to multiplication—then we divide out the common factors and multiply:

$$\frac{a+b}{2} \div \frac{(a+b)^2}{6} = \frac{\cancel{a+b}}{\cancel{2}} \cdot \frac{\overset{3}{\cancel{6}}}{\underset{a+b}{\cancel{(a+b)^2}}}$$

$$= \frac{3}{a+b}.$$

EXAMPLE 9

$$\frac{x^2 + 5x + 6}{x^2 - 4} \div \frac{x^2 + 4x + 4}{x^2 - 4x + 4}$$

SOLUTION

We begin by performing the division operation. Next, we factor each numerator and denominator and divide out the common factors:

$$\frac{x^2 + 5x + 6}{x^2 - 4} \div \frac{x^2 + 4x + 4}{x^2 - 4x + 4} = \frac{x^2 + 5x + 6}{x^2 - 4} \cdot \frac{x^2 - 4x + 4}{x^2 + 4x + 4}$$

$$= \frac{\cancel{(x + 2)}(x + 3)}{(x + 2)\cancel{(x - 2)}} \cdot \frac{\cancel{(x - 2)}(x - 2)}{\cancel{(x + 2)}(x + 2)}$$

$$= \frac{(x + 3)(x - 2)}{(x + 2)(x + 2)}.$$

EXAMPLE 10

$$\frac{w^2 - w}{z^2 - z} \cdot \frac{z^2w - zw}{w - 1} \div \frac{w^2}{w - 1}$$

SOLUTION

We perform the division operation, factor where possible, and divide out the common factors:

$$\frac{w^2 - w}{z^2 - z} \cdot \frac{z^2w - zw}{w - 1} \div \frac{w^2}{w - 1} = \frac{w^2 - w}{z^2 - z} \cdot \frac{z^2w - zw}{w - 1} \cdot \frac{w - 1}{w^2}$$

$$= \frac{\cancel{w}\cancel{(w - 1)}}{\cancel{z}\cancel{(z - 1)}} \cdot \frac{z\cancel{w}\cancel{(z - 1)}}{\cancel{w - 1}} \cdot \frac{w - 1}{\cancel{w^2}}$$

$$= \frac{w - 1}{1}$$

$$= w - 1.$$

EXAMPLE 11

$$\frac{2t - 1}{2t^2 + 2t} \div \left[\frac{6t^4 - 6}{4t^2 + t - 3} \cdot \frac{8t^2 - 10t + 3}{4t^3 - 4t} \right]$$

SOLUTION

First we perform the multiplication operation inside the brackets. Then we perform the division operation and divide out the common factors:

$$\frac{6t^4 - 6}{4t^2 + t - 3} \cdot \frac{8t^2 - 10t + 3}{4t^3 - 4t} = \frac{\overset{3}{\cancel{6}}\cancel{(t^2 - 1)}(t^2 + 1)}{(t + 1)\cancel{(4t - 3)}} \cdot \frac{\cancel{(4t - 3)}(2t - 1)}{\underset{2}{\cancel{4}t\cancel{(t^2 - 1)}}}$$

$$= \frac{3(t^2 + 1)(2t - 1)}{2t(t + 1)}.$$

Thus,

$$\frac{2t - 1}{2t^2 + 2t} \div \left[\frac{6t^4 - 6}{4t^2 + t - 3} \cdot \frac{8t^2 - 10t + 3}{4t^3 - 4t}\right] = \frac{2t - 1}{2t^2 + 2t} \div \frac{3(t^2 + 1)(2t - 1)}{2t(t + 1)}$$

$$= \frac{2t - 1}{2t(t + 1)} \cdot \frac{2t(t + 1)}{3(t^2 + 1)(2t - 1)}$$

$$= \frac{1}{3(t^2 + 1)}.$$

PROBLEM SET 3.2

In problems 1–22 perform each multiplication and reduce the answer to lowest terms.

1. $\dfrac{12}{13} \cdot \dfrac{39}{60}$

2. $\dfrac{10}{17} \cdot \dfrac{34}{25}$

3. $\dfrac{7x^3}{8y^4} \cdot \dfrac{16y}{21x^2}$

4. $\dfrac{3uv^3}{7u^2v} \cdot \dfrac{14u^5}{9uv^6}$

5. $\dfrac{5x^2y}{3t^2s} \cdot \dfrac{6ts}{10x^2}$

6. $\dfrac{15xyz}{16a^2} \cdot \dfrac{12a^3}{25x^2yz^2}$

7. $\dfrac{t + 4}{t^2} \cdot \dfrac{5t}{3t + 12}$

8. $\dfrac{m^2 + mn}{mn} \cdot \dfrac{5n}{m^2 - n^2}$

9. $\dfrac{3x + 6}{5x + 5} \cdot \dfrac{10x + 10}{x^2 - 6x - 16}$

10. $\dfrac{c + 2}{c^2 + 8c - 9} \cdot \dfrac{2c + 18}{2c^2 - 8}$

11. $\dfrac{1 - a^2}{a + 1} \cdot \dfrac{7a^2 - 5a - 2}{a^2 - 2a + 1}$

12. $\dfrac{3y^2 - 12}{3y^2 - 3} \cdot \dfrac{y - 1}{2y + 4}$

13. $\dfrac{v^2 - 1}{v - 3} \cdot \dfrac{3v^2 - 8v - 3}{v^2 - 10v + 9}$

14. $\dfrac{a^2 - 9b^2}{a^2 - b^2} \cdot \dfrac{5a - 5b}{a^2 + 6ab + 9b^2}$

15. $\dfrac{x^2 - 144}{x + 4} \cdot \dfrac{x^2 - 16}{x + 12}$

16. $\dfrac{t^2 + 7t + 10}{t^2 + 10t + 25} \cdot \dfrac{t + 5}{t + 2}$

17. $\dfrac{3y^2 - y - 2}{3y^2 + y - 2} \cdot \dfrac{3y^2 - 5y + 2}{3y^2 + 5y + 2}$

18. $\dfrac{u^2v^2 - 9}{w^2 - w - 2} \cdot \dfrac{w^3 - w^2 - 2w}{u^2v^2 - 6uv + 9}$

19. $\dfrac{3a^2 - 12b^2}{(a + b)^2} \cdot \dfrac{2a^2 - 2b^2}{6a^2 - 24ab + 24b^2}$

20. $\dfrac{2x^3 + 16}{5x^3 - 135} \cdot \dfrac{x^2 - 9}{x^2 - 2x + 4}$

21. $\dfrac{x^3 - y^3}{3x + 3y} \cdot \dfrac{6x^2 + 12xy + 6y^2}{2x^2 + 2xy + 2y^2}$

22. $\dfrac{8m^3 + n^3}{3m - 5n} \cdot \dfrac{9m^2 - 25n^2}{4m^2 - 2mn + n^2}$

In problems 23–42, perform each division and reduce the answer to lowest terms:

23. $\dfrac{9}{15} \div \dfrac{27}{25}$

24. $\dfrac{4}{11} \div \dfrac{24}{55}$

25. $\dfrac{4x^3}{y} \div \dfrac{2x}{3y^2}$

26. $\dfrac{5uv^3}{9a^2b} \div \dfrac{25u^3v}{18a^3b^2}$

27. $\dfrac{5t + 10}{t^3} \div \dfrac{t + 2}{t^4}$

28. $\dfrac{2x - y}{x^2 - y^2} \div \dfrac{4x^2 - y^2}{x + y}$

29. $\dfrac{12 - 6y}{7y - 21} \div \dfrac{2y - 4}{y^2 - 9}$

30. $\dfrac{9a^2 - 1}{a + 1} \div \dfrac{6a + 2}{2a + 2}$

31. $\dfrac{x^3 + 3x}{2x - 1} \div \dfrac{x^2 + 3}{x + 1}$

32. $\dfrac{2c}{c^3d^2 - c^2d^3} \div \dfrac{4c}{c^2 - d^2}$

33. $\dfrac{a^2 - 4a + 4}{a + 2} \div \dfrac{a^2 - 4}{3a + 6}$

34. $\dfrac{x^3 - 4x^2 + 3x}{x + 2} \div (x^2 - 3x)$

35. $\dfrac{15u + 15v}{u^2 + 6uv + 9v^2} \div \dfrac{5u^2 - 5v^2}{u^2 - 9v^2}$

36. $\dfrac{p^2 - 3}{p^3 - 4p} \div \dfrac{p^4 - 9}{p^2 - 4}$

37. $\dfrac{2x^2 - 5x - 3}{3x^2 - 5x - 2} \div \dfrac{2x^2 + 11x + 5}{x^2 + 3x - 10}$

38. $\dfrac{a^2 + 8a + 16}{a^2 - 8a + 16} \div \dfrac{a^3 + 4a^2}{a^2 - 16}$

39. $\dfrac{6y^2 + 7y - 3}{9y^2 - 25} \div \dfrac{12y^2 - y - 1}{12y^2 + 20y}$

40. $\dfrac{6x^2 + 11x - 10}{3x^2 - 5x - 12} \div \dfrac{2x^2 + 9x + 10}{3x^2 + 10x + 8}$

41. $\dfrac{8x^3 - 27}{9x^2 - 3x + 1} \div \dfrac{4x^2 + 6x + 9}{27x^3 + 1}$

42. $\dfrac{u^4 - 8u}{3u^2 + 6u + 12} \div \dfrac{u^4 - 4u^2}{u^2 + 4u + 4}$

In problems 43–50, perform the indicated operations and reduce the answer to lowest terms.

43. $\left[\dfrac{x - 1}{x^2 - 4} \cdot \dfrac{2x + 4}{x^2 - 1}\right] \div \dfrac{2x + 2}{x^2 - 4x + 4}$

44. $\left[\dfrac{t - 3}{t^2 + 2t - 3} \cdot \dfrac{t^2 - 2t + 1}{t^2 - 2t - 3}\right] \div \dfrac{t^2 - 9}{t^2 - 1}$

45. $\left[\dfrac{b - 1}{21 - 4a - a^2} \cdot \dfrac{b - 2}{b - b^3}\right] \div \dfrac{2 - b}{a^2 + 6a - 7}$

46. $\left[\left(\dfrac{x^2 - 1}{x^2}\right)^2 \cdot \dfrac{2}{x - 1}\right] \div \dfrac{x^2 + 2x + 1}{x^3}$

47. $\dfrac{7}{5v^2(v + 3)} \div \left[\dfrac{v^2 - 5v + 6}{8v^2} \cdot \dfrac{21}{5v^2 - 45}\right]$

48. $\dfrac{x^3 - 16x}{x^2 - 3x - 10} \div \left[\dfrac{x^3 + x^2 - 12x}{x^2 + 5x + 6} \cdot \dfrac{x^2 - 4}{x^2 - 9}\right]$

49. $\dfrac{2y^2 + 5y - 3}{6y^2 - 5y - 6} \div \left[\dfrac{2y^2 + 9y - 5}{12y^2 - y - 6} \cdot \dfrac{4y^2 + 9y - 9}{2y^2 + y - 6}\right]$

50. $\dfrac{3m^2 + m - 10}{8m^2 - 10m - 3} \div \left[\dfrac{2m^2 + 7m + 6}{8m^2 - 2m - 1} \div \dfrac{4m^2 - 9}{3m^2 + 10m - 25}\right]$

3.3 Addition and Subtraction of Fractions

In arithmetic, when we add (or subtract) numerical fractions having the same denominators, we add (or subtract) the numerator and keep the denominator. If the denominators of the fractions are different, we replace the fractions with equivalent fractions that do have the same denominators, and then we add or subtract the numerators. The same procedures are used for rational expressions.

To **add** (or **subtract**) fractions with the same denominators, add (or subtract) the numerators and keep the common denominator. Thus, if $Q \neq 0$, then:

$$\text{(i)}\quad \frac{P}{Q} + \frac{R}{Q} = \frac{P + R}{Q} \qquad\qquad \text{(ii)}\quad \frac{P}{Q} - \frac{R}{Q} = \frac{P - R}{Q}$$

In Examples 1–6, perform each operation and simplify the result.

EXAMPLE **1** $\dfrac{3}{7} + \dfrac{2}{7}$

SOLUTION

$$\frac{3}{7} + \frac{2}{7} = \frac{3 + 2}{7} = \frac{5}{7}$$

equal denominators

EXAMPLE **2** $\dfrac{7}{9} - \dfrac{4}{9}$

SOLUTION

$$\frac{7}{9} - \frac{4}{9} = \frac{7 - 4}{9} = \frac{3}{9} = \frac{1}{3}$$

equal denominators

EXAMPLE **3** $\dfrac{7}{c^2} + \dfrac{13c}{c^2}$

SOLUTION

$$\frac{7}{c^2} + \frac{13c}{c^2} = \frac{7 + 13c}{c^2}$$

Note that the fraction $(7 + 13c)/c^2$ cannot be reduced because we cannot factor the numerator.

EXAMPLE 4 $\dfrac{25}{5 - y} - \dfrac{y^2}{5 - y}$

SOLUTION

$$\frac{25}{5 - y} - \frac{y^2}{5 - y} = \frac{25 - y^2}{5 - y} = \frac{(5 - y)(5 + y)}{5 - y} = 5 + y$$

EXAMPLE 5 $\dfrac{3x - 1}{x^2 + x - 12} + \dfrac{2x - 14}{x^2 + x - 12}$

SOLUTION

$$\frac{3x - 1}{x^2 + x - 12} + \frac{2x - 14}{x^2 + x - 12} = \frac{3x - 1 + 2x - 14}{x^2 + x - 12} = \frac{5x - 15}{x^2 + x - 12}$$

$$= \frac{5(x - 3)}{(x - 3)(x + 4)} = \frac{5}{x + 4}$$

EXAMPLE 6 $\dfrac{8t + 5}{6t^2 - 11t + 3} - \dfrac{5t + 6}{6t^2 - 11t + 3}$

SOLUTION

$$\frac{8t + 5}{6t^2 - 11t + 3} - \frac{5t + 6}{6t^2 - 11t + 3} = \frac{8t + 5 - (5t + 6)}{6t^2 - 11t + 3} = \frac{8t + 5 - 5t - 6}{6t^2 - 11t + 3}$$

$$= \frac{3t - 1}{(3t - 1)(2t - 3)} = \frac{1}{2t - 3}$$

To add (or subtract) fractions with different denominators, we change the fractions to equivalent fractions that do have the same denominators (fractions with a **common denominator**). We then follow the rule for adding fractions with like denominators. For example:

$$\frac{3}{4} + \frac{5}{7} = \frac{3 \cdot 7}{4 \cdot 7} + \frac{5 \cdot 4}{7 \cdot 4} = \frac{21}{28} + \frac{20}{28} = \frac{21 + 20}{28} = \frac{41}{28}$$

and

$$\frac{3}{4} - \frac{2}{5} = \frac{3 \cdot 5}{4 \cdot 5} - \frac{2 \cdot 4}{5 \cdot 4} = \frac{15}{20} - \frac{8}{20} = \frac{15 - 8}{20} = \frac{7}{20}.$$

In general, if $Q \neq S$, we can add and subtract the fractions P/Q and R/S as follows:

Rule 1. $\dfrac{P}{Q} + \dfrac{R}{S} = \dfrac{PS + RQ}{QS}$ Rule 2. $\dfrac{P}{Q} - \dfrac{R}{S} = \dfrac{PS - RQ}{QS}$

In Examples 7–9 perform each operation.

EXAMPLE 7 $\dfrac{2}{5} + \dfrac{5}{6}$

SOLUTION   Following Rule 1 of addition, we have

$$\frac{2}{5} + \frac{5}{6} = \frac{2(6) + 5(5)}{5(6)} = \frac{12 + 25}{30} = \frac{37}{30}.$$

EXAMPLE 8 $\dfrac{5}{x} - \dfrac{3}{y}$

SOLUTION   Following Rule 2 of subtraction, we have

$$\frac{5}{x} - \frac{3}{y} = \frac{5y - 3x}{xy}.$$

EXAMPLE 9 $\dfrac{7}{10x} + \dfrac{8}{15x}$

SOLUTION   Following Rule 1 of addition, we have

$$\frac{7}{10x} + \frac{8}{15x} = \frac{7(15x) + 8(10x)}{(10x)(15x)}$$

$$= \frac{105x + 80x}{150x^2} = \frac{185x}{150x^2} = \frac{\cancel{5}(37)\cancel{x}}{\cancel{5}(30)\cancel{x^2}\,_x} = \frac{37}{30x}.$$

We could have saved steps in Example 9 by changing each fraction to an equivalent fraction which had the **least common denominator** (LCD) of the original denominators. We can determine the LCD of two fractions by using the following procedure:

Finding the LCD

Step 1. Factor each denominator completely into a product of prime factors or into a product of prime factors and -1.

Step 2. List each prime factor the greatest number of times it appears in any one factored denominator. The product of the listed factors is the least common denominator, or LCD.

Returning to Example 9,

$$\frac{7}{10x} + \frac{8}{15x},$$

we determine the LCD by factoring the first denominator, $10x$, as

$$2 \cdot 5 \cdot x,$$

and the second denominator, $15x$, as

$$3 \cdot 5 \cdot x.$$

We list each factor the greatest number of times it appears in any one factored denominator to obtain the LCD:

$$2 \cdot 5 \cdot x \cdot 3 \quad \text{or} \quad 30x.$$

We must now multiply the numerator and the denominator of each of the original fractions by an expression that will make each denominator equal to the LCD:

$$\frac{7}{10x} + \frac{8}{15x} = \frac{7(3)}{10x(3)} + \frac{8(2)}{15x(2)} = \frac{21}{30x} + \frac{16}{30x}$$

$$= \frac{21 + 16}{30x} = \frac{37}{30x}.$$

In Examples 10–16 perform each operation.

EXAMPLE 10 $\dfrac{3}{4x} + \dfrac{5}{6xy}$

SOLUTION   First we find the LCD of the fractions. The following table is helpful in constructing the LCD:

Denominators	Powers of Prime Factors			
$4x$	2^2		x	
$6xy$	2	3	x	y
LCD	2^2	3	x	y

Thus the LCD is

$$2^2 \cdot 3 \cdot x \cdot y = 12xy.$$

Now we multiply the numerator and the denominator of each fraction by an expression that will make each denominator equal to the LCD:

$$\frac{3}{4x} + \frac{5}{6xy} = \frac{3(3y)}{4x(3y)} + \frac{5(2)}{6xy(2)}$$

$$= \frac{9y}{12xy} + \frac{10}{12xy}$$

$$= \frac{9y + 10}{12xy}.$$

EXAMPLE 11 $\dfrac{5}{a^2b} - \dfrac{3}{ab^3}$

SOLUTION

Note that the denominators are in factored form. Therefore, the LCD is

$$a^2b^3,$$

which is the product of the highest power of a and the highest power of b that occur in either denominator. Now we have

$$\frac{5}{a^2b} - \frac{3}{ab^3} = \frac{5(b^2)}{a^2b(b^2)} - \frac{3(a)}{ab^3(a)}$$

$$= \frac{5b^2 - 3a}{a^2b^3}.$$

EXAMPLE 12 $\dfrac{5}{y^2 - y} - \dfrac{4}{y^2 - 1}$

SOLUTION

The following table shows how the LCD of the fractions is obtained:

Denominators	Prime Factors		
$y^2 - y$	y	$y - 1$	
$y^2 - 1$		$y - 1$	$y + 1$
LCD	y	$y - 1$	$y + 1$

Thus, the LCD is

$$y(y - 1)(y + 1),$$

and we have

$$\frac{5}{y^2 - y} - \frac{4}{y^2 - 1} = \frac{5}{y(y - 1)} - \frac{4}{(y - 1)(y + 1)}$$

$$= \frac{5(y + 1)}{y(y - 1)(y + 1)} - \frac{4y}{(y - 1)(y + 1)y}$$

$$= \frac{5(y + 1) - 4y}{y(y - 1)(y + 1)}$$

$$= \frac{5y + 5 - 4y}{y(y - 1)(y + 1)}$$

$$= \frac{y + 5}{y(y - 1)(y + 1)}.$$

EXAMPLE **13**

$$\frac{6}{w^2 - 2w - 8} + \frac{5}{w^2 + 2w}$$

SOLUTION   The following table shows how the LCD is obtained:

Denominators	Prime Factors		
$w^2 - 2w - 8$	$w + 2$	$w - 4$	
$w^2 + 2w$	$w + 2$		w
LCD	$w + 2$	$w - 4$	w

Thus, the LCD is

$$(w + 2)(w - 4)w,$$

and we have

$$\frac{6}{w^2 - 2w - 8} + \frac{5}{w^2 + 2w} = \frac{6}{(w + 2)(w - 4)} + \frac{5}{w(w + 2)}$$

$$= \frac{6w}{(w + 2)(w - 4)w} + \frac{5(w - 4)}{w(w + 2)(w - 4)}$$

$$= \frac{6w + 5(w - 4)}{(w + 2)(w - 4)w}$$

$$= \frac{6w + 5w - 20}{(w + 2)(w - 4)w}$$

$$= \frac{11w - 20}{(w + 2)(w - 4)w}.$$

EXAMPLE **14** $\dfrac{5}{a-3} - \dfrac{5}{3-a}$

SOLUTION Here, we observe that the denominators are negatives of each other:

$$a - 3 = -(3 - a).$$

Therefore, we can change one denominator into the other by applying Rule (i), page 88, which states that

$$-\frac{P}{Q} = \frac{P}{-Q}$$

$$\frac{5}{a-3} - \frac{5}{3-a} = \frac{5}{a-3} + \frac{5}{-(3-a)}$$

$$= \frac{5}{a-3} + \frac{5}{a-3}$$

$$= \frac{5+5}{a-3} = \frac{10}{a-3}.$$

EXAMPLE **15** $\dfrac{x}{x+1} - \dfrac{x}{x-1} + \dfrac{2}{x^2-1}$

SOLUTION The following table shows how the LCD is obtained:

Denominators	Prime Factors	
$x + 1$	$x + 1$	
$x - 1$		$x - 1$
$x^2 - 1$	$x + 1$	$x - 1$
LCD	$x + 1$	$x - 1$

Therefore, the LCD is

$$(x - 1)(x + 1),$$

and we have

$$\frac{x}{x+1} - \frac{x}{x-1} + \frac{2}{x^2-1} = \frac{x}{x+1} - \frac{x}{x-1} + \frac{2}{(x-1)(x+1)}$$

$$= \frac{x(x-1)}{(x+1)(x-1)} - \frac{x(x+1)}{(x-1)(x+1)} + \frac{2}{(x-1)(x+1)}$$

$$= \frac{x(x-1) - x(x+1) + 2}{(x+1)(x-1)}$$

$$= \frac{x^2 - x - x^2 - x + 2}{(x + 1)(x - 1)}$$

$$= \frac{-2x + 2}{(x + 1)(x - 1)}$$

$$= \frac{-2(x - 1)}{(x + 1)(x - 1)}$$

$$= \frac{-2}{x + 1}.$$

EXAMPLE **16** $\dfrac{p}{p^2 + 6p + 5} - \dfrac{2}{p^2 + 4p - 5} + \dfrac{3}{p^2 - 1}$

SOLUTION   The following table shows how the LCD of the fractions is obtained:

Denominators	Prime Factors		
$p^2 + 6p + 5$	$p + 1$	$p + 5$	
$p^2 + 4p - 5$		$p + 5$	$p - 1$
$p^2 - 1$	$p + 1$		$p - 1$
LCD	$p + 1$	$p + 5$	$p - 1$

Thus, the LCD is
$$(p + 1)(p + 5)(p - 1),$$
and we have

$$\frac{p}{p^2 + 6p + 5} - \frac{2}{p^2 + 4p - 5} + \frac{3}{p^2 - 1}$$

$$= \frac{p}{(p + 1)(p + 5)} - \frac{2}{(p + 5)(p - 1)} + \frac{3}{(p + 1)(p - 1)}$$

$$= \frac{p(p - 1)}{(p + 1)(p + 5)(p - 1)} - \frac{2(p + 1)}{(p + 5)(p - 1)(p + 1)} + \frac{3(p + 5)}{(p + 1)(p - 1)(p + 5)}$$

$$= \frac{p(p - 1) - 2(p + 1) + 3(p + 5)}{(p + 1)(p + 5)(p - 1)}$$

$$= \frac{p^2 - p - 2p - 2 + 3p + 15}{(p + 1)(p + 5)(p - 1)}$$

$$= \frac{p^2 + 13}{(p + 1)(p + 5)(p - 1)}.$$

PROBLEM SET 3.3

In problems 1–20, perform each operation and simplify the result.

1. $\dfrac{5}{8} + \dfrac{1}{8}$

2. $\dfrac{7}{12} + \dfrac{5}{12}$

3. $\dfrac{6}{5x} + \dfrac{14}{5x}$

4. $\dfrac{3}{2y^2} + \dfrac{7}{2y^2}$

5. $\dfrac{7'}{12} - \dfrac{3}{12}$

6. $\dfrac{9}{14} - \dfrac{2}{14}$

7. $\dfrac{t-1}{4} + \dfrac{t+1}{4}$

8. $\dfrac{3x-1}{5} + \dfrac{2x+1}{5}$

9. $\dfrac{2x}{x^2-4} - \dfrac{4}{x^2-4}$

10. $\dfrac{2t}{4t^2-9} - \dfrac{-3}{4t^2-9}$

11. $\dfrac{36}{6-v} + \dfrac{-v^2}{6-v}$

12. $\dfrac{4}{16-x^2} + \dfrac{x}{16-x^2}$

13. $\dfrac{3m}{m^2+m-2} + \dfrac{6}{m^2+m-2}$

14. $\dfrac{12y}{2y^2+y-1} - \dfrac{6}{2y^2+y-1}$

15. $\dfrac{3x}{3x^2-10x-8} - \dfrac{12}{3x^2-10x-8}$

16. $\dfrac{4t}{4t^2+5t-6} + \dfrac{8}{4t^2+5t-6}$

17. $\dfrac{15}{8u^2-6u-5} - \dfrac{12u}{8u^2-6u-5}$

18. $\dfrac{4y}{21x^2+32xy-5y^2} - \dfrac{28x}{21x^2+32xy-5y^2}$

19. $\dfrac{8u^3}{4u^2+6uv+9v^2} - \dfrac{27v^3}{4u^2+6uv+9v^2}$

20. $\dfrac{16t^2}{64t^3+27s^3} - \dfrac{12ts-9s^2}{64t^3+27s^3}$

In problems 21–34, perform each operation by using Rule 1 or 2 from page 98. Simplify the result.

21. $\dfrac{5}{8} + \dfrac{3}{7}$

22. $\dfrac{2}{3} + \dfrac{5}{8}$

23. $\dfrac{7}{12} - \dfrac{1}{5}$

24. $\dfrac{15}{16} - \dfrac{3}{7}$

25. $\dfrac{2x}{3} + \dfrac{9x}{10}$

26. $\dfrac{3t}{4} + \dfrac{2t}{5}$

27. $\dfrac{5u}{7} - \dfrac{2u}{3}$

28. $\dfrac{3-x}{4} - \dfrac{x+1}{2}$

29. $\dfrac{9}{y-5} + \dfrac{6}{y-3}$

30. $\dfrac{2a}{a-3} + \dfrac{3}{a+7}$

31. $\dfrac{m}{m-3} - \dfrac{2}{m+3}$

32. $\dfrac{2y}{y-2} - \dfrac{y}{y+1}$

33. $\dfrac{x}{3x+2} + \dfrac{1}{x-4}$

34. $\dfrac{2u}{4u+3} + \dfrac{u}{2u+5}$

In problems 35–60, perform each operation and simplify the result.

35. $\dfrac{5}{t^2s} + \dfrac{3}{2ts^3}$

36. $\dfrac{7}{4x^3y^2} + \dfrac{5}{12xy^4}$

37. $\dfrac{12}{5m^3n} - \dfrac{4}{3m^2n^5}$

38. $\dfrac{9}{4u^4v^2} - \dfrac{3}{8v^5w^3}$

39. $\dfrac{3}{x^2 + x} + \dfrac{2}{x^2 - 1}$

40. $\dfrac{3v}{4v^2 - 1} + \dfrac{1}{2v + 1}$

41. $\dfrac{c}{c^2 - 9} - \dfrac{c - 1}{c^2 - 5c + 6}$

42. $\dfrac{t}{t^2 + 5t - 6} + \dfrac{3}{t + 6}$

43. $\dfrac{m - 5}{m^2 - 5m - 6} + \dfrac{m + 4}{m^2 - 6m}$

44. $\dfrac{2}{a^2 - 4} + \dfrac{7}{a^2 - 4a - 12}$

45. $\dfrac{x - 2}{x^2 + 10x + 16} + \dfrac{x + 1}{x^2 + 9x + 14}$

46. $\dfrac{u}{u^2 - 2uv + v^2} - \dfrac{v}{u^2 - v^2}$

47. $\dfrac{3y}{y^2 + 7y + 10} - \dfrac{y}{y^2 + y - 20}$

48. $\dfrac{1}{3x^2 - x - 2} - \dfrac{1}{2x^2 - x - 1}$

49. $\dfrac{4}{3t - 2} + \dfrac{1}{2 - 3t}$

50. $\dfrac{12}{5y - 7} - \dfrac{3}{7 - 5y}$

51. $\dfrac{3x + 1}{2x^2 - 5x - 12} - \dfrac{x + 4}{6x^2 + 7x - 3}$

52. $\dfrac{2t + 3}{3t^2 + 2t - 8} + \dfrac{3t + 4}{2t^2 + t - 6}$

53. $\dfrac{v}{v + 2} - \dfrac{v}{v - 2} - \dfrac{v^2}{v^2 - 4}$

54. $\dfrac{x - 3}{x + 3} - \dfrac{x + 3}{3 - x} + \dfrac{x^2}{9 - x^2}$

55. $\dfrac{2t^2 - t}{3t^2 - 27} - \dfrac{t - 3}{3t - 9} + \dfrac{6t^2}{9 - t^2}$

56. $\dfrac{4}{a^2 - 4} + \dfrac{2}{a^2 - 4a + 4} + \dfrac{1}{a^2 + 4a + 4}$

57. $\dfrac{1}{x^2 - x - 6} + \dfrac{2}{x^2 - 6x + 9} - \dfrac{1}{x^2 + 4x + 4}$

58. $\dfrac{2x^2}{x^4 - 1} + \dfrac{1}{x^2 + 1} + \dfrac{1}{x^2 - 1}$

59. $\dfrac{2y - 1}{3y^2 - y - 2} + \dfrac{y + 2}{y^2 + 2y - 3} - \dfrac{y - 3}{3y^2 + 11y + 6}$

60. $\dfrac{15c}{8c^2 - 26c + 15} - \dfrac{2c}{6c^2 - 13c - 5} - \dfrac{c}{12c^2 - 5c - 3}$

3.4 Complex Fractions

We have worked with fractions of the form P/Q, $Q \neq 0$, where P and Q are polynomial expressions (called simple fractions). In this section, we look at **complex fractions**—fractions in which the numerator or the denominator (or both) contain fractions. For example,

$$\frac{\dfrac{5}{7}}{3}, \quad \frac{\dfrac{2}{x}}{\dfrac{11}{x}}, \quad \frac{3 + \dfrac{3}{y}}{9}, \quad \frac{9}{\dfrac{w}{w - 5}}, \quad \frac{\dfrac{5}{t}}{7 + \dfrac{2}{t}}, \quad \text{and} \quad \frac{p + \dfrac{3}{p}}{\dfrac{p}{2 - p} - \dfrac{p}{2 - p}}$$

are complex fractions.

To simplify a complex fraction, we express it as a simple fraction in lowest terms. We do this by using one of the following methods:

> Method 1. Express the complex fraction as a quotient of simple fractions: add and subtract as indicated in the numerator and denominator of the complex fraction; then divide and simplify the result.
>
> Method 2. Multiply the numerator and the denominator of the complex fraction by the least common denominator (LCD) of all the fractions in the numerator and the denominator. Simplify the result.

In Examples 1–3, express each of the following fractions as a simple fraction and simplify the result.

EXAMPLE 1

$$\frac{\dfrac{7}{11}}{\dfrac{5}{44}}$$

SOLUTION   Method 1:

$$\frac{\dfrac{7}{11}}{\dfrac{5}{44}} = \frac{7}{11} \div \frac{5}{44} = \frac{7}{\cancel{11}} \cdot \frac{\overset{4}{\cancel{44}}}{5} = \frac{28}{5}$$

Method 2: The LCD is 44. Therefore,

$$\frac{\dfrac{7}{11}}{\dfrac{5}{44}} = \frac{\dfrac{7}{11} \cdot 44}{\dfrac{5}{44} \cdot 44} = \frac{7 \cdot 4}{5 \cdot 1} = \frac{28}{5}$$

EXAMPLE 2

$$\frac{\dfrac{9}{xy^2}}{\dfrac{12}{x^2 y}}$$

SOLUTION   Method 1:

$$\frac{\dfrac{9}{xy^2}}{\dfrac{12}{x^2 y}} = \frac{9}{xy^2} \div \frac{12}{x^2 y}$$

$$= \frac{\overset{3}{\cancel{9}}}{\cancel{xy^2}_{y}} \cdot \frac{\cancel{x^2 y}^{x}}{\cancel{12}_{4}} = \frac{3x}{4y}$$

Method 2: The LCD is $x^2 y^2$, so

$$\frac{\dfrac{9}{xy^2}}{\dfrac{12}{x^2 y}} = \frac{\dfrac{9}{xy^2} \cdot x^2 y^2}{\dfrac{12}{x^2 y} \cdot x^2 y^2}$$

$$= \frac{\overset{3}{\cancel{9}x}}{\cancel{12}y_{4}} = \frac{3x}{4y}$$

EXAMPLE 3

$$\frac{8 + \dfrac{4}{t}}{\dfrac{2t + 1}{12}}$$

SOLUTION

Method 1:

$$\frac{8 + \dfrac{4}{t}}{\dfrac{2t + 1}{12}} = \frac{\dfrac{8t + 4}{t}}{\dfrac{2t + 1}{12}}$$

$$= \frac{8t + 4}{t} \div \frac{2t + 1}{12}$$

$$= \frac{8t + 4}{t} \cdot \frac{12}{2t + 1}$$

$$= \frac{4(2t + 1)}{t} \cdot \frac{12}{2t + 1}$$

$$= \frac{48}{t}$$

Method 2: The LCD is $12t$. So we have

$$\frac{8 + \dfrac{4}{t}}{\dfrac{2t + 1}{12}} = \frac{\left(8 + \dfrac{4}{t}\right) \cdot 12t}{\left(\dfrac{2t + 1}{12}\right) \cdot 12t}$$

$$= \frac{96t + 48}{2t^2 + t} = \frac{48(2t + 1)}{t(2t + 1)}$$

$$= \frac{48}{t}$$

In the following examples, we use only Method 2. This method is usually easier to apply to more complicated problems.

In Examples 4–6, simplify each complex fraction.

EXAMPLE 4

$$\frac{c - \dfrac{1}{d}}{1 - \dfrac{c}{d}}$$

SOLUTION

We multiply both the numerator and the denominator of the fraction by d (their LCD):

$$\frac{c - \dfrac{1}{d}}{1 - \dfrac{c}{d}} = \frac{d\left(c - \dfrac{1}{d}\right)}{d\left(1 - \dfrac{c}{d}\right)} = \frac{d(c) - d\left(\dfrac{1}{d}\right)}{d(1) - d\left(\dfrac{c}{d}\right)}$$

$$= \frac{cd - 1}{d - c}.$$

EXAMPLE 5

$$\frac{\dfrac{a}{b} - \dfrac{b}{a}}{\dfrac{1}{a} + \dfrac{1}{b}}$$

SOLUTION

The LCD is ab, so we have

$$\frac{\dfrac{a}{b} - \dfrac{b}{a}}{\dfrac{1}{a} + \dfrac{1}{b}} = \frac{\left(\dfrac{a}{b} - \dfrac{b}{a}\right)ab}{\left(\dfrac{1}{a} + \dfrac{1}{b}\right)ab} = \frac{\left(\dfrac{a}{b}\right)ab - \left(\dfrac{b}{a}\right)ab}{\left(\dfrac{1}{a}\right)ab + \left(\dfrac{1}{b}\right)ab}$$

$$= \frac{a^2 - b^2}{b + a} = \frac{(a - b)(a + b)}{b + a} = a - b.$$

EXAMPLE 6

$$\frac{\dfrac{1}{t + 1} - 1}{1 - \dfrac{1}{t}}$$

SOLUTION

The LCD is $t(t + 1)$, so we have

$$\frac{\dfrac{1}{t + 1} - 1}{1 - \dfrac{1}{t}} = \frac{\left(\dfrac{1}{t + 1} - 1\right)t(t + 1)}{\left(1 - \dfrac{1}{t}\right)t(t + 1)} = \frac{\left(\dfrac{1}{t + 1}\right)t(t + 1) - (1)t(t + 1)}{(1)t(t + 1) - \left(\dfrac{1}{t}\right)t(t + 1)}$$

$$= \frac{t - t(t + 1)}{t(t + 1) - (t + 1)} = \frac{t - t^2 - t}{t^2 + t - t - 1} = \frac{-t^2}{t^2 - 1} = \frac{t^2}{1 - t^2}.$$

PROBLEM SET 3.4

In problems 1–24, simplify each complex fraction.

1. $\dfrac{\dfrac{2}{3}}{\dfrac{4}{5}}$

2. $\dfrac{\dfrac{17}{6}}{\dfrac{34}{9}}$

3. $\dfrac{\dfrac{35x}{24y}}{\dfrac{7x}{9y^2}}$

4. $\dfrac{\dfrac{a^2b^2}{8}}{\dfrac{5ab^3}{16}}$

5. $\dfrac{3 + \dfrac{1}{c}}{2 - \dfrac{3}{c}}$

6. $\dfrac{5 + \dfrac{1}{x}}{1 - \dfrac{1}{x}}$

7. $\dfrac{\dfrac{1}{y}}{2 - \dfrac{1}{y}}$

8. $\dfrac{3 - \dfrac{x}{5}}{5 - \dfrac{x}{3}}$

9. $\dfrac{\dfrac{2}{1+v}}{3+\dfrac{1}{1+v}}$

10. $\dfrac{\dfrac{1}{t+1}-2}{3-\dfrac{1}{t+1}}$

11. $\dfrac{1+\dfrac{2}{m}}{1-\dfrac{4}{m^2}}$

12. $\dfrac{\dfrac{3}{y}-\dfrac{x}{y}}{\dfrac{3}{y}-1}$

13. $\dfrac{\dfrac{x}{y}-\dfrac{y}{x}}{\dfrac{x}{y}+\dfrac{y}{x}}$

14. $\dfrac{\dfrac{u}{v}+1+\dfrac{v}{u}}{\dfrac{u^3-v^3}{uv}}$

15. $\dfrac{\dfrac{1}{a+b}-\dfrac{1}{a-b}}{\dfrac{4}{a^2-b^2}}$

16. $\dfrac{\dfrac{a}{b}-2+\dfrac{b}{a}}{\dfrac{a}{b}+2+\dfrac{b}{a}}$

17. $\dfrac{\dfrac{3}{1-x}+\dfrac{x}{x-1}}{\dfrac{1}{1-x}}$

18. $\dfrac{\dfrac{c-3}{c-2}+\dfrac{c+1}{c+2}}{\dfrac{c^2-3c+6}{c^2-4}}$

19. $\dfrac{\dfrac{2}{t}+\dfrac{1}{t+1}}{\dfrac{3}{t+1}-\dfrac{1}{t}}$

20. $\dfrac{\dfrac{x^2-y^2}{x+y}}{\dfrac{x}{y}+1}$

21. $\dfrac{\dfrac{x-1}{x+1}-\dfrac{x+1}{x-1}}{\dfrac{x-1}{x+1}+\dfrac{x+1}{x-1}}$

22. $\dfrac{m+\dfrac{n}{n-m}}{n+\dfrac{m}{m-n}}$

23. $\dfrac{\dfrac{a}{b}-1+\dfrac{b}{a}}{\dfrac{a^3+b^3}{a^2b+ab^2}}$

24. $\dfrac{\dfrac{2u-v}{2u+v}+\dfrac{2u+v}{2u-v}}{\dfrac{2u-v}{2u+v}-\dfrac{2u+v}{2u-v}}$

REVIEW PROBLEM SET

In problems 1–4, determine all values of the variable for which the fraction is not defined.

1. $\dfrac{8+t}{t-5}$

2. $\dfrac{3x}{x+11}$

3. $\dfrac{3x}{(x+2)(x-7)}$

4. $\dfrac{y^2+1}{(y-9)(y+4)}$

In problems 5–10, determine whether the given pair of fractions are equivalent.

5. $\dfrac{3}{uv}$ and $\dfrac{6u}{2u^2v}$

6. $\dfrac{9a}{11b^2}$ and $\dfrac{27a^3b^2}{33a^2b^4}$

7. $\dfrac{c-4}{c^2-16}$ and $\dfrac{1}{c+4}$

8. $\dfrac{2x-6}{2}$ and $\dfrac{x^2-5x+6}{x-3}$

9. $\dfrac{y+3}{2y+2}$ and $\dfrac{y-3}{2y-2}$

10. $\dfrac{3a-b}{3}$ and $\dfrac{a^3-b^3}{a^2+ab+b^2}$

In problems 11–20, reduce each fraction to lowest terms.

11. $\dfrac{12u^3v^5}{18u^7v^2}$

12. $\dfrac{28ts^3}{35t^4s}$

13. $\dfrac{3m+3n}{9m^2-9n^2}$

14. $\dfrac{4y^2-9}{2y^2-3y}$

15. $\dfrac{x^2-7x+12}{x^2+3x-18}$

16. $\dfrac{2a^2-2b^2}{5a^2+10ab+5b^2}$

17. $\dfrac{2t^2-7t-15}{4t^2-4t-15}$

18. $\dfrac{3y^2-17y-28}{3y^2+10y+8}$

19. $\dfrac{x^4-8x}{x^4+2x^3+4x^2}$

20. $\dfrac{ax+ay-bx-by}{bx-by-ax+ay}$

In problems 21–28, find the missing numerator that will make the fractions equivalent.

21. $\dfrac{7x}{9y} = \dfrac{?}{27xy^3}$

22. $\dfrac{5u^2}{11vw} = \dfrac{?}{33v^2w^4}$

23. $\dfrac{3a}{2a - 2b} = \dfrac{?}{2a^2 - 2b^2}$

24. $\dfrac{2c}{3c + 1} = \dfrac{?}{3c^2 - 14c - 5}$

25. $\dfrac{2z + 1}{z - 2} = \dfrac{?}{3z^2 - 4z - 4}$

26. $\dfrac{xy}{x - 2y} = \dfrac{?}{2x^3 - 3x^2y - 2xy^2}$

27. $\dfrac{w}{w + 3} = \dfrac{?}{w^3 + 27}$

28. $\dfrac{3}{t - 2} = \dfrac{?}{3t^2 - 3t - 6}$

In problems 29–32, write an equivalent fraction with denominator $b - a$.

29. $\dfrac{4b}{a - b}$

30. $\dfrac{-5}{a - b}$

31. $-\dfrac{-3b}{a - b}$

32. $\dfrac{2b - a}{a - b}$

In problems 33–58, perform the indicated operations and simplify the result.

33. $\dfrac{5x^3y}{12uv^4} \cdot \dfrac{6u^3v}{25xy^2}$

34. $\dfrac{14m^3n^5}{9ab^4} \cdot \dfrac{3a^4b}{7mn^7}$

35. $\dfrac{b - 1}{b - 5} \cdot \dfrac{3}{2b - 2}$

36. $\dfrac{t^2 - 4}{5t + 15} \cdot \dfrac{5}{t + 2}$

37. $\dfrac{9 - y^2}{x^3 - x} \cdot \dfrac{x - 1}{y + 3}$

38. $\dfrac{a^2 + 8a + 16}{a^2 - 9} \cdot \dfrac{a - 3}{a + 4}$

39. $\dfrac{u^2 - v^2}{u^2 + 2uv + v^2} \cdot \dfrac{3u + 3v}{6u}$

40. $\dfrac{4x^2 - 64}{2x^2 - 8x} \cdot \dfrac{x - 4}{x + 4}$

41. $\dfrac{x^2 + 5x + 6}{x^2 + x - 2} \cdot \dfrac{x^2 + 3x - 4}{x^2 + 7x + 12}$

42. $\dfrac{4x^2 - 11x - 3}{6x^2 - 5x - 6} \cdot \dfrac{6x^2 - 13x + 6}{4x^2 + 13x + 3}$

43. $\dfrac{8t^3 + 1}{t^3 - 27} \cdot \dfrac{t^2 + 3t + 9}{4t^2 - 2t + 1}$

44. $\dfrac{a^3 - b^3}{a^2 - 4b^2} \cdot \dfrac{a^2 - ab - 2b^2}{a^2 + ab + b^2}$

45. $\dfrac{14x^2}{5b^2} \div \dfrac{21x^4}{15b}$

46. $\dfrac{5a}{12yz^2} \div \dfrac{25a^3}{18y^2z^3}$

47. $\dfrac{v + 7}{v + 2} \div \dfrac{v^2 - 49}{v^2 - 4}$

48. $\dfrac{4x^2 - 9}{9x^2 - 4} \div \dfrac{2x - 3}{3x + 2}$

49. $\dfrac{m^2 - m - 2}{m^2 - m - 6} \div \dfrac{m^2 - 2m}{2m + m^2}$

50. $\dfrac{y^2 - 5y + 6}{y^2 + y - 2} \div \dfrac{y^2 + y - 12}{y^2 + 3y - 4}$

51. $\dfrac{x^2 + 3xy + 2y^2}{x^2 - 2xy} \div \dfrac{x^2 + 4xy + 3y^2}{x^2 + xy - 6y^2}$

52. $\dfrac{t^3 + 1}{t^2 - 4u^2} \div \dfrac{t^2 - t + 1}{t - 2u}$

53. $\dfrac{2w^2 - w - 6}{3w^2 - 11w - 4} \div \dfrac{2w^2 + 5w + 3}{3w^2 + 7w + 2}$

54. $\dfrac{8x^2 + 18x + 9}{6x^2 - 7x + 2} \div \dfrac{4x^2 + 7x + 3}{2x^2 + 9x - 5}$

55. $\left[\dfrac{x^2 - 1}{x^2 + 4x + 4} \cdot \dfrac{x^2 - x - 6}{x^2 - 2x + 1} \right] \div \dfrac{x^2 + x}{x^2 - x}$

56. $\left[\dfrac{2y^2 - 3y - 2}{9y^2 - 1} \cdot \dfrac{3y^2 - 13y + 4}{2y^2 - y - 6} \right] \div \dfrac{2y^2 - 7y - 4}{3y^2 - 5y - 2}$

57. $\dfrac{25u^2 - 9}{2u^2 + 3u - 14} \div \left[\dfrac{5u^2 + 22u - 15}{u^2 + 7u + 10} \cdot \dfrac{5u^2 + 3u}{2u^2 + 7u} \right]$

58. $\dfrac{30x^2 + 6x}{40x^2 - 8x} \div \left[\dfrac{25x^2 + 10x + 1}{25x^2 - 10x + 1} \cdot \dfrac{25x^2 - 1}{20x + 4} \right]$

In problems 59–80, perform the indicated operations and simplify the result.

59. $\dfrac{5}{7x} + \dfrac{9}{7x}$

60. $\dfrac{3}{15y^2} + \dfrac{2}{15y^2}$

61. $\dfrac{3}{9 - t^2} - \dfrac{t}{9 - t^2}$

62. $\dfrac{3a}{a^2 - 4b^2} - \dfrac{6b}{a^2 - 4b^2}$

63. $\dfrac{3y}{y^2 - y - 2} + \dfrac{3}{y^2 - y - 2}$

64. $\dfrac{15t}{6t^2 - 5t - 6} + \dfrac{10}{6t^2 - 5t - 6}$

65. $\dfrac{y}{x - y} + \dfrac{x}{x + y}$

66. $\dfrac{t + 1}{2t - 1} - \dfrac{t - 1}{2t + 1}$

67. $\dfrac{2}{3 - t} - \dfrac{5}{t - 3}$

68. $\dfrac{x}{x + 2} + \dfrac{x}{x + 3}$

69. $\dfrac{y + 2}{y + 4} - \dfrac{y - 1}{y + 6}$

70. $\dfrac{z + 1}{3z + 2} - \dfrac{z - 3}{2z - 5}$

71. $\dfrac{4}{5m^2n} + \dfrac{3}{4mn^3}$

72. $\dfrac{3}{14x^3y^2} + \dfrac{5}{21x^2y^5}$

73. $\dfrac{3v}{4v^2 - 9} - \dfrac{v}{2v^2 + v - 6}$

74. $\dfrac{5}{x^2 + 8x + 15} - \dfrac{4}{x^2 + 2x - 3}$

75. $\dfrac{5}{x^2 - 7x + 12} + \dfrac{3}{2x^2 - 10x + 8}$

76. $\dfrac{7}{w^2 - z^2} + \dfrac{2}{w^2 - 2wz + z^2}$

77. $\dfrac{2}{2m^2 + m - 3} + \dfrac{4}{m^2 - 1} - \dfrac{1}{2m^2 + 5m + 3}$

78. $\dfrac{4c}{3c^2 + 5c - 2} - \dfrac{c}{2c^2 + 5c + 2} - \dfrac{2}{6c^2 + c - 1}$

79. $\dfrac{y}{x^2 - 9y^2} - \dfrac{x}{x^2 - 2xy - 3y^2} + \dfrac{x}{x^2 + 4xy + 3y^2}$

80. $\dfrac{3u}{2u^2 + 9u + 4} + \dfrac{2u}{4u^2 - 1} + \dfrac{u}{2u^2 + 7u - 4}$

In problems 81–90, simplify each complex fraction.

81. $\dfrac{\dfrac{34a}{4x^2y}}{\dfrac{17a^3}{2xy}}$

82. $\dfrac{\dfrac{25x^2}{9y^2z}}{\dfrac{100x^3}{27yz^2}}$

83. $\dfrac{\dfrac{3}{y} + \dfrac{x}{y}}{\dfrac{4}{y} + 2}$

84. $\dfrac{\dfrac{m}{n} + 1}{\dfrac{m^2}{n} - n}$

85. $\dfrac{2 + \dfrac{2}{y - 1}}{\dfrac{2}{y - 1}}$

86. $\dfrac{\dfrac{x^2 + y^2}{y} - 2z}{\dfrac{1}{y} - \dfrac{1}{x}}$

87. $\dfrac{1 - \dfrac{1}{t}}{t - 2 + \dfrac{1}{t}}$

88. $\dfrac{\dfrac{1}{a} - \dfrac{1}{b}}{\dfrac{a^2 - b^2}{ab}}$

89. $\dfrac{\dfrac{1}{2x - 3} - \dfrac{1}{2x + 3}}{\dfrac{x}{4x^2 - 9}}$

90. $\dfrac{\dfrac{y}{y^2 - 1} - \dfrac{1}{y + 1}}{\dfrac{y}{y - 1} + \dfrac{1}{y + 1}}$

4

Linear Equations
and Inequalities

Equations and inequalities are often used to describe situations and to solve problems. In this chapter, we apply the algebraic skills we have developed in the preceding chapters in order to solve algebraic equations and inequalities.

4.1 Equations

The following are examples of equations in one variable:

$$-2x + 37 = 39 \qquad 6y + 5 = 3(2y - 4) \qquad \sqrt{p - 2} = 5$$

$$u^2 - u - 6 = 0 \qquad \frac{2}{t} + \frac{1}{3} = \frac{2}{5} \qquad 2 + \frac{3}{3w + 1} = \frac{-4}{3w}.$$

An **equation** is a statement that two mathematical expressions (each representing a real number) are equal. The expressions on either side of the equals sign of an equation are called the **sides** or **members** of the equation. An equation will produce a statement which is either true or false when a particular number is substituted for the variable. If a true statement results when we substitute a number for the variable, we say the number substituted **satisfies** the equation. Take, for instance, the equation $3x + 6 = 18$; if we substitute $x = 4$, we have a true statement: $12 + 6 = 18$. Thus, $x = 4$ satisfies the equation $3x + 6 = 18$, and we say that the number 4 is a **solution** or a **root** of the equation. To **solve** an equation we find all of its solutions.

An **identity** is an equation that is satisfied by *every* real number substituted for the variable in which both sides of the equation are defined. For example, the equation

$$(x + 2)^2 = x^2 + 4x + 4$$

is an identity. An equation that is not an identity is called a **conditional equation.** Both sides of a conditional equation are defined, but they are not equal when at least one real number is substituted for the variable. For instance, $7x = 14$ is a conditional equation because there is at least one substitution (say, $x = 5$) that produces a false statement.

Two equations are **equivalent** if they have exactly the same solutions. Thus, the equation $5x - 15 = 0$ is equivalent to $5x = 15$ because both equations have the same solution, $x = 3$. You can change an equation into an equivalent one by performing any of the following operations:

(i) Addition and Subtraction Properties

Add or subtract the same quantity on both sides of the equation. That is,

$$\text{if } P = Q, \text{ then } P + R = Q + R \text{ and } P - R = Q - R.$$

(ii) Multiplication and Division Properties

Multiply or divide both sides of the equation by the same nonzero quantity. That is,

$$\text{if } P = Q, \text{ then } PR = QR \text{ and } \frac{P}{R} = \frac{Q}{R}, \text{ where } R \neq 0.$$

(iii) Interchanging Property

Interchange the two sides of the equation. That is,

$$\text{if } P = Q, \text{ then } Q = P.$$

In this section, we will use these properties to solve first-degree equations in one variable.

First-Degree or Linear Equations

Equations of the form

$$2x - 1 = 5, \qquad 3t - 5 = 7(1 + t), \qquad \text{and} \qquad 3u + u = 7u - 4,$$

are called **first-degree** or **linear equations.** These first-degree equations have a single variable which always has an exponent of 1. The most common technique for **solving** a first-degree (linear) equation is to write a sequence of equations (starting with the given equation) in which each equation is equivalent to the previous one. When we write this sequence of equations, we collect all the variable terms on one side of the equation, and the constant terms on the other side. Then we simplify each side of the equation, and we divide both sides by the coefficient of the variable to obtain the solution.

For example, to solve the equation

$$7x - 28 = 0,$$

we begin by adding 28 to both sides of the equation to obtain the equivalent equation:

$$7x - 28 + 28 = 0 + 28$$

or $\quad 7x = 28.$

We then divide both sides of the equation by 7 to get the equivalent equation

$$\frac{7x}{7} = \frac{28}{7}$$

or $\quad x = 4.$

Thus the solution is 4.

In Examples 1–7, solve each equation.

EXAMPLE 1 $\quad x + 7 = 13$

SOLUTION

$$x + 7 = 13$$
$$x + 7 - 7 = 13 - 7 \qquad \text{(We subtracted 7 from both sides.)}$$
$$x = 6.$$

Check We substitute 6 for x in the original equation:

$$6 + 7 \overset{?}{=} 13 \qquad \text{or} \qquad 13 = 13.$$

Thus, 6 is the solution.

EXAMPLE 2 $\quad 5t = 30$

SOLUTION

$$5t = 30$$
$$\frac{5t}{5} = \frac{30}{5} \qquad \text{(We divided both sides by 5.)}$$
$$t = 6.$$

Check We substitute $t = 6$ in the original equation:

$$5(6) \overset{?}{=} 30$$
$$30 = 30.$$

Therefore, the solution is 6.

EXAMPLE 3 $7y - 3 = 28$

SOLUTION
$$7y - 3 = 28$$
$$7y - 3 + 3 = 28 + 3 \qquad \text{(We added 3 to both sides.)}$$
$$7y = 31$$
$$\frac{7y}{7} = \frac{31}{7} \qquad \text{(We divided both sides by 7.)}$$
$$y = \frac{31}{7}.$$

Check Substitute $y = \frac{31}{7}$ in the original equation:
$$7\left(\frac{31}{7}\right) - 3 \overset{?}{=} 28$$
$$31 - 3 \overset{?}{=} 28$$
$$28 = 28.$$

Therefore, $\frac{31}{7}$ is the solution.

EXAMPLE 4 $5 - 4w = 21$

SOLUTION
$$5 - 4w = 21$$
$$5 - 4w - 5 = 21 - 5 \qquad \text{(We subtracted 5 from both sides.)}$$
$$-4w = 16$$
$$\frac{-4w}{-4} = \frac{16}{-4} \qquad \text{(We divided both sides by } -4.)$$
$$w = -4.$$

Check $5 - 4(-4) \overset{?}{=} 21$
$$5 + 16 \overset{?}{=} 21$$
$$21 = 21.$$

Therefore, -4 is the solution.

EXAMPLE 5 $12m + 1 = 25 - 12m$

SOLUTION

$$12m + 1 = 25 - 12m$$

$$12m + 1 + 12m = 25 - 12m + 12m$$ (We added $12m$ to both sides.)

$$24m + 1 = 25$$

$$24m + 1 - 1 = 25 - 1$$ (We subtracted 1 from both sides.)

$$24m = 24$$

$$\frac{24m}{24} = \frac{24}{24}$$ (We divided both sides by 24.)

$$m = 1.$$

Check Substitute $m = 1$ in the original equation:

$$12(1) + 1 \overset{?}{=} 25 - 12(1)$$

$$12 + 1 \overset{?}{=} 25 = 12$$

$$13 = 13.$$

Therefore, 1 is the solution.

EXAMPLE **6** $4(u + 1) = 21 + 2u$

SOLUTION

$$4(u + 1) = 21 + 2u$$

$$4u + 4 = 21 + 2u$$ (We removed parentheses on the left side.)

$$4u + 4 - 2u = 21 + 2u - 2u$$ (We subtracted $2u$ from both sides.)

$$2u + 4 = 21$$

$$2u + 4 - 4 = 21 - 4$$ (We subtracted 4 from both sides.)

$$2u = 17$$

$$\frac{2u}{2} = \frac{17}{2}$$ (We divided both sides by 2.)

$$u = \frac{17}{2}.$$

Check Substitute $u = \frac{17}{2}$ in the original equation:

$$4\left(\frac{17}{2} + 1\right) \overset{?}{=} 21 + 2\left(\frac{17}{2}\right)$$

$$4\left(\frac{19}{2}\right) \overset{?}{=} 21 + \frac{34}{2}$$

$$38 = 38.$$

Therefore, $\frac{17}{2}$ is the solution.

EXAMPLE 7 $5 + 8(x + 2) = 23 - 2(2x - 5)$

SOLUTION

$$5 + 8(x + 2) = 23 - 2(2x - 5)$$

$$5 + 8x + 16 = 23 - 4x + 10 \qquad \text{(We removed the parentheses on both sides.)}$$

$$4x + 5 + 8x + 16 = 4x + 23 - 4x + 10 \qquad \text{(We added } 4x \text{ to both sides.)}$$

$$12x + 21 = 33 \qquad \text{(We combined like terms on both sides.)}$$

$$12x + 21 - 21 = 33 - 21 \qquad \text{(We subtracted 21 from both sides.)}$$

$$12x = 12$$

$$\frac{12x}{12} = \frac{12}{12} \qquad \text{(We divided both sides by 12.)}$$

$$x = 1.$$

Therefore, 1 is the solution. You can check this by substituting $x = 1$ in the original equation.

The procedure for solving first-degree or linear equations with one variable consists of one or all of the following steps:

Procedure for Solving Linear Equations

Step 1. Remove all parentheses or grouping symbols on each side of the equation by using the distributive property.

Step 2. Combine all like terms on each side of the equation.

Step 3. Convert the equation to an equivalent equation in which all the variable terms are on one side of the equation and all the constant terms are on the other side. Combine like terms after each operation.

Step 4. Complete the solution by applying the multiplication or division property to bring the coefficient of the variable to 1.

The following example illustrates an interesting application of linear equations.

EXAMPLE 8   Express the repeating decimal $0.\overline{61}$ as a quotient of integers.

SOLUTION   Let $x = 0.\overline{61}$. Then $100x = 61.\overline{61}$. If we subtract $0.\overline{61}$ from $61.\overline{61}$, the repeating portion of the decimals cancels out:

$$100x - x = 61.\overline{61} - 0.\overline{61} = 61$$

$$99x = 61$$

$$x = \frac{61}{99}.$$

Therefore, $0.\overline{61} = 61/99$.

PROBLEM SET 4.1

In problems 1–50, solve each equation.

1. $x + 3 = 10$	**2.** $t + 2 = 8$	**3.** $u - 4 = 8$
4. $y - 5 = 2$	**5.** $w + 11 = 17$	**6.** $z + 12 = 18$
7. $8t = 24$	**8.** $7y = 35$	**9.** $-4u = 12$
10. $6x = 42$	**11.** $10c = -18$	**12.** $-8y = -32$
13. $15t = 75$	**14.** $14w = 42$	**15.** $16b = 30$
16. $9t = 54$	**17.** $-12b = 8$	**18.** $-3z = 1$
19. $3x - 6 = 15$	**20.** $7 - 5x = 11$	**21.** $6y + 7 = 31$
22. $7t - 3 = 18$	**23.** $5 + 4u = 17$	**24.** $3w - 2 = 10$
25. $3t - 5 = 20$	**26.** $-8t + 1 = 17$	**27.** $12u + 25 = 40$
28. $10w - 9 = 21$	**29.** $6x - 8 = 7 - x$	**30.** $18 - x = 3 + 4x$
31. $10 - 2m = 3m + 25$	**32.** $2u + 1 = 5u + 8$	**33.** $5 - 9z = -8z + 3$
34. $7c + 4 = c - 8$	**35.** $5y - 1 = 2y + 8$	**36.** $7y - 16 = y - 10$
37. $12t + 1 = 25 - 12t$	**38.** $17w - 2 = 12w + 8$	**39.** $3x - 2(x + 1) = 2(x - 1)$
40. $8(5x - 1) + 36 = -3(x + 5)$	**41.** $1 - 2(5 - 2y) = 26 - 3y$	**42.** $2(1 - 2y) = 3(2y - 4) + 94$

43. $7t - 3(9 - 5t) = 4t - 9t$ **44.** $7(w - 3) = 4(w + 5) - 47$

45. $6(c - 10) + 3(2c - 7) = -45$ **46.** $16 - 9(3 - u) + 4u = 15$

47. $13 - 5(2 - x) - 18 = 0$ **48.** $3(z - 2) + 5(z + 1) = 4(z - 1)$

49. $34 - 3y = 8(7 - y) + 23$ **50.** $11 - 7(1 - 2v) = 9(v + 1)$

In problems 51–58, express each repeating decimal as a quotient of integers.

51. $0.\overline{5}$	**52.** $0.\overline{46}$	**53.** $1.\overline{3}$	**54.** $0.0\overline{53}$
55. $0.4\overline{9}$	**56.** $-7.\overline{362}$	**57.** $-3.\overline{128}$	**58.** $1.5\overline{821}$

© In problems 59–62, solve each equation with the aid of a calculator.

59. $41.03x + 49.37 = 0$ **60.** $271.73t + 839.41 = 972.82$

61. $0.1347y - 6.738 = 0.2814y - 1.813$ **62.** $2.719u - 3.482 = 6.432u - 1.713$

4.2 Equations Involving Fractions

We solved first-degree (linear) equations in Section 4.1. Now we will solve equations involving fractions that can be changed to linear equations. We will consider two types of equations. One type has fractions in which the denominators are constants. Examples of such equations are:

$$\frac{x}{5} = \frac{2}{3},$$

$$\frac{3t + 1}{10} + \frac{1}{2} = \frac{7}{8},$$

and $$\frac{u + 7}{10} - \frac{u}{25} = 1.$$

The other type of equations has fractions in which at least one denominator contains the variable. Examples of this type of equations are:

$$\frac{1}{3x} + \frac{1}{x} = \frac{1}{9},$$

$$\frac{1}{2y} - \frac{1}{6} = -\frac{1}{3y},$$

and $$\frac{4}{10 + w} + \frac{4}{10 - w} = \frac{5}{6}.$$

We must be careful when the variable appears in the denominator of a fraction. Do not assign a value to the variable that would produce a zero in any denominator, because division by zero is undefined.

Procedure for Solving Equations Involving Fractions

Step 1. Determine the LCD of the fractions.

Step 2. Multiply both sides of the equation by the LCD in order to produce an equation that contains no fractions.

Step 3. Solve the resulting equation.

Step 4. If the original equation contains a variable in any denominator, you must check the proposed solution to see whether it should be accepted or rejected.

In Examples 1–6, solve each equation

EXAMPLE 1 $\dfrac{2x}{3} + \dfrac{1}{2} = \dfrac{5}{6}$

SOLUTION The LCD of the fractions in the equation is 6. In order to produce an equation that contains no fractions, we multiply both sides of the equation by 6:

$$6\left(\dfrac{2x}{3} + \dfrac{1}{2}\right) = 6\left(\dfrac{5}{6}\right)$$

$$6\left(\dfrac{2x}{3}\right) + 6\left(\dfrac{1}{2}\right) = 6\left(\dfrac{5}{6}\right) \qquad \text{(We used the distributive property.)}$$

$$4x + 3 = 5$$

$$4x + 3 - 3 = 5 - 3 \qquad \text{(We subtracted 3 from both sides.)}$$

$$4x = 2$$

$$x = \dfrac{2}{4} \qquad \text{(We divided both sides by 4.)}$$

$$x = \dfrac{1}{2}.$$

Therefore, $\frac{1}{2}$ is the solution.

EXAMPLE 2 $\dfrac{u - 4}{3} = \dfrac{u}{5} + 2$

SOLUTION We multiply both sides of the equation by 15, the LCD of the fractions:

$$15\left(\dfrac{u - 4}{3}\right) = 15\left(\dfrac{u}{5} + 2\right)$$

$$5(u - 4) = 15\left(\dfrac{u}{5}\right) + 15(2) \qquad \text{(We applied the distributive property to the right side.)}$$

$$5u - 20 = 3u + 30 \qquad \text{(We applied the distributive property to the left side.)}$$

$$5u - 3u = 30 + 20 \qquad \text{(We subtracted } 3u \text{ from both sides and added 20 to both sides.)}$$

$$2u = 50$$

$$u = \dfrac{50}{2} \qquad \text{(We divided both sides by 2.)}$$

$$u = 25.$$

Therefore, 25 is the solution.

EXAMPLE 3 $\dfrac{y-3}{4} - \dfrac{y-2}{3} = \dfrac{y-11}{12}$

SOLUTION   We multiply both sides of the equation by 12, the LCD of the fractions:

$$12\left(\dfrac{y-3}{4} - \dfrac{y-2}{3}\right) = 12\left(\dfrac{y-11}{12}\right)$$

$$12\left(\dfrac{y-3}{4}\right) - 12\left(\dfrac{y-2}{3}\right) = 12\left(\dfrac{y-11}{12}\right) \qquad \text{(We used the distributive property.)}$$

$$3(y-3) - 4(y-2) = y - 11$$

$$3y - 9 - 4y + 8 = y - 11 \qquad \text{(We used the distributive property.)}$$

$$-y - 1 = y - 11 \qquad \text{(We collected like terms.)}$$

$$-2y = -10 \qquad \text{(We subtracted } y \text{ from both sides, and added 1 to both sides.)}$$

$$y = \dfrac{-10}{-2} \qquad \text{(We divided both sides by } -2.)$$

$$y = 5.$$

Therefore, 5 is the solution.

EXAMPLE 4 $\dfrac{3}{5t} + 1 = \dfrac{4}{t} + \dfrac{18}{35}$

SOLUTION   We multiply both sides of the equation by $35t$, the LCD of the fractions:

$$35t\left(\dfrac{3}{5t} + 1\right) = 35t\left(\dfrac{4}{t} + \dfrac{18}{35}\right)$$

$$35t\left(\dfrac{3}{5t}\right) + 35t(1) = 35t\left(\dfrac{4}{t}\right) + 35t\left(\dfrac{18}{35}\right) \qquad \text{(We used the distributive property.)}$$

$$21 + 35t = 140 + 18t$$

$$35t - 18t = 140 - 21 \qquad \text{(We subtracted } 18t \text{ and 21 from both sides.)}$$

$$17t = 119 \qquad \text{(We collected like terms.)}$$

$$t = \dfrac{119}{17} \qquad \text{(We divided both sides by 17.)}$$

$$t = 7.$$

The original equation contains a variable in the denominator; therefore, we must check the solution.

Check If we substitute $t = 7$ in the original equation, we have

$$\frac{3}{5(7)} + 1 \stackrel{?}{=} \frac{4}{7} + \frac{18}{35}$$

$$\frac{3}{35} + 1 \stackrel{?}{=} \frac{4}{7} + \frac{18}{35}$$

$$\frac{38}{35} = \frac{38}{35}.$$

Therefore, 7 is the solution.

EXAMPLE 5 $\qquad \dfrac{2}{r-1} + \dfrac{6}{r} = \dfrac{5}{r-1}$

SOLUTION We multiply both sides of the equation by $r(r-1)$, the LCD of the fractions:

$$r(r-1)\left(\frac{2}{r-1} + \frac{6}{r}\right) = r(r-1)\left(\frac{5}{r-1}\right)$$

$$r(r-1)\left(\frac{2}{r-1}\right) + r(r-1)\left(\frac{6}{r}\right) = r(r-1)\left(\frac{5}{r-1}\right) \qquad \text{(We used the distributive property.)}$$

$$2r + 6(r-1) = 5r$$

$$2r + 6r - 6 = 5r \qquad \text{(We used the distributive property.)}$$

$$8r - 6 = 5r \qquad \text{(We collected like terms.)}$$

$$3r = 6 \qquad \text{(We added 6 to both sides, and subtracted } 5r \text{ from both sides.)}$$

$$r = \frac{6}{3}$$

$$r = 2.$$

Check If we substitute $r = 2$ in the original equation, we have

$$\frac{2}{2-1} + \frac{6}{2} \stackrel{?}{=} \frac{5}{2-1}$$

$$\frac{2}{1} + 3 \stackrel{?}{=} \frac{5}{1}$$

$$5 = 5.$$

Therefore, 2 is the solution.

EXAMPLE **6** $\dfrac{3}{w-3} + 4 = \dfrac{w}{w-3}$

SOLUTION   We multiply both sides of the equation by $w - 3$, the LCD of the fractions:

$$(w-3)\left(\dfrac{3}{w-3} + 4\right) = (w-3)\left(\dfrac{w}{w-3}\right)$$

$$(w-3)\left(\dfrac{3}{w-3}\right) + (w-3)4 = (w-3)\left(\dfrac{w}{w-3}\right) \quad \text{(We used the distributive property.)}$$

$$3 + (w-3)4 = w$$

$$3 + 4w - 12 = w \quad \text{(We used the distributive property.)}$$

$$4w - 9 = w$$

$$3w = 9 \quad \text{(We subtracted } w \text{ from both sides and added 9 to both sides.)}$$

$$w = 3.$$

Check If we substitute $w = 3$ in the original equation, we have

$$\dfrac{3}{3-3} + 4 \overset{?}{=} \dfrac{3}{3-3}$$

$$\dfrac{3}{0} + 4 \overset{?}{=} \dfrac{3}{0}.$$

Because we cannot divide by zero, two of the terms are undefined. Therefore, we cannot substitute 3 for w—this does not produce a true statement—and there is *no* solution for this equation. We call 3 an **extraneous root.**

PROBLEM SET 4.2

In problems 1–42, solve each equation.

1. $\dfrac{2}{3} - \dfrac{5x}{3} = \dfrac{17}{3}$

2. $\dfrac{2x}{5} - \dfrac{4}{5} = \dfrac{9}{5}$

3. $\dfrac{y}{6} - \dfrac{1}{2} = \dfrac{2}{3}$

4. $\dfrac{5u}{4} + \dfrac{3}{16} = \dfrac{1}{2}$

5. $\dfrac{t}{6} - \dfrac{t}{7} = \dfrac{1}{42}$

6. $\dfrac{4z}{3} - \dfrac{5z}{6} = \dfrac{3}{4}$

7. $\dfrac{5w}{6} = \dfrac{52}{2} - \dfrac{w}{4}$

8. $\dfrac{3y}{7} - \dfrac{10}{7} = \dfrac{-2y}{3}$

9. $\dfrac{u - 1}{2} + \dfrac{u}{7} = \dfrac{11}{14}$

10. $\dfrac{9z + 1}{4} = z + \dfrac{1}{3}$

11. $\dfrac{5x - 15}{7} - \dfrac{x}{3} = \dfrac{2}{5}$

12. $\dfrac{3w}{5} - \dfrac{13}{15} = \dfrac{w + 2}{12}$

13. $\dfrac{5u}{4} - 1 = \dfrac{3u}{4} + \dfrac{1}{2}$

14. $z + \dfrac{16}{3} = \dfrac{3z}{2} + \dfrac{25}{6}$

15. $c - 1 = \dfrac{2c}{5} - \dfrac{7}{5}$

16. $\dfrac{2b + 1}{3} - \dfrac{b}{2} = \dfrac{b}{5}$

17. $\dfrac{1}{3}(3x - 2) + \dfrac{1}{2}(x - 3) = \dfrac{5}{6}$

18. $\dfrac{x - 14}{5} + 4 = \dfrac{x + 16}{10}$

19. $\dfrac{y + 9}{4} - \dfrac{6y - 9}{14} = 2$

20. $\dfrac{3u - 6}{4} - \dfrac{u + 6}{6} + \dfrac{2u}{3} = 5$

21. $\dfrac{z - 2}{3} - \dfrac{z - 3}{5} = \dfrac{13}{15}$

22. $\dfrac{8w + 10}{5} - \dfrac{6w + 1}{4} = \dfrac{3}{20}$

23. $\dfrac{1}{2x} + \dfrac{8}{5} = \dfrac{3}{x}$

24. $\dfrac{5}{x} + \dfrac{3}{8} = \dfrac{7}{16}$

25. $\dfrac{1}{y} + \dfrac{2}{y} = 3 - \dfrac{3}{y}$

26. $\dfrac{2}{3u} + \dfrac{1}{6u} = \dfrac{1}{4}$

27. $\dfrac{3}{8u} - \dfrac{1}{5u} = \dfrac{7}{10}$

28. $\dfrac{3}{4c} - \dfrac{1}{6} = \dfrac{4}{8c} + \dfrac{1}{2}$

29. $\dfrac{x}{x - 1} - \dfrac{3}{x + 1} = 1$

30. $\dfrac{8}{x - 3} = \dfrac{12}{x + 3}$

31. $\dfrac{1}{y} + \dfrac{1}{y - 1} = \dfrac{5}{y - 1}$

32. $\dfrac{t}{t + 1} + 2 = \dfrac{3t}{t + 2}$

33. $\dfrac{-4}{3u} = \dfrac{3}{3u + 1} + \dfrac{2}{u}$

34. $\dfrac{1}{y^2 - 2y} - \dfrac{1}{y} = \dfrac{1}{1 - y}$

35. $\dfrac{5}{3b + 1} - \dfrac{2}{2b - 1} = \dfrac{1}{2b - 1}$

36. $\dfrac{u + 1}{u - 4} - \dfrac{u}{u - 2} = \dfrac{3}{u - 6}$

37. $\dfrac{4}{y - 2} = \dfrac{5y}{y^2 - 4} - \dfrac{y + 3}{y^2 - 2y}$

38. $\dfrac{1}{2t + 5} - \dfrac{4}{2t - 1} = \dfrac{4t + 4}{(2t + 5)(2t - 1)}$

39. $\dfrac{1}{y - 3} - \dfrac{1}{3 - y} = \dfrac{1}{y^2 - 9}$

40. $\dfrac{2}{u - 2} + \dfrac{1}{u + 1} = \dfrac{1}{(u - 2)(u + 1)}$

41. $\dfrac{1}{t(t - 1)} - \dfrac{1}{t} = \dfrac{1}{t - 1}$

42. $\dfrac{5}{y - 5} = \dfrac{y}{y - 5} - 4$

4.3 Literal Equations and Formulas

The equations you have worked with so far have contained one variable or unknown (represented by a letter) and the constants have been known numbers. In this section we examine **literal equations**—equations containing more than one letter symbol in which one of the letter symbols represents the variable or unknown and the others represent constants. For instance, consider the literal equation

$$ax = b$$

in which x is the unknown and a and b represent unspecified constants. You may be asked to *solve* a literal equation for one of the letters. In this case, you express that unknown in terms of the other letters. Thus, to solve the equation $ax = b$ for x, you divide both sides of the equation by a $(a \neq 0)$ to produce the equivalent equation

$$x = \frac{b}{a}.$$

In applied work, letters other than x are often used, because certain quantities are designated by conventional symbols. In each case, we specify which unknown we are solving for. For example, to solve the equation

$$3r + t = 6t + 5$$

for r, we can apply the procedure used in Section 4.1:

$$3r = 6t + 5 - t \qquad \text{(We subtracted } t \text{ from each side.)}$$

$$3r = 5t + 5 \qquad \text{(We collected like terms.)}$$

$$r = \frac{5t + 5}{3}. \qquad \text{(We divided both sides by 3.)}$$

In Examples 1–4, solve each equation for the indicated unknown.

EXAMPLE 1 $4x - a = x + 8a$, for x.

SOLUTION

$$4x - a = x + 8a$$

$$4x - x = 8a + a \qquad \text{(We subtracted } x \text{ from both sides and added } a \text{ to both sides.)}$$

$$3x = 9a \qquad \text{(We collected like terms.)}$$

$$x = \frac{9a}{3} \qquad \text{(We divided both sides by 3.)}$$

$$x = 3a.$$

Check Substitute $x = 3a$ in the original equation:

$$4(3a) - a \stackrel{?}{=} 3a + 8a$$

$$12a - a \stackrel{?}{=} 3a + 8a$$

$$11a = 11a.$$

Therefore, the solution is $3a$.

EXAMPLE **2** $c(t - c) = d(t - d)$, for t.

SOLUTION

$$c(t - c) = d(t - d)$$

$$ct - c^2 = dt - d^2 \qquad \text{(We used the distributive property.)}$$

$$ct - dt = c^2 - d^2 \qquad \text{(We added } -dt \text{ and } c^2 \text{ to both sides.)}$$

$$t(c - d) = (c - d)(c + d) \qquad \text{(We factored both sides.)}$$

$$t = \frac{(c - d)(c + d)}{c - d} \qquad \text{(We divided both sides by } c - d.\text{)}$$

$$t = c + d.$$

EXAMPLE **3** $\dfrac{y - 3a}{b} = \dfrac{2a}{b} + y$, for y.

SOLUTION   We multiply both sides of the equation by b, the LCD:

$$b\left(\frac{y - 3a}{b}\right) = b\left(\frac{2a}{b} + y\right)$$

$$y - 3a = b\left(\frac{2a}{b}\right) + by$$

$$y - 3a = 2a + by$$

$$y - by = 2a + 3a$$

$$y(1 - b) = 5a$$

$$y = \frac{5a}{1 - b}.$$

EXAMPLE **4** $\dfrac{2x}{x - a} = 3 - \dfrac{x - a}{x}$, for x.

SOLUTION   We multiply both sides of the equation by $x(x - a)$, the LCD:

$$x(x - a)\left(\frac{2x}{x - a}\right) = x(x - a)\left(3 - \frac{x - a}{x}\right)$$

$$x(2x) = 3x(x - a) - (x - a)^2$$

$$2x^2 = 3x^2 - 3ax - (x^2 - 2ax + a^2)$$

$$2x^2 = 3x^2 - 3ax - x^2 + 2ax - a^2$$

$$2x^2 = 2x^2 - ax - a^2$$

$$ax = -a^2$$

$$x = -a.$$

In many fields, algebraic formulas are used to express relations between various quantities. These formulas often contain more than one unknown. For example, in physics, the formula

$$d = rt$$

gives the distance d in terms of the speed r and the time t. If we are asked to solve for t in terms of d and r, we divide both sides of the equation by r:

$$\frac{d}{r} = \frac{rt}{r}$$

or

$$\frac{d}{r} = t.$$

EXAMPLE 5   The formula

$$A = P + Prt$$

gives the total amount of money due at the end of t years if P is the amount of money invested at a simple interest rate r. Solve for P.

SOLUTION

$$A = P + Prt$$

$$A = P(1 + rt) \qquad \text{(We factored out } P \text{ on the right side.)}$$

$$\frac{A}{1 + rt} = P \qquad \text{(We divided both sides by } 1 + rt.\text{)}$$

EXAMPLE 6   The formula

$$F = \frac{9}{5}C + 32$$

expresses the temperature F in degrees Fahrenheit in terms of the temperature C in degrees Celsius. Solve for C.

SOLUTION

$$F = \frac{9}{5}C + 32$$

$$5F = 5\left(\frac{9}{5}C + 32\right) \qquad \text{(We multiplied both sides by 5, the LCD.)}$$

$$5F = 9C + 160 \qquad \text{(We used the distributive property.)}$$

$$5F - 160 = 9C \qquad \text{(We subtracted 160 from both sides.)}$$

$$\frac{5F - 160}{9} = C \qquad \text{(We divided both sides by 9.)}$$

or

$$C = \frac{5}{9}(F - 32)$$

EXAMPLE 7 The formula

$$\frac{1}{R} = \frac{1}{R_1} + \frac{1}{R_2}$$

relates the values of the three resistances R, R_1, and R_2 in a certain type of electrical circuit. Solve for R_1.

SOLUTION

$$\frac{1}{R} = \frac{1}{R_1} + \frac{1}{R_2}$$

$$RR_1R_2\left(\frac{1}{R}\right) = RR_1R_2\left(\frac{1}{R_1} + \frac{1}{R_2}\right) \qquad \text{(We multiplied both sides by } RR_1R_2, \text{ the LCD.)}$$

$$R_1R_2 = RR_2 + RR_1 \qquad \text{(We used the distributive property.)}$$

$$R_1R_2 - RR_1 = RR_2 \qquad \text{(We subtracted } RR_1 \text{ from both sides.)}$$

$$R_1(R_2 - R) = RR_2 \qquad \text{(We factored out } R_1 \text{ on the left side.)}$$

$$R_1 = \frac{RR_2}{R_2 - R} \qquad \text{(We divided both sides by } R_2 - R.)$$

PROBLEM SET 4.3

In problems 1–34, solve each equation for the indicated unknown.

1. $6x + 7c = 37c$, for x.

2. $4u - 19a = 5u$, for u.

3. $at + b = c$, for t.

4. $ay - b = c$, for y.

5. $12z - 4b = 6z - 7b$, for z.

6. $13f + 6y = 8f - 9y$, for y.

7. $34c + 11n = 7c + 17n$, for n.

8. $18u + 11d = 14u - 19d$, for u.

9. $4x - 3a - (10x + 7a) = 0$, for x.

10. $27t - 4b - (15t - 6b) = 0$, for t.

11. $9z + 7h - (11h - 13z) = 6z$, for z.

12. $4(2k - 3m) - 3(5k - 7m) = 0$, for m.

13. $5(4r - 3c) - 2(7r - 9c) = 0$, for r.

14. $5(mu - 2d) - 3m(u - 4d) = 8d$, for u.

15. $8(w - 2b) - 3(5w + 11b) = 0$, for w.

16. $(ah + 7)(ah - 3) = ah(ah + 1)$, for h.

17. $3(a - 2b) + 4(b + a) = 5$, for a.

18. $a(y - a) = ab + 2b(y - b)$, for y.

19. $\dfrac{ay}{b} = c + \dfrac{d}{b}$, for y.

20. $\dfrac{ax}{7} - \dfrac{bc}{3} = \dfrac{2x}{3b^2}$, for x.

21. $\dfrac{b - x}{3} = \dfrac{2a - b}{4} - \dfrac{3x}{5}$, for x.

22. $\dfrac{y}{b^4} - \dfrac{3}{2b^4} = 2$, for y.

23. $\dfrac{u + 2c}{3} + \dfrac{u - 3c}{2} = \dfrac{5}{12}$, for u.

24. $\dfrac{m^2 t}{4p} = \dfrac{2m^2 t - 9}{5p}$, for t.

25. $\dfrac{2t + a}{4} - \dfrac{6t + 3a}{7} = \dfrac{15a}{28}$, for t.

26. $\dfrac{a^2 r + 8}{b} = \dfrac{a^2 r + 10}{3b}$, for r.

27. $\dfrac{a + b}{x} + \dfrac{a - b}{x} = 2a$, for x.

28. $\dfrac{8c^4 - 3y}{2y} + \dfrac{7}{2} = 0$, for y.

29. $\dfrac{3}{x} - \dfrac{4}{b} = \dfrac{5}{3b}$, for x.

30. $\dfrac{b}{x} - \dfrac{a}{3} = \dfrac{b - a}{3x}$, for x.

31. $\dfrac{3}{a - x} + \dfrac{a}{a + x} = \dfrac{1}{a^2 - x^2}$, for x.

32. $\dfrac{x + a}{2x - b} = \dfrac{x + b}{2x - a}$, for x.

33. $\dfrac{1}{y} + \dfrac{2}{y + a} = \dfrac{3}{y - a}$, for y.

34. $\dfrac{t - 2a}{2t + a} = \dfrac{2t - 7a}{4t - 3a}$, for t.

In problems 35–56, solve each formula for the indicated unknown.

35. $A = \frac{1}{2}bh$, for h. (area of a triangle)

36. $E = IR$, for I. (physics)

37. $V = lwh$, for l. (volume of a box)

38. $V = \pi r^2 h$, for h. (volume of a cylinder)

39. $C = 2\pi r$, for r. (circumference of a circle)

40. $S = 2\pi rh$, for r. (surface area of a cylinder)

41. $V = gt$, for t. (physics)

42. $V = \frac{1}{3}\pi r^2 h$, for h. (volume of a cone)

43. $I = Prt$, for t. (simple interest)

44. $F = \dfrac{W}{g} a$, for a. (physics)

45. $F = mx + b$, for m. (analytic geometry)

46. $E = I(R + r)$, for R. (physics)

47. $P = 2l + 2w$, for w. (perimeter of a rectangle)

48. $S = gt^2 + vt$, for v. (physics)

49. $S = \dfrac{n}{2}(a + l)$, for a.

50. $S = \dfrac{a}{1 - r}$, for r.

51. $S = \dfrac{n}{2}[2a + (n - 1)d]$, for d.

52. $S = \dfrac{a - rl}{l - r}$, for r.

53. $L = a + (m - 1)d$, for d.

54. $\dfrac{1}{u} + \dfrac{1}{v} = \dfrac{1}{f}$, for f.

55. $pv = k\left(1 + \dfrac{t}{m}\right)$, for t.

56. $I = \dfrac{E}{R + nr}$, for R.

4.4 Applications of Linear Equations—Word Problems

When we describe problems of any kind, whether they have to do with geometry, business, the social sciences, or the physical sciences, we typically ask questions and supply facts in the form of words rather than letters or symbols. These "word problems" or "story problems" often give rise to first-degree or linear equations. The most challenging step in solving these problems is translating the situation described in the problem into an appropriate algebraic form. This translation process requires a kind of ability that you can only acquire with practice. The following step-by-step procedure provides some useful guidelines:

Step 1. Read the problem carefully, and clearly identify the question or questions you must answer. Draw a diagram whenever possible to help interpret the given information.

Step 2. List all the unknown quantities involved in the problem, and represent them in terms of an algebraic symbol (say, x, y, etc.).

Step 3. Use the information given in the problem to write algebraic relationships among the quantities identified in step 2.

Step 4. Combine the algebraic relationships into a single equation.

Step 5. Solve the equation for the unknown.

Step 6. Check your answer to see if it agrees with the facts in the problem.

Number Problems

EXAMPLE 1   The sum of two numbers is 94, and the larger number is 5 less than twice the smaller number. Find the numbers.

SOLUTION   We follow the outlined procedure:

Step 1. Question: What are the two numbers?

Step 2. Unknown quantities:
Let
$$x = \text{the smaller number.}$$
Then
$$94 - x = \text{the larger number.}$$

Step 3. Information given:

the larger number = twice the smaller number -5
$$= 2x - 5.$$

Step 4. Equation: The relationship can be written as

$$94 - x = 2x - 5.$$

Step 5. We solve the equation as follows:

$$94 - x = 2x - 5$$
$$-x - 2x = -5 - 94$$
$$-3x = -99$$
$$x = 33.$$

Thus, the smaller number is 33 and the larger number is $94 - 33 = 61$.

Step 6. Check: Indeed,

$$33 + 61 = 94.$$

Is it also true that

$$94 - 33 = 2(33) - 5?$$

Yes,

$$61 = 66 - 5.$$

Age Problems

Many word problems involving age can be worked by solving first-degree equations. If x is the age of a person now, then the person's age after 5 years is $x + 5$; and the person's age 5 years ago is $x - 5$. The following problems give information about two people's ages at different times.

EXAMPLE 2   Janice is 7 years older than her brother, and 5 years from now the sum of their ages will be 63 years. How old is each now?

SOLUTION   Step 1. Question: What are the ages of Janice and her brother now?

Step 2. Unknown quantities:
Let

$$x = \text{Janice's age now.}$$

Then

$$x - 7 = \text{her brother's age now.}$$

Step 3. Information given:

	Janice	Brother
Ages now	x	$x - 7$
Ages 5 years from now	$x + 5$	$(x - 7) + 5$

Janice's age 5 years from now + her brother's age 5 years from now = 63.

Step 4. Equation:

$$x + 5 + [(x - 7) + 5] = 63.$$

Step 5. We solve the equation as follows:

$$x + 5 + [(x - 7) + 5] = 63$$
$$x + 5 + x - 7 + 5 = 63$$
$$2x + 3 = 63$$
$$2x = 60$$
$$x = 30.$$

Therefore, Janice's age now is 30, and her brother's age is $30 - 7 = 23$.

Step 6. Check:

$$(30 + 5) + (23 + 5) = 63.$$

Money Value Problems

Linear equations are often used to solve problems involving a specific number of items, in which each item has a particular value. For example, in coin problems we consider the number and the value of each type of coin. For instance

$$1 \text{ nickel} = 5\text{¢} = \$0.05$$
$$1 \text{ dime} = 10\text{¢} = \$0.10$$
$$1 \text{ quarter} = 25\text{¢} = \$0.25$$

and

the total value of a quantity of each type of coin
= (the number of coins) × (the value of the coin).

EXAMPLE 3   Pedro has $6.80 in nickels, dimes, and quarters. He has the same number of coins of each kind. How many coins of each kind does he have?

SOLUTION   Let

$$d = \text{the number of coins of each type Pedro has.}$$

Information given:

Coins	Number of Coins	Individual Value (in dollars)	Total Value (in dollars)
Nickels	d	0.05	$0.05d$
Dimes	d	0.10	$0.10d$
Quarters	d	0.25	$0.25d$

Because the total value of all kinds of coins is $6.80, the equation and its solution are:

$$0.05d + 0.10d + 0.25d = 6.80$$
$$5d + 10d + 25d = 680 \qquad \text{(We multiplied each side by 100.)}$$
$$40d = 680$$
$$d = 17.$$

Therefore, Pedro has 17 nickels, 17 dimes, and 17 quarters.

Check

$$0.05(17) + 0.10(17) + 0.25(17)$$
$$= 0.85 + 1.70 + 4.25$$
$$= 6.80.$$

Finance and Investment Problems

Some problems deal with the sale of an item at a price that has been discounted from the original price. We may use the following formula to solve such problems:

$$\text{sale price} = \text{original price} - \text{discount}$$
$$S = P - D.$$

Other problems involve the investment of a sum of money (the principal) at a specified interest rate, over a given period of time. We may use the following formula to solve this type of problem:

$$\text{simple interest} = \text{principal} \times \text{rate} \times \text{time}$$

$$I = Prt.$$

EXAMPLE 4   A clothing store has discounted the price of a topcoat by 40%. If the sale price of the topcoat is $180, what was the original price?

SOLUTION   Let

$$x = \text{the original price of the topcoat.}$$

Then

$$\text{original price} - \text{discount} = \text{sale price} = \$180,$$

that is,

$$x - \text{discount} = 180$$

and

$$\text{discount} = 40\% \text{ of original price} = 0.40x.$$

The equation and its solution are:

$$x - 0.40x = 180$$
$$0.60x = 180$$
$$x = 300.$$

Therefore, the original price of the topcoat was $300.

Check
$$40\% \text{ of } \$300 = 0.40(300)$$
$$= \$120 \text{ discount}$$

and
$$\$300 - \$120 = \$180 \text{ sale price.}$$

EXAMPLE 5   A businesswoman treated a client to a dinner and spent $76.16, which included the 4% tax and a 15% tip. What was the amount of the bill before tax and tip?

SOLUTION   Let

$$x = \text{the original amount of the bill before tax and tip.}$$

The following table summarizes the information given:

Amount of Bill Before Tax and Tip (in dollars)	Tax (in dollars)	Tip (in dollars)
x	$0.04x$	$0.15x$

Because the total bill is $76.16, the equation and its solution are:

$$x + 0.04x + 0.15x = 76.16$$
$$1.19x = 76.16$$
$$x = 64.$$

Therefore, the amount of the bill before tax and tip was $64.00.

Check

$$\text{bill before tax and tip} = \$64$$
$$\text{tax: 4\% of } \$64 = \$\ 2.56$$
$$\text{tip: 15\% of } \$64 = \$\ 9.60$$
$$\text{total} = \$76.16.$$

EXAMPLE 6   A small company had $24,000 to invest. It invested some of the money in a bank that paid 8% annual simple interest. The rest of the money was invested in stocks that paid dividends equivalent to 11% annual simple interest. At the end of 1 year, the combined income from these investments was $2,340. How much money was originally invested in stocks?

SOLUTION   In solving this problem, we use the formula

$$I = Prt.$$

Let

$$x = \text{amount of money (in dollars) invested in stocks.}$$

Since a total of $24,000 was invested,

$$24,000 - x = \text{the amount of money (in dollars) invested in the bank.}$$

The following table summarizes the information given about the quantities in the formula $I = Prt$:

Investment	Principal (in dollars)	Rate	Time (in years)	Simple Interest (in dollars)
Stocks	x	0.11	1	$0.11x$
Bank	$24,000 - x$	0.08	1	$0.08(24,000 - x)$

The combined simple interest is $2,340. The equation and its solution are as follows:

$$0.11x + 0.08(24,000 - x) = 2,340$$
$$11x + 8(24,000 - x) = 234,000$$
$$11x + 192,000 - 8x = 234,000$$
$$3x = 42,000$$
$$x = 14,000.$$

Therefore, $14,000 was invested in stocks.

Check

$$0.11(14,000) + 0.08(10,000) = 1,540 + 800$$
$$= 2,340.$$

Geometric Problems

Many problems involve perimeters and areas of squares and rectangles. To solve these problems, we use the following formulas for areas and perimeters:

	Length	Width	Area A	Perimeter P
Rectangle	l	w	$A = lw$	$P = 2l + 2w$
Square	l	l	$A = l^2$	$P = 4l$

EXAMPLE 7   A 130-meter length of fence is used to enclose a rectangular garden. The length of the garden is 5 meters more than its width. Find the length and the width of the garden to be enclosed.

SOLUTION   Let

$$x = \text{the width of the garden (in meters).}$$

Then

$$x + 5 = \text{the length of the garden (in meters).}$$

Thus

$$2(\text{length}) + 2(\text{width}) = \text{length of fence} = \text{perimeter of garden.}$$

Figure 1

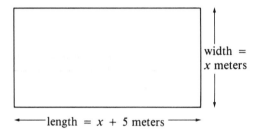

width = x meters

length = x + 5 meters

The equation and its solution are:

$$2x + 2(x + 5) = 130$$
$$2x + 2x + 10 = 130$$
$$4x + 10 = 130$$
$$4x = 120$$
$$x = 30.$$

Therefore, the width of the garden is 30 meters and the length is 35 meters.

Check

$$2(\text{length}) + 2(\text{width}) = 2(35) + 2(30)$$
$$= 70 + 60$$
$$= 130 = \text{length of fence.}$$

Motion Problems

Problems involving motion may also lead to first-degree equations. For example, if an object moves a distance d at a constant rate r (also called the **speed**) in t units of time, then

$$d = rt.$$

EXAMPLE **8**   A bicycle rider travels 18 miles in the same amount of time that it takes a jogger to travel 10 miles. If the bicyclist goes 9.6 miles per hour faster than the jogger, how fast does each person travel?

SOLUTION   Let

$$r = \text{the jogger's rate (in miles per hour).}$$

Then

$$r + 9.6 = \text{the bicyclist's rate (in miles per hour).}$$

The following table summarizes the given information:

	Rate (in miles per hour)	Distance (in miles)	Time (in hours) = $\dfrac{\text{Distance}}{\text{Rate}}$
Jogger	r	10	$\dfrac{10}{r}$
Bicycle rider	$r + 9.6$	18	$\dfrac{18}{r + 9.6}$

We are told that

the time for the jogger = the time for the bicycle rider.

Therefore, the equation is

$$\frac{10}{r} = \frac{18}{r + 9.6}.$$

We solve the equation:

$$r(r + 9.6)\left(\frac{10}{r}\right) = r(r + 9.6)\left(\frac{18}{r + 9.6}\right)$$

$$10r + 96 = 18r$$

$$-8r = -96$$

$$r = 12.$$

Therefore, the speed of the jogger is 12 miles per hour and the speed of the bicycle rider is $12 + 9.6 = 21.6$ miles per hour.

Check

$$\frac{10}{12} = \frac{18}{21.6} = \frac{5}{6}.$$

Mixture Problems

Problems involving mixtures of substances can often be worked by solving first-degree equations.

EXAMPLE 9   A chemist has one solution containing a 16% concentration of acid and a second solution containing a 26% concentration of acid. How many milliliters of each should be mixed to obtain 30 milliliters of a solution containing an 18% concentration of acid?

SOLUTION   Let

$$x = \text{the number of milliliters of the first solution.}$$

Then

$$30 - x = \text{the number of milliliters of the second solution.}$$

The following table summarizes the given information:

	Milliliters of Solution	Acid Concentration	Milliliters of Acid in Solution
First solution	x	0.16	$0.16x$
Second solution	$30 - x$	0.26	$0.26(30 - x)$
Mixture	30	0.18	$0.18(30) = 5.4$

Because the amount of acid in the mixture is the sum of the amounts of acid in the two solutions, the equation and its solution are:

$$0.16x + 0.26(30 - x) = 5.4$$
$$0.16x + 7.8 - 0.26x = 5.4$$
$$-0.10x = -2.4$$
$$x = 24.$$

Therefore, 24 milliliters of the first solution and $30 - 24 = 6$ milliliters of the second solution should be mixed.

Check

$$16\% \text{ of } 24 = 0.16(24) = 3.84$$
$$26\% \text{ of } 6 = 0.26(6) = 1.56$$

so that

$$3.84 + 1.56 = 5.4 = 0.18(30).$$

Work Problems

We may come across problems concerning a job that is done at a constant rate. These problems can often be solved by using the following principle: If a job can be done in t hours, then $1/t$ of the job can be done in 1 hour.

EXAMPLE 10 Two cranes, operating together, can unload a cargo ship in 4 hours. If it takes one crane twice as long as the other to unload the ship, how long would it take each crane to unload the ship by itself?

SOLUTION Let

x = the number of hours for the faster crane to unload the ship.

Then

$2x$ = the number of hours for the slower crane to unload the ship.

The given information is displayed in the following table:

Cranes	Part of Job Done in 1 Hour	Number of Hours	Part of Job Done in 4 Hours
Faster crane	$\dfrac{1}{x}$	4	$\dfrac{4}{x}$
Slower crane	$\dfrac{1}{2x}$	4	$\dfrac{4}{2x}$

When the two cranes operate together,

$$\begin{pmatrix} \text{part of job done} \\ \text{by faster crane} \end{pmatrix} + \begin{pmatrix} \text{part of job done} \\ \text{by slower crane} \end{pmatrix} = \begin{pmatrix} \text{part of job done} \\ \text{by both cranes in 4 hours} \end{pmatrix}$$

$$= 1 \qquad \text{(the entire job)}.$$

The equation is:

$$\frac{4}{x} + \frac{4}{2x} = 1.$$

To solve the equation, multiply both sides by $2x$, the LCD:

$$2x\left(\frac{4}{x} + \frac{4}{2x}\right) = 2x(1)$$

$$8 + 4 = 2x$$

$$2x = 12$$

$$x = 6.$$

Therefore, it would take the faster crane 6 hours to unload the ship by itself, and it would take $2 \cdot 6 = 12$ hours for the slower crane to unload the ship by itself.

Check

$$\frac{4}{6} + \frac{4}{12} = \frac{2}{3} + \frac{1}{3}$$

$$= 1.$$

PROBLEM SET 4.4

© In some of the following problems, a calculator may be useful to speed up the arithmetic.

1. Three more than twice a certain number is 57. Find the number.
2. Find two numbers whose sum is 18, if one number is 8 larger than the other.
3. Find two consecutive even integers such that seven times the first exceeds five times the second by 54.
4. Find three consecutive even integers such that the first plus twice the second plus four times the third equals 174.
5. One-fourth of a number is 3 greater than one-sixth of it. Find the number.
6. Two-thirds of a number plus five-sixths of the same number is equal to 42. What is the number?
7. Gus is 5 years younger than his brother, and 8 years from now he will be four-fifths as old as his brother is then. How old is each now?
8. Wendy's mother is three times as old as Wendy, and 14 years from now she will be twice as old as Wendy is then. How old is each now?
9. Raul is 3 years older than his brother, and 4 years from now the sum of their ages will be 33 years. How old is each now?
10. Jose is 5 years younger than his brother, and 3 years ago the sum of their ages was 23. How old is each now?
11. At a Christmas party, there are five times as many men as women. If 12 more women arrive, there will be only twice as many men as women. How many men are at the party?
12. Psychologists define an individual's intelligence quotient (IQ) to be 100 times the person's mental age divided by his or her chronological age. What is the chronological age of a person with an IQ of 160 and a mental age of 16?
13. A pay-phone slot receives quarters, dimes, and nickels. When the phone box was emptied, it yielded $6.50 in coins. If there were four more dimes than quarters and three times as many nickels as dimes, find the number of coins of each kind.
14. A cashier has four times as many nickels as quarters. She has $3.60. How many of each coin does she have?

15. A parking-meter slot receives dimes and nickels. When emptied, the box produced 70 coins worth $4.85. How many nickels and how many dimes were there?

16. A bank teller has $75 in $1 bills and $5 bills. He has three more $1 bills than $5 bills. How many bills of each kind does he have?

17. A vending machine receives dimes and quarters. $12.50 worth of coins is found when the machine is emptied. If the number of dimes is 20 more than the number of quarters, find the number of each kind of coin.

18. Two electricians worked a total of 12 hours one day. One electrician earns $15 per hour and the other earns $18 per hour. If that day's payroll is $195, how many hours did each electrician work?

19. A pants suit on sale is discounted by 30%. If the selling price is $84, what was its original price?

20. A racquetball racket is on sale at 20% off. If the saving is $7.80, find the original price of the racket.

21. Joe has received an 11% increase in salary. His new weekly earnings are $222. What was his previous weekly salary?

22. Mary received an 18% increase in salary. If the increase was $54 per week, what was her weekly salary before the increase?

23. At the end of the year, a car dealer advertises that the list prices on all of last year's models have been discounted by 15%. What was the original price of a car that now carries a discounted price of $6,849.30?

24. Harry's stocks show a 38% increase in value for the year. If the stocks are now worth $27,000, what were they worth one year ago?

25. On his first dinner date with Lucy, Jack spent $19.04, which included the 4% tax and a 15% tip on the original bill. What was the amount of the bill before tax and tip?

26. John's new annual salary is $23,976, which includes a 7% pay raise and the addition of a 4% cost-of-living allowance. What was his original salary?

27. A tire was sold for $35.65, which included the 4% state sales tax and an 11% federal tax. What was the original price of the tire before the taxes?

28. A man has the first $1,200 exempted from his income tax, but he pays a 20% tax on the remainder of his income. If the total taxes he pays after the exemption are $5,440, find his total income.

29. A businesswoman had $18,000 to invest. She invested some of it in a bank certificate that paid 13% annual simple interest, and the rest in another certificate that paid 14% annual simple interest. At the end of 1 year, the combined income from these investments was $2,395. How much money did she originally invest at 13% interest?

30. A family invests a total of $85,000 in two taxfree municipal bonds to reduce its income tax. One bond pays 8% taxfree simple annual interest, and the other bond pays 8.5%. The total nontaxable income from both investments at the end of 1 year is $7,072. How much did the family invest in each bond?

31. A person invests part of $62,000 in a certificate that yields 13.2% simple annual interest, and the rest of the money in a certificate that yields 13.7% simple annual interest. At the end of 1 year, the combined interest on the two certificates is $8,354. How much money did the person invest in each certificate?

32. A retail store invested $25,000 in two kinds of toys. Over the course of 1 year, it made a profit of 15% from the first kind but lost 5% on the second kind. If the income from the two investments was a return of 8% on the entire amount invested, how much had the store invested in each kind of toy?

33. A certain amount of money is invested in a passbook savings account at 7% simple annual interest. In addition, $8,000 is invested in a certificate that pays 14% simple annual interest. The income from both investments amounts to 11.5% of their total. How much money is invested at 7%?

34. A small company takes advantage of a state income tax credit that provides 15% of the cost of installing solar-heating equipment, and 8% of the cost of upgrading insulation in a building. After spending a total of $6,510 on insulation and solar heating, the company receives a state income tax credit of $854. How much money was spent on solar heating?

35. The length of a rectangular house lot is twice the width, and the difference between the length and the width is 32 meters. What are the dimensions of the lot?

36. The length of a rectangular rug is 6 feet more than the width. The perimeter is 40 feet. Find the length and the width of the rug.

37. The length of a rectangle is four times its width. What are its dimensions if its perimeter is 150 meters?

38. The length of a rectangle is 17 centimeters less than three times the width, and the perimeter is 238 centimeters. Find the dimensions of the rectangle.

39. The length and width of a square are increased by 6 feet and 8 feet, respectively. The result is a rectangle whose area is 188 square feet more than the area of the square. Find the length of a side of the square.

40. The sides of two squares differ by 6 centimeters and their areas differ by 468 square centimeters. Find the lengths of the sides of the squares.

41. Carlos can walk 8 miles in the same time it takes him to jog 12 miles. His jogging rate is 5 miles per hour faster than his walking rate. At what rate does he walk? How long does it take him to walk 8 miles?

42. A bicycle rider travels 8 miles per hour faster than a jogger. It takes the bicycle rider half as much time as it takes the jogger to travel 16 miles. Find the jogger's speed.

43. If a freight train traveling at 30 miles per hour is 300 miles ahead of an express train traveling at 55 miles per hour, how long will it take the express train to catch up with the freight train?

44. An airplane travels 1,620 kilometers in the same time it takes a train to travel 180 kilometers. If the airplane goes 480 kilometers per hour faster than the train, find the rate of each.

45. Maria jogs 15 miles and bicycles back. The total trip requires 3 hours. If she bicycles twice as fast as she jogs, how fast does she bicycle?

46. Joshua rides his bicycle 5 miles from his home to the school bus stop at a rate of 8 miles per hour. He arrives in time to catch the bus, which travels at 25 miles per hour. If he spends $1\frac{1}{2}$ hours traveling from home to school, how far does he travel on the bus?

47. A chemist has one solution containing a 10% concentration of acid and a second solution containing a 15% concentration of acid. How many milliliters of each should be mixed in order to obtain 10 milliliters of a solution containing a 12% concentration of acid?

48. A chemist has 10 milliliters of a solution which contains a 30% concentration of acid. How many milliliters of pure acid must be added in order to increase the concentration to 50%?

49. A petroleum distributor has two gasohol storage tanks. The first contains 9% alcohol and the second contains 12% alcohol. The distributor receives an order for 300,000 gallons of gasohol containing 10% alcohol. How can this order be filled by mixing gasohol from the two storage tanks?

50. A car radiator contains 8 quarts of a mixture of water and antifreeze. If 40% of the mixture is antifreeze, how much of the mixture should be drained and replaced by pure antifreeze so that the resultant mixture will contain 60% antifreeze?

51. A grocer mixes two kinds of coffee to form a blend that sells for $3.85 per pound. He mixes coffee selling for $3.80 per pound with coffee selling for $4.00 per pound to get 100 pounds of the blend. How many pounds of each kind of coffee does he use?

52. A grocer has 100 pounds of candy worth $4.80 per pound. How many pounds of a different type of candy worth $5.20 per pound should he mix with the 100 pounds in order to obtain a mixture worth $5.00 per pound?

53. At a factory, smokestack A pollutes the air 1.25 times faster than smokestack B. How long would it take smokestack B, operating alone, to pollute the air by as much as both smokestacks do in 20 hours?

54. Two plumbers can do a job together in 6 days. If the first plumber can do the job alone in 10 days, how long will it take the second plumber to do the job?

55. A computer can do a payroll in 12 hours. A second computer can do the payroll in 6 hours. How long will it take to do the payroll if both computers operate at the same time?

56. One pipe can fill a tank in 18 minutes, and another pipe can fill it in 24 minutes. The drain pipe can empty the tank in 15 minutes. With all pipes open, how long will it take to fill the tank?

57. John can mow a lawn in 1 hour and 20 minutes. Tom can mow the same lawn in 2 hours. How long would it take John and Tom together to mow the lawn?

58. Jamal can fill the vending machines in 45 minutes. However, if his brother Gus helps, it takes them only 20 minutes. How long would it take Gus to fill the machines by himself?

4.5 **Linear Inequalities**

Statements such as

$$2x + 5 < 3 \qquad\qquad 3y + 7 > 10$$
$$4x - 3 \leq 7x + 1 \qquad 7t - 13 \geq 4t + 2$$

are examples of inequalities. An **inequality** is a statement that an expression representing some real number is *greater than* (or *less than*) another expression representing a real number.

If the point with coordinate a lies to the left of the point with coordinate b on the number line (Figure 1), we say that a is **less than** b (or, equivalently, that b is **greater than** a), and we write $a < b$ (or $b > a$).

Figure 1

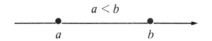

More formally, we have the following definition:

DEFINITION   **Inequality**

> $a < b$ (or $b > a$) means that $b - a$ is a positive number.
> A statement of the form $a < b$ (or $b > a$) is called an **inequality.**

For example, $3 < 7$ (or $7 > 3$), because $7 - 3 = 4$ (a positive number). Also, $-5 < -3$ (or $-3 > -5$), because $-3 - (-5) = 2$ (a positive number). Other inequality signs are $\leq$, which means *less than or equal to*, and $\geq$, which means *greater than or equal to*. By definition,

$$a \leq b \qquad \text{if either} \qquad a < b \qquad \text{or} \qquad a = b$$

and

$$a \geq b \qquad \text{if either} \qquad a > b \qquad \text{or} \qquad a = b.$$

By writing one of the symbols $<$, $\leq$, $>$, or $\geq$ between two expressions, we obtain an inequality. The two expressions are called the **sides** or **members** of the inequality.

An inequality containing a variable will produce a statement that is either true or false when a particular number is substituted for the variable. If we substitute a particular number for the variable in an inequality and we obtain a true statement, we say that the number substituted **satisfies** the inequality, and that the number is a **solution** of the inequality. The set of all solutions of an inequality is called the **solution set.**

To **graph** an inequality on a number line we sketch the graph of its solution set.

EXAMPLE 1   Sketch the graph of each inequality on a number line.

(a) $x < \frac{3}{2}$

(b) $x \geq 4$

(c) $x \leq -2$

(d) $x > -\frac{2}{3}$

(e) $-1 \leq x \leq 2$

SOLUTION   (a) $x < \frac{3}{2}$ consists of all numbers less than $\frac{3}{2}$. The fact that $\frac{3}{2}$ is *excluded* from the solution set is shown on the graph by a *parenthesis* at the point with the coordinate $\frac{3}{2}$ (Figure 2a).

(b) $x \geq 4$ consists of all numbers greater than or equal to 4. The fact that 4 is *included* in the solution set is shown on the graph by a *bracket* at the point with the coordinate 4 (Figure 2b).

(c) $x \leq -2$ consists of all numbers less than and including -2 (Figure 2c).

(d) $x > -\frac{2}{3}$ consists of all real numbers greater than and excluding $-\frac{2}{3}$ (Figure 2d).

(e) $-1 \leq x \leq 2$ means that $-1 \leq x$ and $x \leq 2$. The solution set of the combined inequalities is the set of all real numbers in which x is between and including the numbers -1 and 2 (Figure 2e).

Figure 2

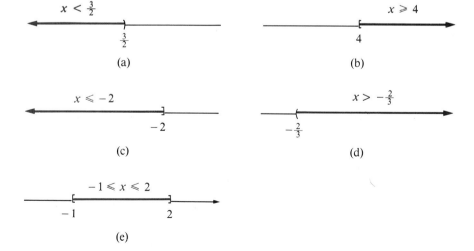

To **solve** a linear inequality—that is, to find its solution set—we proceed in much the same way as we do in solving equations. The only exceptions occur when we multiply (or divide) by a negative number. The general process for solving linear inequalities parallels that for solving linear equations. The procedure is based on the following properties:

Properties of Inequalities

Let a, b, and c be real numbers. Then:

(1) **Addition Property**
   If $a < b$, then $a + c < b + c$ for any number c.

(2) **Multiplication Properties**
   (i) If $a < b$ and $c > 0$, then $ac < bc$.
   (ii) If $a < b$ and $c < 0$, then $ac > bc$.

(3) **Division Properties**
   (i) If $a < b$ and $c > 0$, then $\dfrac{a}{c} < \dfrac{b}{c}$.

   (ii) If $a < b$ and $c < 0$, then $\dfrac{a}{c} > \dfrac{b}{c}$.

(4) **Transitive Property**
   If $a < b$ and $b < c$, then $a < c$.

The following graphs illustrate some of the properties of inequalities:
In Figure 3, we illustrate Property (1), by letting $a = 2$, $b = 4$, and $c = 1$, so that

$$2 + 1 < 4 + 1.$$

Figure 3

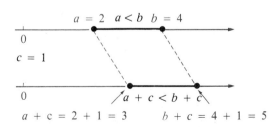

In Figure 4, we illustrate Properties (2i) and (2ii), by letting $a = 3$, $b = 6$, $c = 2$ (Figure 4a) and $c = -2$ (Figure 4b).

Figure 4

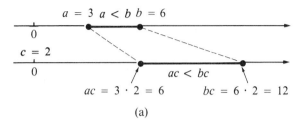

(a)

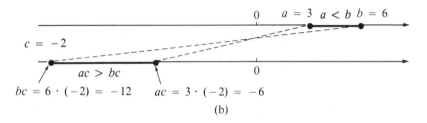

(b)

EXAMPLE **2**   State the property of the inequality that justifies each statement.

(a) $2 < 3$, so it follows that $2 + 5 < 3 + 5$.

(b) $-2 < -1$, so it follows that $-2 + (-3) < -1 + (-3)$.

(c) $3 < 7$, so it follows that $3 \cdot 4 < 7 \cdot 4$.

(d) $x < y$, so it follows that $\dfrac{x}{-3} > \dfrac{y}{-3}$.

(e) $-2 < -1$ and $-1 < 5$, so it follows that $-2 < 5$.

SOLUTIONS   (a) The addition property   (b) The addition property

(c) The multiplication property   (d) The division property

(e) The transitive property

   In solving linear inequalities, we use a technique similar to the step by step procedure used to solve equations (see page 118). However, we must be careful when using the multiplication (or division) property since multiplying or dividing both sides of an inequality by a negative number *always reverses* the sign of the inequality. It is important also to note that the properties of inequalities also hold when the inequality symbol $<$ is replaced by $>$, $\le$, or $\ge$. For example, the addition property could be stated as "If $a \ge b$, then $a + c \ge b + c$ for any number c."

In Examples 3–9, solve each inequality and sketch the graph of its solution set on a number line.

EXAMPLE 3 $x + 2 < 4$

SOLUTION

$$x + 2 < 4$$
$$x < 4 - 2 \qquad \text{(We subtracted 2 from both sides.)}$$
$$x < 2$$

Thus, the solution set consists of all numbers x such that $x < 2$ (Figure 5).

Figure 5

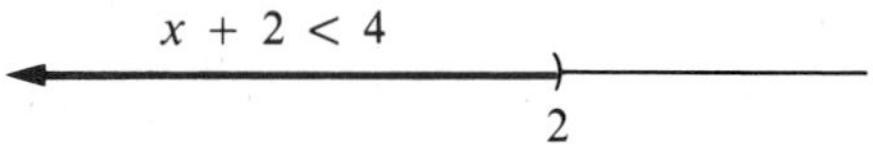

EXAMPLE 4 $3x \geq 12$

SOLUTION

$$3x \geq 12$$
$$x \geq \frac{12}{3} \qquad \text{(We divided both sides by 3, a positive number.)}$$
$$x \geq 4$$

Thus, the solution set consists of all numbers x such that $x \geq 4$ (Figure 6).

Figure 6

EXAMPLE 5 $-5x \geq 15$

SOLUTION

$$-5x \geq 15$$
$$x \leq \frac{15}{-5} \qquad \text{(We divided both sides by } -5 \text{ and reversed the inequality because } -5 \text{ is a negative number.)}$$
$$x \leq -3$$

Thus, the solution set consists of all numbers x such that $x \leq -3$ (Figure 7).

Figure 7

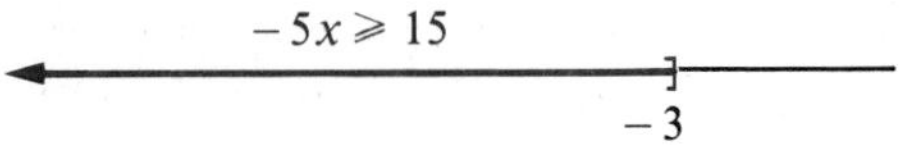

EXAMPLE 6 $4x + 3 < 11$

SOLUTION

$$4x + 3 < 11$$

$$4x < 8 \qquad \text{(We subtracted 3 from both sides.)}$$

$$x < \frac{8}{4} \qquad \text{(We divided both sides by 4.)}$$

$$x < 2.$$

Thus, the solution set consists of all numbers x such that $x < 2$ (Figure 8).

Figure 8

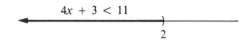

EXAMPLE 7 $\quad 4(x - 3) \geq 3(x - 2)$

SOLUTION

$$4(x - 3) \geq 3(x - 2)$$

$$4x - 12 \geq 3x - 6 \qquad \text{(We used the distributive property.)}$$

$$4x - 3x \geq -6 + 12$$

$$x \geq 6.$$

Thus, the solution set consists of all numbers x such that $x \geq 6$ (Figure 9).

Figure 9
$$\overset{4(x - 3) \geqslant 3(x - 2)}{\underset{6}{\longrightarrow}}$$

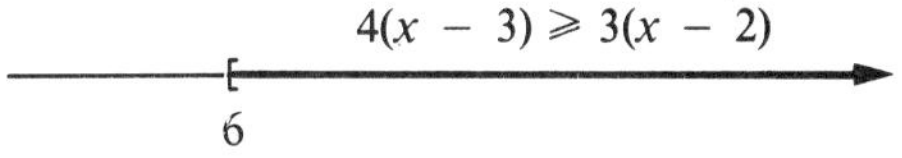

EXAMPLE **8** $\quad \frac{1}{2}(x + 1) \leq \frac{1}{3}(x - 5)$

SOLUTION

We begin by multiplying both sides by 6, the LCD of the fractional coefficients:

$$6(\tfrac{1}{2})(x + 1) \leq 6(\tfrac{1}{3})(x - 5)$$

$$3(x + 1) \leq 2(x - 5)$$

$$3x + 3 \leq 2x - 10 \qquad \text{(We used the distributive property.)}$$

$$3x - 2x \leq -10 - 3$$

$$x \leq -13.$$

The solution set consists of all numbers x such that $x \leq -13$ (Figure 10).

Figure 10
$$\overset{\frac{1}{2}(x + 1) \leqslant \frac{1}{3}(x - 5)}{\underset{-13}{\longleftarrow}}$$

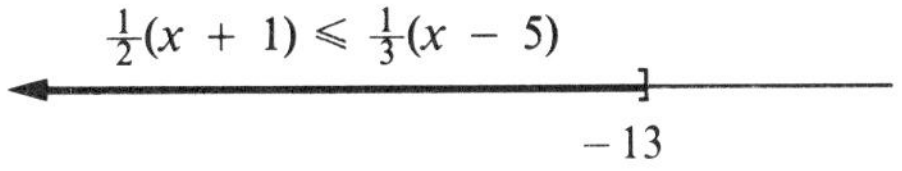

EXAMPLE 9 $-11 \leq 2x - 3 \leq 7$

SOLUTION   This is an example of a compound inequality—an abbreviation for the two simple inequalities

$$-11 \leq 2x - 3 \quad \text{and} \quad 2x - 3 \leq 7.$$

We solve the compound inequality by applying the properties of inequalities to all three members of the inequality. We begin by adding 3 to each member of the inequality to help isolate x in the middle:

$$-11 \leq 2x - 3 \leq 7$$
$$-11 + 3 \leq 2x - 3 + 3 \leq 7 + 3$$
$$-8 \leq 2x \leq 10$$
$$-4 \leq x \leq 5. \qquad \text{(We divided all members by 2.)}$$

Therefore, the solution set consists of all numbers x so that $-4 \leq x \leq 5$ (Figure 11).

Figure 11

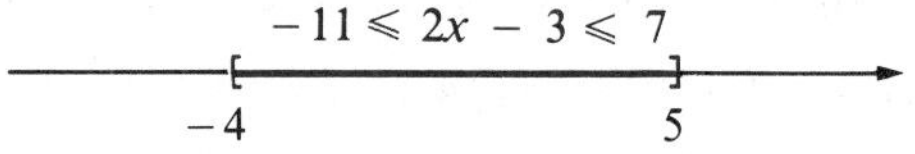

$-11 \leqslant 2x - 3 \leqslant 7$

$-4 \qquad 5$

A practical application of inequalities is illustrated below:

EXAMPLE 10   A telephone company offered its customers a choice between two schedules of billing for its services: a fixed $25.00 monthly charge for unlimited local calls, or a base rate of $7.00 per month plus 6¢ per message unit. Above what level of usage (number of message units) does it cost less to choose the unlimited service?

SOLUTION   Let

$x =$ the number of message units above
which the fixed monthly charge is preferable.

The first choice is to pay $25.00 a month. The second choice is to pay $7.00 plus 6¢ ($ = \$0.06$) for each message unit, that is,

the second choice $= 7 + 0.06x$.

To determine when the first choice produces a lower cost than the second choice, we must solve the inequality

$$7 + 0.06x > 25.$$

We have

$$0.06x > 18$$

$$x > \frac{18}{0.06}$$

$$x > 300.$$

The customer should make the first choice if the anticipated use exceeds 300 message units.

PROBLEM SET 4.5

In problems 1–6, sketch the graph of each inequality on a number line.

1. $x < 1$ **2.** $x \geq -3$ **3.** $x \geq -\frac{3}{4}$
4. $x < \frac{5}{2}$ **5.** $-2 \leq x \leq \frac{3}{5}$ **6.** $-1 < x < \frac{12}{7}$

In problems 7–16, state the property of inequalities that justifies each statement.

7. $2 < 4$, so it follows that $2 + 3 < 4 + 3$.
8. $-3 < 0$, so it follows that $-3 + (-1) < 0 + (-1)$.
9. $-5 < -3$ and $-3 < 4$, so it follows that $-5 < 4$.
10. $7 > -2$ and $-2 > -4$, so it follows that $7 > -4$.

11. If $x > -4$, then $-5x < 20$. **12.** If $t < 10$, than $\dfrac{t}{-2} > -5$.

13. If $a - 3 > 2$, then $a > 5$. **14.** If $y - 2 < 7$, then $y < 9$.

15. If $-\dfrac{x}{3} \leq -5$, then $x \geq 15$. **16.** If $a > 1$, then $a^3 > a$.

In problems 17–62, solve each inequality and sketch the graph of its solution set on a number line.

17. $x - 2 < 4$ **18.** $3 - x < 2$
19. $x + 2 \geq 3$ **20.** $x - 4 \geq -2$
21. $4 + x \leq -1$ **22.** $5 - x \leq 7$
23. $2x \geq 6$ **24.** $3x \leq 9$
25. $-2x < -3$ **26.** $-4x > 6$
27. $5x > -15$ **28.** $-7x \leq 21$
29. $7x \leq -2$ **30.** $9x \geq -1$
31. $4x - 1 \geq 11$ **32.** $5 + 3x \leq 8$
33. $3x + 4 < 7$ **34.** $2x - 1 \leq 5$
35. $-5x + 2 > 12$ **36.** $-7x - 1 > 13$
37. $3x - 4 \geq 6x$ **38.** $3x - 2 + x \leq 5x$

39. $5 - 3x \geq 7$

40. $x + 6 \leq 4 - 3x$

41. $5 + x < -x + 3$

42. $3x - 4 > 2x - 9$

43. $-4x > -21 + 3x$

44. $2x + 1.3 > -4.1$

45. $0 \leq 2x + 7 - 9x$

46. $2 - 8x \leq 5 + x$

47. $4(3 - x) \geq 2(x - 1)$

48. $5x \geq -3(x - 2)$

49. $-3(x + 1) < -4(2x - 1)$

50. $4(-x + 2) - (1 - 5x) \geq -8$

51. $7(x - 3) \leq 4(x + 5) - 47$

52. $6(x - 10) + 3(2x - 7) < -45$

53. $\frac{1}{4}x \leq 2$

54. $\frac{3}{5}x \leq \frac{8}{10}$

55. $\dfrac{3x}{2} > -6 - \dfrac{x}{2}$

56. $\dfrac{x}{3} + 2 < \dfrac{x}{4} - 2x$

57. $\frac{1}{3}(4x - 3) \geq 5$

58. $\frac{1}{3}(x - 3) \leq -15$

59. $\frac{1}{3}(2x + 3) \leq \frac{3}{4}x$

60. $\frac{2}{3}(2x - 1) - \frac{2}{3}x \leq 4$

61. $\frac{2}{3}(3x - 2) > \frac{1}{10}(6x + 7)$

62. $\frac{1}{6}(2x - 7) > \frac{1}{2}(x + 1)$

In problems 63 and 64, solve each compound inequality and show the solution on a number line.

63. $5 \leq 4x + 1 \leq 17$

64. $2 < \dfrac{x + 10}{5} \leq 3$

65. Tom has purchased shares of stock for $60. The stock pays him an 11% dividend every full year. In how many full years will the stock have paid him more than $58 in dividends?

66. A bank teller is entitled to a 2-week vacation for the first year of employment. Thereafter, she is entitled to a 3-week vacation for each year she works. How many years must she work without taking a vacation to entitle her to a vacation of 30 weeks or more?

67. A taxpayer has the following choices: either to pay a 30% tax on his gross income, or to pay a 35% tax on the difference between his gross income and $5,000. The taxpayer should elect to pay at the 30% rate when his income is above what level?

68. An investor had $10,000 to invest. She invested some of the money in a savings account at a simple annual rate of 5% and the rest in a commercial paper that paid interest at a simple annual rate of 15%. If her income from both investments for 1 year was at least $700, how much did she invest in the savings account?

69. Marguerite sold five times as many tickets as Rosalind at a fund-raising gathering, but Marguerite could not have sold more than 35 tickets. How many tickets might Rosalind have sold?

70. The formula $F = \frac{9}{5}C + 32$ expresses the temperature F (in degrees Fahrenheit) in terms of the temperature C (in degrees Celsius). If the temperature range on a certain day is 66° to 84° Fahrenheit (that is, $66 \leq F \leq 84$), what is the temperature range in degrees Celsius?

4.6 Equations and Inequalities Involving Absolute Value

We learned in Chapter 1, Section 1.3, that the absolute value notation $|x|$ is used to represent the distance between x and 0. On a number line, $|x|$ is the number of units of distance between the point with coordinate x and the origin, regardless of whether the point is to the right of the origin or to the left of the origin (Figure 1).

Figure 1

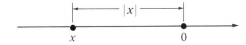

More formally, we have the following definition:

DEFINITION   **Absolute Value**

> If x is a real number, then the **absolute value of x**, represented by $|x|$, is defined by
>
> $$|x| = \begin{cases} x & \text{if } x \geq 0 \\ -x & \text{if } x < 0. \end{cases}$$

For instance,

$$|3| = 3 \quad \text{since} \quad 3 \geq 0,$$

$$|0| = 0 \quad \text{because} \quad 0 \geq 0,$$

and

$$|-3| = -(-3) = 3 \quad \text{since} \quad -3 < 0.$$

Notice that $|x| \geq 0$ is always true.

Equations Involving Absolute Value

Geometrically, the equation $|x| = 3$ means that the point with coordinate x is 3 units from 0 on the number line. Obviously, there are two points that are 3 units from the origin—one to the right of the origin and the other to the left (Figure 2).

Figure 2

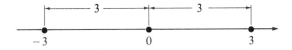

Figure 2 illustrates the following property:

Property 1

Let a be a positive real number, then the equation
$$|u| = a$$
is equivalent to the statement $u = -a$ or $u = a$.

In Examples 1–5, solve each equation.

EXAMPLE **1** $|3x| = 12$.

SOLUTION   Using Property 1 with $u = 3x$, we have

$$
\begin{array}{ccc}
u = -12 & \text{or} & u = 12 \\
3x = -12 & \text{or} & 3x = 12 \\
x = -4 & \text{or} & x = 4.
\end{array}
$$

EXAMPLE **2** $|4t| - 1 = 7$

SOLUTION   The equation is equivalent to $|4t| = 8$. Using Property 1 with $u = 4t$, we have

$$
\begin{array}{ccc}
u = -8 & \text{or} & u = 8 \\
4t = -8 & \text{or} & 4t = 8 \\
t = -2 & \text{or} & t = 2.
\end{array}
$$

EXAMPLE **3** $3|y - 2| = 15$

SOLUTION   We divide both sides of the equation by 3:
$$|y - 2| = 5,$$
which is equivalent to $|u| = 5$ in which $u = y - 2$.

$$
\begin{array}{ccc}
u = -5 & \text{or} & u = 5 \\
y - 2 = -5 & \text{or} & y - 2 = 5 \\
y = -3 & \text{or} & y = 7.
\end{array}
$$

EXAMPLE **4** $|3 - 2t| = 15$

SOLUTION   Let $u = 3 - 2t$:

$$u = -15 \qquad \text{or} \qquad u = 15$$
$$3 - 2t = -15 \qquad \text{or} \qquad 3 - 2t = 15$$
$$-2t = -18 \qquad \text{or} \qquad -2t = 12$$
$$t = 9 \qquad \text{or} \qquad t = -6.$$

EXAMPLE 5 $\left| 1 - \tfrac{2}{5}w \right| = -7$

SOLUTION   Let $u = 1 - \tfrac{2}{5}w$, so that $|u| = -7$. Since $|u|$ must be ≥ 0, the equation has *no* solution.

Inequalities Involving Absolute Value

Examining the number line in Figure 3, we see that the inequality

$$|x| < 3$$

describes all real numbers whose distance from the origin is less than 3 (Figure 3). These are the numbers between -3 and 3. Thus,

$$|x| < 3 \qquad \text{is equivalent to} \qquad -3 < x < 3.$$

($-3 < x < 3$ is a shorthand notation for $-3 < x$ and $x < 3$.)

Figure 3
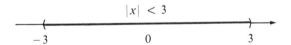

Figure 3 illustrates the following property:

Property 2

If $a > 0$, then

(i) $|u| < a$ is equivalent to $-a < u < a$.
(ii) $|u| \leq a$ is equivalent to $-a \leq u \leq a$.

In Examples 6 and 7, solve each inequality and sketch the graph of its solution set on a number line.

EXAMPLE 6 $|2x| < 10$

SOLUTION   We use Property 2(i) with $u = 2x$: We have

$$-10 < u < 10$$

$$-10 < 2x < 10$$

$$-5 < x < 5. \qquad \text{(We divided all three members by 2.)}$$

Thus, the solution set consists of all real numbers x such that $-5 < x < 5$ (Figure 4).

Figure 4

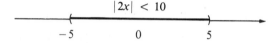

EXAMPLE 7 $|7x - 2| \leq 9$

SOLUTION   We use Property 2(ii) with $u = 7x - 2$: We have

$$-9 \leq u \leq 9$$

$$-9 \leq 7x - 2 \leq 9$$

$$-7 \leq 7x \leq 11 \qquad \text{(We added 2 to all members.)}$$

$$-1 \leq x \leq \frac{11}{7}. \qquad \text{(We divided all members by 7.)}$$

Thus, the solution set consists of all real numbers x such that $-1 \leq x \leq \frac{11}{7}$ (Figure 5).

Figure 5

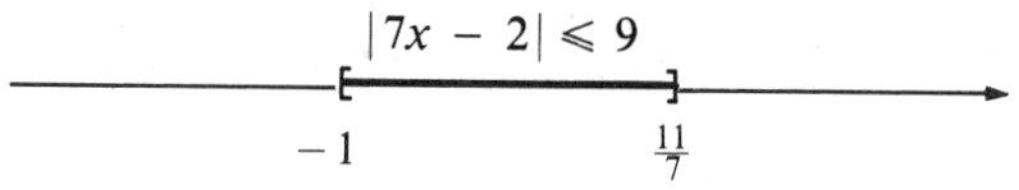

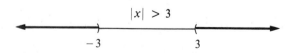

On the basis of the discussion above, you may conclude that the graph of the inequality

$$|x| > 3,$$

on a number line, consists of all real numbers whose distance from the origin is greater than 3 (Figure 6). These numbers are either less than -3 or greater than 3. That is,

$$|x| > 3 \qquad \text{is equivalent to} \qquad x < -3 \qquad \text{or} \qquad x > 3 \text{ (Figure 6)}.$$

Figure 6

It is a *common error* to write $x < -3$ or $x > 3$ as $-3 > x > 3$. The statement $-3 > x > 3$ says that $-3 > 3$, which is not correct.

We can state the following property:

Property 3

If $a > 0$, then

(i) $|u| > a$ is equivalent to $u < -a$ or $u > a$.
(ii) $|u| \geq a$ is equivalent to $u \leq -a$ or $u \geq a$.

In Examples 8–10, solve each inequality and sketch its solution set on a number line.

EXAMPLE 8 $|7x| + 1 > 15$

SOLUTION   The inequality is equivalent to $|7x| > 14$. We use Property 3(i) with $u = 7x$. We have

$$u < -14 \qquad \text{or} \qquad u > 14$$
$$7x < -14 \qquad \text{or} \qquad 7x > 14$$
$$x < -2 \qquad \text{or} \qquad x > 2.$$

Therefore, the solution set consists of all real numbers x for which $x < -2$ or $x > 2$ (Figure 7).

Figure 7

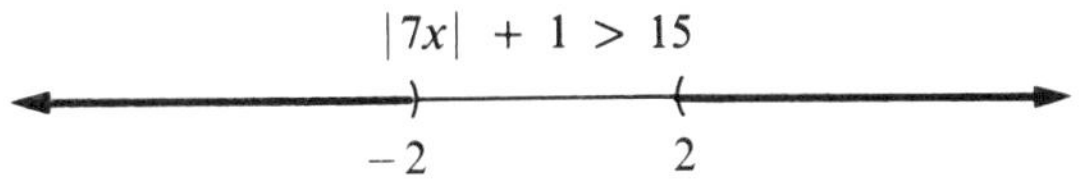

$$|7x| + 1 > 15$$

$-2 \qquad 2$

EXAMPLE 9 $|2x + 7| \geq 11$

SOLUTION   We use Property 3(ii) with $u = 2x + 7$. We have

$$u \leq -11 \qquad \text{or} \qquad u \geq 11$$
$$2x + 7 \leq -11 \qquad \text{or} \qquad 2x + 7 \geq 11$$
$$2x \leq -18 \qquad \text{or} \qquad 2x \geq 4$$
$$x \leq -9 \qquad \text{or} \qquad x \geq 2.$$

Figure 8

$$|2x + 7| \geq 11$$

$-9 \qquad 2$

Therefore, the solution set consists of all real numbers x for which $x \leq -9$ or $x \geq 2$ (Figure 8).

EXAMPLE 10 $|2 - 3x| > 5$

SOLUTION   We use Property 3(i) with $u = 2 - 3x$. We have

Figure 9

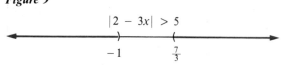

$$|2 - 3x| > 5$$

$$
\begin{array}{ccc}
u < -5 & \text{or} & u > 5 \\
2 - 3x < -5 & \text{or} & 2 - 3x > 5 \\
-3x < -7 & \text{or} & -3x > 3 \\
x > \dfrac{7}{3} & \text{or} & x < -1.
\end{array}
$$

Therefore, the solution set consists of all real numbers x for which $x < -1$ or $x > \frac{7}{3}$ (Figure 9).

PROBLEM SET 4.6

In problems 1–24, solve each equation.

1. $|5x| = 15$
2. $|-4t| = 16$
3. $|-7y| = 14$
4. $3|2x| = 12$

5. $3|6w| = 24$
6. $-4|5c| = 2$
7. $|3z| - 1 = 14$
8. $|5w| - 2 = 13$

9. $5 - |2u| = 3$
10. $8 + |3w| = 17$
11. $|2y| = -4$
12. $|-5t| = |-20|$

13. $|x - 2| = 3$
14. $|p - 3| = 3$
15. $|5 - q| = 6$
16. $|3 - y| = 4$

17. $3|2w - 3| = 7$
18. $2|5 - 2t| = 1$
19. $4|2 - 7x| = 16$
20. $3|3 - 2y| = 7$

21. $|3w + 2| = -1$
22. $|\frac{1}{3} - 2x| = -\frac{2}{3}$
23. $|3t + 2| = 0$
24. $|5 - 4x| = 0$

In problems 25–50, solve each inequality and sketch the graph of its solution set on a number line.

25. $|3x| < 15$
26. $|4x| < |-8|$
27. $|5x| \leq 20$
28. $|-7x| \leq 14$

29. $|\frac{1}{3}x| \leq 2$
30. $|-\frac{1}{5}x| < 1$
31. $|4x| > 12$
32. $|5x| > 25$

33. $|-6x| \geq 18$
34. $|-9x| \geq 27$
35. $|3x| \geq |-12|$
36. $|7x| > |-28|$

37. $|x - 1| < 3$
38. $|x + 2| < 5$
39. $|2x - 1| \leq 3$
40. $|5x - 3| > 12$

41. $|4x - 1| \geq 11$
42. $|3 - 2x| < 5$
43. $|\frac{2}{3} - x| \leq \frac{1}{3}$
44. $|5x + 1| \geq -2$

45. $\left|\dfrac{1}{3} - \dfrac{x}{6}\right| \geq \dfrac{1}{2}$
46. $\left|\dfrac{3x}{7} + \dfrac{2}{21}\right| < 0$
47. $|5x + 3| + 2 > 8$
48. $|1 - 2x| - 5 \leq 7$

49. $\frac{1}{4}|2x - 1| \leq 3$
50. $\frac{1}{2}|3x + 2| \geq 4$

REVIEW PROBLEM SET

In problems 1–34, solve each equation.

1. $y + 1 = 3$
2. $z - 11 = 4$
3. $u + 8 = 21$
4. $m - 7 = 53$

5. $-17u = 51$
6. $-8t = 24$
7. $2y + 13 = -1$
8. $10 = 2 - 3m$

9. $-4 + 5z = 16$
10. $5x + 2 = 32$
11. $-4x - 7 = 21$
12. $4y - 16 = 6y$

13. $3w - 4 = w - 12$
14. $17 - 3c = c + 5$

15. $10 - 8(3 - y) = 0$
16. $3(3t - 1) + 2 = 4(2 - t)$

17. $2(x - 1) = 3 - 3(x - 5)$
18. $12(y - 2) + 8 = 5(y - 1) + 2y$

19. $z - 7(4 + z) = 5z - 6(3 - 4z)$

20. $5(w + 2) - 3(w - 1) = 4(w + 1) + 15$

21. $2(u - 1) + 3(u - 2) = -4(u - 3)$

22. $m - 2(3 - m) = 2(m + 3) - (m - 2)$

23. $\dfrac{2z}{3} + z = 5$

24. $\dfrac{3t}{6} - \dfrac{2t}{8} = 8$

25. $\dfrac{x}{3} - \dfrac{x}{6} = 1$

26. $\dfrac{2x - 1}{3} = \dfrac{x}{5} + \dfrac{2}{15}$

27. $\dfrac{5}{y} + \dfrac{3}{8} = \dfrac{7}{16}$

28. $\dfrac{t + 4}{3} - \dfrac{t + 2}{2} = \dfrac{t - 2}{5}$

29. $\dfrac{13}{14z} - \dfrac{1}{2z} = \dfrac{1}{7}$

30. $\dfrac{2}{w} + \dfrac{w - 1}{3w} = \dfrac{2}{5}$

31. $\dfrac{3}{y - 2} = \dfrac{5}{y + 2}$

32. $\dfrac{10 - w}{w} + \dfrac{3w + 3}{3w} = 3$

33. $\dfrac{t + 1}{t - 1} - \dfrac{t}{t + 1} = \dfrac{5 - t}{t^2 - 1}$

34. $\dfrac{12}{c^2 - 25} = \dfrac{1}{c + 5} + \dfrac{2}{c - 5}$

In problems 35–38, express each repeating decimal as a quotient of integers.

35. $0.\overline{37}$

36. $-0.0\overline{3}$

37. $0.04\overline{8}$

38. $1.\overline{431}$

In problems 39–52, solve each equation for the indicated unknown.

39. $a - 2x = 5a$, for x.

40. $5u - 17b = 39u$, for u.

41. $a - bt = 3a$, for t.

42. $a - y = a(y - 1)$, for y.

43. $az = b + cz$, for z.

44. $y - a = m(x - b)$, for x.

45. $6(a - bc) = a^3 - 6a$, for c.

46. $a^2t - 2at = 1 - t$, for t.

47. $3(ax - 2b) = 4(2b - ax)$, for a.

48. $7(a^2b + c) - 4a^2(b + 3a^2) = 10c$, for b.

49. $\dfrac{x - 1}{a} + \dfrac{2}{a} = 3$, for x.

50. $\dfrac{u}{5b} - \dfrac{u}{7b} = 3$, for u.

51. $\dfrac{c}{a} - \dfrac{3}{a - b} = 1$, for b.

52. $\dfrac{2w - a}{5} + \dfrac{w}{3} - 3a = 1$, for w.

In problems 53–60, solve each formula for the indicated unknown.

53. $E = mc^2$, for m.

54. $S = 2A + ph$, for A.

55. $S = 2\pi rh + 2\pi r^2$, for h.

56. $A = \pi r^2 + 2\pi rs$, for s.

57. $G = \dfrac{m}{r^3}$, for m.

58. $S = \dfrac{ax^n - a}{x - 1}$, for a.

59. $g = \dfrac{v^2}{R + h}$, for h.

60. $En = RI + \dfrac{nr\ell}{m}$, for ℓ.

61. In a Red Cross drive, Dawn contributed twice as much as Joe, and Mike contributed $4 more than Dawn. How much did each person give if their donations totaled $79?

62. A billfold contains $223 in $10, $5, and $1 bills. There are 47 bills in all, and there are five more $5 bills than $10 bills. How many bills of each kind are there?

63. A woman used 420 meters of wire to fence a rectangular garden. If the length of the garden is four times the width, what are the length and the width of the garden?

64. How many pounds of water must be evaporated from 50 pounds of a 3% salt solution so that the remaining portion will be a 5% solution?

65. A retail store invested $25,000 in two kinds of radios. All the radios had been sold and the store made a profit of 15% from the first type, but lost 5% on the second kind. If the two investments provided a return of 10% on the entire amount invested, how much had the store invested in each kind of radio?

66. A chemist has a gallon containing a 4% solution of a certain chemical. How much water should she add so that the resulting mixture will be a 1% solution?

67. A jogger traveling at 10 miles per hour leaves a starting point. Twenty minutes later a bicycle rider traveling at 12 miles per hour leaves the same starting point and follows the same route. How long does it take the bicyclist to overtake the jogger?

68. In a predator-prey model, the total population is 34,100, and the number of the prey is 10 times the number of predators. What is the predator population?

69. If a taxicab fare is $0.80 for the first mile and $0.50 for each additional mile, the cost C for n miles is represented by the formula

$$C = 80 + 50(n - 1).$$

How many miles are traveled in a trip that cost $3.75?

70. A rocket is fired straight up at an initial velocity v feet per second. The rocket's distance of s feet above the ground t seconds after being fired is given by the formula $s = vt - 16t^2$. Solve this formula for v.

In problems 71–82, solve each inequality and graph the solution set on a number line.

71. $3x > 6$

72. $-3x \leq 12$

73. $2x + 1 > 13$

74. $3 - 2x \leq 9$

75. $3x + 5 < 2x + 1$

76. $3(x - 4) < 2(4 - x)$

77. $2(1 - 2x) \geq 3(4 - x)$

78. $5x + 3(3x - 1) \geq -1 - 5(x - 3)$

79. $\frac{1}{19}(3x - 5) \geq 1$

80. $\frac{1}{4}(2x + 3) \geq 1 - x$

81. $\frac{1}{3}(1 - 3x) \leq \frac{1}{4}(1 + 2x)$

82. $\frac{1}{5}(2x + 1) - 2 \leq \frac{1}{2}(3x + 1)$

83. A postal service charges a fixed fee of $0.70 plus an additional $0.10 for every ounce over the first pound. What restriction in weight is there on a package that is to be delivered for less than $4.35?

84. Judy asks for a $7 weekly payroll deduction from her salary in order to save money for a new camera. How many weeks will it take for her payroll deductions to add up to at least $437 (the amount she needs for the camera)?

In problems 85–94, solve each equation.

85. $|7x| = 35$

86. $|-11x| = 55$

87. $|y + 2| = 31$

88. $|4 - w| = 13$

89. $|4 - 2z| = -2$

90. $|3t + 17| = -1$

91. $|4u + 3| = 13$

92. $|2(y - 3)| = 8$

93. $|7 - 11x| = 29$

94. $|3t - 1| + 4 = 8$

In problems 95–104, solve each inequality and graph the solution set on a number line.

95. $|x| < 11$

96. $|x| \geq 12$

97. $|13x| \geq 52$

98. $|17x| < 51$

99. $|x + 40| \leq 43$

100. $|8 - 3x| \leq 6$

101. $|4x + 7| \geq 9$

102. $\left|\dfrac{3x}{4} - 1\right| > 2$

103. $\left|\frac{1}{5}(1 - x)\right| < 3$

104. $\left|\frac{1}{3}(2 - x)\right| \leq 5$

5 Exponents, Radicals, and Complex Numbers

Exponents and radicals are often used in applied mathematics and the sciences. In dealing with very large or very small numbers on computers or calculators, for example, these numbers are typically expressed in exponential notation. Scientific problems are commonly formulated as mathematical equations which involve exponents and/or radicals.

We begin this chapter by extending the properties of exponents, discussed in Chapter 2, beyond the positive integers. We then translate the properties of exponents to radicals. The chapter concludes with a discussion of complex numbers.

5.1 Zero and Negative Integer Exponents

In Chapter 2, Section 2.3, we established the properties of exponents that follow from the definition

$$a^n = \overbrace{a \cdot a \cdot a \cdots a,}^{n \text{ factors}}$$

where n is a positive integer. We now extend this definition so that zero and the negative integers can also be used as exponents, and do so in such a way that the properties of positive exponents (discussed in Section 2.3) continue to hold.

We begin with a question: How can we define a^0? If the multiplication property, $a^m \cdot a^n = a^{m+n}$, continues to hold for $m = 0$, we have

$$a^0 \cdot a^n = a^{0+n}$$
$$= a^n.$$

This equation is true only if $a \neq 0$ and if $a^0 = 1$. This leads us to the following definition.

DEFINITION 1   **Zero as an Exponent**

> If a is any nonzero real number, we define $a^0 = 1$.

For example

$$3^0 = 1, \qquad \left(-\frac{4}{7}\right)^0 = 1, \qquad \text{and} \qquad (1 + y^2)^0 = 1.$$

Notice that 0^0 is not defined.

To define a^{-n}, when n is a positive integer, we also assume that $a^m \cdot a^n = a^{m+n}$ continues to hold for $m = -n$. Thus, we have

$$a^{-n} \cdot a^n = a^{-n+n} = a^0 = 1, \qquad \text{for } a \neq 0.$$

This equation is true only if

$$a^{-n} = \frac{1}{a^n},$$

which leads to the following definition:

DEFINITION 2   **Negative Integer Exponents**

> If a is a nonzero real number and n is any positive integer, then
>
> $$a^{-n} = \frac{1}{a^n}.$$

For instance,

$$10^{-4} = \frac{1}{10^4}, \qquad 3^{-4} = \frac{1}{3^4}, \qquad \text{and} \qquad b^{-1} = \frac{1}{b}, \qquad \text{for} \qquad b \neq 0.$$

We obtain the following properties from Definition 2:

> If $a \neq 0$ and $b \neq 0$, then
>
> (1) $\left(\dfrac{a}{b}\right)^{-n} = \left(\dfrac{b}{a}\right)^n.$
>
> (2) $\dfrac{a^{-n}}{b^{-m}} = \dfrac{b^m}{a^n}.$

The proofs of these properties are straightforward. (Problem 68, Problem Set 5.1.)

In Examples 1 and 2, rewrite each expression so that it contains only positive exponents, and simplify the results.

EXAMPLE **1**   (a) 6^{-2} (b) $(2x)^{-3}$ (c) $(3y^0)^{-4}$

(d) $\dfrac{a^{-1}}{c^{-3}}$ (e) $\dfrac{1}{7(w+4)^{-3}}$ (f) $\left(\dfrac{p^2}{q^3}\right)^{-1}$

(g) $5t^{-4}$

SOLUTION   (a) $6^{-2} = \dfrac{1}{6^2} = \dfrac{1}{36}$ (by Definition 2)

(b) $(2x)^{-3} = \dfrac{1}{(2x)^3}$ (by Definition 2)

$= \dfrac{1}{2^3 \cdot x^3} = \dfrac{1}{8x^3}$ (why?)

(c) $(3y^0)^{-4} = (3 \cdot 1)^{-4}$ (by Definition 1)

$= 3^{-4} = \dfrac{1}{3^4} = \dfrac{1}{81}$ (by Definition 2)

(d) $\dfrac{a^{-1}}{c^{-3}} = \dfrac{c^3}{a}$ [by Property (2)]

(e) $\dfrac{1}{7(w+4)^{-3}} = \dfrac{(w+4)^3}{7}$ [by Property (2)]

Notice that the 7 remains in the denominator because it was not part of the expression under the exponent.

(f) $\left(\dfrac{p^2}{q^3}\right)^{-1} = \left(\dfrac{q^3}{p^2}\right)^1$ [by Property (1)]

$= \dfrac{q^3}{p^2}$

(g) $5t^{-4} = \dfrac{5}{t^4}$ (by Definition 2)

EXAMPLE **2**   (a) $\dfrac{1}{2^{-2} + 3^{-2}}$ (b) $\dfrac{1}{(2+3)^{-2}}$ (c) $(4 + 4^{-1})^{-1}$

(d) $\dfrac{u^3}{v^{-3}} + \dfrac{v^3}{u^{-3}}$ (e) $\dfrac{1}{p^{-1} + q^{-1}}$ (f) $\dfrac{x^{-1} + y^{-1}}{x + y}$

SOLUTION   (a) $\dfrac{1}{2^{-2} + 3^{-2}} = \dfrac{1}{\dfrac{1}{2^2} + \dfrac{1}{3^2}} = \dfrac{1}{\dfrac{1}{4} + \dfrac{1}{9}}$

$$= \dfrac{1}{\dfrac{9 + 4}{36}} = \dfrac{1}{\dfrac{13}{36}} = \dfrac{36}{13}$$

(b) $\dfrac{1}{(2 + 3)^{-2}} = \dfrac{1}{5^{-2}} = 5^2 = 25$

(c) $(4 + 4^{-1})^{-1} = \left(4 + \dfrac{1}{4}\right)^{-1} = \left(\dfrac{16 + 1}{4}\right)^{-1}$

$$= \left(\dfrac{17}{4}\right)^{-1} = \dfrac{4}{17}$$

(d) $\dfrac{u^3}{v^{-3}} + \dfrac{v^3}{u^{-3}} = u^3 v^3 + v^3 u^3 = 2u^3 v^3$

(e) $\dfrac{1}{p^{-1} + q^{-1}} = \dfrac{1}{\dfrac{1}{p} + \dfrac{1}{q}} = \dfrac{1}{\dfrac{q + p}{pq}} = \dfrac{pq}{q + p}$

(f) $\dfrac{x^{-1} + y^{-1}}{x + y} = \dfrac{\dfrac{1}{x} + \dfrac{1}{y}}{x + y} = \dfrac{\dfrac{y + x}{xy}}{x + y} = \dfrac{y + x}{xy} \cdot \dfrac{1}{x + y} = \dfrac{1}{xy}$

Definitions 1 and 2 enable you to use any *integer*—positive, negative or zero—as an exponent, with the exception that 0^n is defined only when n is positive. We can verify that Properties (i) through (v) listed below hold for *all* integer exponents.

Properties of Integer Exponents

Let a and b be real numbers, and let m and n be integers. Then each of the following is true for all values of a and b for which both sides of the equation are defined:

(i) $a^m \cdot a^n = a^{m+n}$ (ii) $(a^m)^n = a^{mn}$ (iii) $(ab)^m = a^m b^m$

(iv) $\left(\dfrac{a}{b}\right)^m = \dfrac{a^m}{b^m}$ (v) $\dfrac{a^m}{a^n} = a^{m-n}$

In Examples 3 and 4, use the properties of exponents to rewrite each expression so that it contains only positive exponents, and simplify the results.

EXAMPLE 3 (a) $x^6 \cdot x^{-2}$ (b) $(p^4)^{-3}$

(c) $(6y^3)^{-2}$ (d) $\left(\dfrac{3}{5}\right)^{-4}$

(e) $\dfrac{w^3}{w^{-6}}$ (f) $(x^2y^{-3})^{-1}$

SOLUTION   (a) $x^6 \cdot x^{-2} = x^{6-2}$ [by Property (i)]
$$= x^4$$

(b) $(p^4)^{-3} = p^{4(-3)}$ [by Property (ii)]

$$= p^{-12} = \frac{1}{p^{12}}$$

(c) $(6y^3)^{-2} = 6^{-2} \cdot (y^3)^{-2}$ [by Property (iii)]

$$= \frac{1}{6^2} \cdot \frac{1}{(y^3)^2} = \frac{1}{36} \cdot \frac{1}{y^6} = \frac{1}{36y^6}$$

(d) $\left(\dfrac{3}{5}\right)^{-4} = \dfrac{3^{-4}}{5^{-4}}$ [by Property (iv)]

$$= \frac{5^4}{3^4}$$ $\left(\text{by the Property } \dfrac{a^{-n}}{b^{-m}} = \dfrac{b^m}{a^n}\right)$

$$= \frac{625}{81}$$

(e) $\dfrac{w^3}{w^{-6}} = w^{3-(-6)}$ [by Property (v)]

$$= w^{3+6} = w^9$$

(f) $(x^2y^{-3})^{-1} = (x^2)^{-1} \cdot (y^{-3})^{-1}$ [by Property (iii)]
$$= x^{-2} \cdot y^3$$ [by Property (ii)]

$$= \frac{y^3}{x^2}$$

EXAMPLE 4 (a) $\dfrac{3ab^{-2}}{c^3d^{-4}}$ (b) $\left[\dfrac{x^{-4}(-3x)}{(-5x)^{-2}}\right]^{-1}$ (c) $\dfrac{5c^{-3}(r+s)^2}{15c^4(r+s)^{-5}}$

SOLUTION

(a) $\dfrac{3ab^{-2}}{c^3d^{-4}} = \dfrac{3a}{c^3}\left(\dfrac{b^{-2}}{d^{-4}}\right)$ (why?)

$= \dfrac{3a}{c^3}\left(\dfrac{d^4}{b^2}\right)$ [by Property (2)]

$= \dfrac{3ad^4}{c^3b^2}$

(b) $\left[\dfrac{x^{-4}(-3x)}{(-5x)^{-2}}\right]^{-1} = \left[\dfrac{x^{-4}}{(-5x)^{-2}}(-3x)\right]^{-1}$

$= \left[\dfrac{(-5x)^2(-3x)}{x^4}\right]^{-1}$ [by Property (2)]

$= \left[\dfrac{25x^2(-3x)}{x^4}\right]^{-1} = \left(\dfrac{-75x^3}{x^4}\right)^{-1}$

$= \left(\dfrac{-75}{x^{4-3}}\right)^{-1} = \left(-\dfrac{75}{x}\right)^{-1} = -\dfrac{x}{75}$

(c) $\dfrac{5c^{-3}(r+s)^2}{15c^4(r+s)^{-5}} = \dfrac{(r+s)^2(r+s)^5}{3c^4c^3}$ (why?)

$= \dfrac{(r+s)^{2+5}}{3c^{4+3}} = \dfrac{(r+s)^7}{3c^7}$ [by Property (i)]

PROBLEM SET 5.1

In problems 1–34, rewrite each expression so that it contains only positive exponents, and simplify the results.

1. 5^{-2}

2. 2^{-4}

3. $\left(\dfrac{1}{3}\right)^{-4}$

4. $(-10)^{-1}$

5. p^{-5}

6. y^{-3}

7. $(5x)^{-2}$

8. $(3u)^{-4}$

9. $u^{-3}v^{-6}$

10. $x^{-4}y^{-7}$

11. $4a^{-7}b^2$

12. $5k^{-2}r^{-3}$

13. $\dfrac{1}{7^{-1}}$

14. $\dfrac{y^{-5}}{7x^{-2}}$

15. $\dfrac{a^{-2}}{8p^{-4}}$

16. $\dfrac{(ab)^{-1}}{a^{-1}b^{-1}}$

17. $\dfrac{2}{-3x^{-4}y^{-5}}$

18. $\dfrac{4}{5m^{-6}n^7}$

19. $\left(\dfrac{p^3}{q^2}\right)^{-1}$

20. $\left(\dfrac{x^4}{y^3}\right)^{-1}$

21. $(-3y^2)^0$

22. $(3 + 5x^2)^0$

23. $(1 + 7y^{-3})^0$

24. $(7^{-1} + 3^{-2})^0$

25. $\dfrac{2^{-2} + 5^{-2}}{10^{-2}}$

26. $\dfrac{a^{-1} + b^{-1}}{(a + b)^{-1}}$

27. $\dfrac{x^{-1} + y^{-1}}{(xy)^{-1}}$

28. $\dfrac{5^{-1} + 3^{-2}}{(45)^{-1}}$

29. $\dfrac{1}{4^{-2} + 3^{-2}}$

30. $\dfrac{1}{(4 + 3)^{-2}}$

31. $\dfrac{1}{u^{-2} + v^{-2}}$

32. $\dfrac{a^2}{b^{-2}} + \dfrac{b^2}{a^{-2}}$

33. $\dfrac{x^{-1}}{y^{-1}} + \left(\dfrac{x}{y}\right)^{-1}$

34. $\left(\dfrac{x}{y}\right)^{-1} + \left(\dfrac{y}{x}\right)^{-1}$

In problems 35–58, use the properties of exponents to rewrite each expression so that it contains only positive exponents, and simplify the results.

35. $7^{-1} \cdot 7^3$

36. $5^{-3} \cdot 5^{-2}$

37. $9^{-3} \cdot 9$

38. $3^{-4} \cdot 3^7$

39. $x^{-3} \cdot x^{-2}$

40. $y^{-7} \cdot y^{-2}$

41. $(2^3)^{-2}$

42. $(5^2)^{-1}$

43. $[(-4)^2]^{-2}$

44. $(p^{-2})^{-4}$

45. $(3x)^{-4}$

46. $(3y^{-4})^2$

47. $(5^{-1}p^4)^{-2}$

48. $(p^{-1}y^{-2})^5$

49. $\left(\dfrac{5}{x}\right)^{-2}$

50. $\left(\dfrac{3}{p^{-1}}\right)^{-1}$

51. $\left(\dfrac{p^2}{q^2}\right)^{-2}$

52. $\left(\dfrac{x^{-3}}{y^{-2}}\right)^{-1}$

53. $\dfrac{8^{-3}}{8^{-5}}$

54. $\dfrac{7^{-3}}{7^{-8}}$

55. $\dfrac{x^{-3}}{x^5}$

56. $\dfrac{y}{y^{-6}}$

57. $\dfrac{xy^{-7}}{xy^{-4}}$

58. $\dfrac{(ab)^{-4}}{(ab)^{-7}}$

In problems 59–67, rewrite each expression without negative exponents, and simplify.

59. $\left(\dfrac{b^{-1}}{3a^{-1}}\right)\left(\dfrac{3a}{b}\right)^{-1}$

60. $\left(\dfrac{a^{-1}c^2}{a^{-2}c^{-1}}\right)^{-2}$

61. $\left(\dfrac{x^{-1}y^{-2}}{x^{-2}y^3}\right)^2$

62. $\left(\dfrac{p^{-2}q^{-1}r^{-3}}{q^3r^{-1}}\right)^{-4}$

63. $\left(\dfrac{a^2c^{-2}b^{-1}}{(abc)^{-1}}\right)^{-3}$

64. $\left(\dfrac{a^{-4}b^2c^{-6}}{(ab)^{-2}(bc)^{-3}}\right)^{-1}$

65. $\left(\dfrac{u^{-4}v^3z^{-3}}{u^3(vz)^{-2}}\right)^{-4}$

66. $\left(\dfrac{(x^{-2})^{-1}(y^{-2})^3}{(x^{-1})^2(y^{-1})^2}\right)^{-2}$

67. $\left[\left(\dfrac{a^{-1}}{b}\right)^{-1} \cdot \left(\dfrac{b^{-1}}{a}\right)^{-1}\right]^{-2}$

68. Show that:

(a) $\left(\dfrac{a}{b}\right)^{-n} = \left(\dfrac{b}{a}\right)^n$

(b) $\dfrac{a^{-n}}{b^{-m}} = \dfrac{b^m}{a^n}$

5.2 Roots and Rational Exponents

The process of finding a *root* of a number is the inverse of the process of raising a number to a power. If we want to find the square of the number 5, we write $5^2 = 25$. The process of finding a number whose square is 25 is called finding the *square root* of 25. That is, because

$$5^2 = 25,$$

we say that 5 is a **square root** of 25.

Similarly, because

$$2^3 = 8,$$

we say that 2 is a **cube root** of 8, and because

$$3^4 = 81,$$

we say that 3 is a **fourth root** of 81.

We have the following definition:

DEFINITION 1   **The *n*th Root of a Number**

> If a and b are real numbers, and n is a positive integer greater than 1, such that
>
> $$b^n = a,$$
>
> then b is called an ***n*th root** of a.

Radical notation is commonly used to indicate a root. We write $\sqrt[n]{a}$ to represent the nth root of a number a.

In the expression $\sqrt[n]{a}$, the symbol $\sqrt{}$ is called the **radical.** The positive integer n is called the **index** (if the index is not written, it is understood to be 2), and the real number under the radical is called the **radicand:**

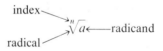

A number may have both a postive and a negative nth root. When this occurs, the *positive* nth root of the number is called the **principal *n*th root,** and is denoted by $\sqrt[n]{a}$. For example, although it is correct to say that both 3 and -3 are square roots of 9, it is *incorrect* to write $\sqrt{9} = -3$ (but *correct* to write $\sqrt{9} = 3$). If we wish to designate the negative square root of 9 using radical notation, we write $-\sqrt{9} = -3$.

Similarly,

$$\sqrt[4]{81} = 3, \qquad \text{since} \qquad 3^4 = 81 \qquad \text{and} \qquad 3 > 0,$$

and

$$\sqrt[6]{64} = 2, \qquad \text{since} \qquad 2^6 = 64 \qquad \text{and} \qquad 2 > 0.$$

If a number a has *only* a negative nth root, the negative root is considered the principal nth root of a. Thus,

$$\sqrt[3]{-8} = -2, \qquad \text{since} \qquad (-2)^3 = -8, \text{ and } -2 \text{ is the only cube root of } -8;$$

and

$$\sqrt[5]{-32} = -2, \qquad \text{since} \qquad (-2)^5 = -32, \text{ and } -2 \text{ is the only fifth root of } -32.$$

Of course, if a number a has *only* a positive nth root, then the positive root is considered to be the principal nth root of a. For example, $\sqrt[3]{27} = 3$, the only cube root of 27.

We can now formulate the following definition:

DEFINITION 2   **The Principal nth Root of a Number**

Let n be a positive integer greater than 1, and let a be a real number. Then $\sqrt[n]{a}$—the **principal nth root** of a—is defined as follows:

(i) If $a > 0$, $\sqrt[n]{a}$ is the positive nth root of a.

(ii) If $a < 0$ and n is odd, $\sqrt[n]{a}$ is the negative nth root of a.

(iii) If $a = 0$, then $\sqrt[n]{0} = 0$.

Notice that $\sqrt[n]{a}$ is not defined when $a < 0$ and n is even. Furthermore, if $\sqrt[n]{a}$ is defined, it represents only one real number.

EXAMPLE 1   Find the principal root (if it is defined).

(a) $\sqrt{36}$ (b) $\sqrt{121}$ (c) $\sqrt[4]{16}$

(d) $\sqrt[3]{-125}$ (e) $\sqrt[4]{-16}$

SOLUTION   (a) $\sqrt{36} = 6$ (b) $\sqrt{121} = 11$

(c) $\sqrt[4]{16} = 2$ (d) $\sqrt[3]{-125} = -5$

(e) $\sqrt[4]{-16}$ is undefined.   (why?)

Rational Exponents

In Section 5.1, we extended the definition of exponents to include negative integers and zero. Now we will expand the definition of exponents to include rational numbers. Radicals can be used to define rational exponents. Let m and n be integers with $n \neq 0$. We wish to define $a^{m/n}$. If Property (ii) of Section 5.1 (page 166) is true, we must have

$$a^{m/n} = a^{(1/n)m} = (a^{1/n})^m.$$

The basic question, therefore, is: How do we define $a^{1/n}$? Again, if Property (ii) is true, then

$$(a^{1/n})^n = a^{(1/n)n} = a^1 = a.$$

In other words, $a^{1/n}$ must be defined as the **principal nth root** of a. Therefore, we have the following definition:

DEFINITION 3   **Rational Exponents**

> Let a be a nonzero real number. Suppose that m and n are integers, that n is positive, and that the fraction m/n is reduced to lowest terms. Then, if $\sqrt[n]{a}$ exists,
>
> (i) $a^{1/n} = \sqrt[n]{a}$ (ii) $a^{m/n} = (a^{1/n})^m = (\sqrt[n]{a})^m = \sqrt[n]{a^m}$.

Notice that $a^{1/n}$ is just an alternative notation for $\sqrt[n]{a}$, the principal nth root of a. For instance,

$$4^{1/2} = \sqrt{4} = 2, \qquad (-8)^{1/3} = \sqrt[3]{-8} = -2, \qquad \text{and} \qquad 0^{1/5} = \sqrt[5]{0} = 0.$$

EXAMPLE 2   Rewrite each expression in terms of rational exponents.

(a) $\sqrt{7}$ (b) $\sqrt[3]{-x^2}$

(c) $\sqrt[5]{11p^4}$ (d) $\sqrt[7]{(a+b)^5}$

SOLUTION   Using Definition 3, we have:

(a) $\sqrt{7} = 7^{1/2}$

(b) $\sqrt[3]{-x^2} = (-x^2)^{1/3} = -(x^2)^{1/3} = -x^{2/3}$ (why?)

(c) $\sqrt[5]{11p^4} = (11p^4)^{1/5} = 11^{1/5}p^{4/5}$

(d) $\sqrt[7]{(a+b)^5} = (a+b)^{5/7}$

EXAMPLE 3   Rewrite each expression in terms of radicals.

(a) $5^{3/4}$ (b) $(6x)^{3/2}$

(c) $(-11y)^{2/3}$ (d) $(8-r)^{2/5}$

SOLUTION   We use $a^{m/n} = \sqrt[n]{a^m}$:

(a) $5^{3/4} = \sqrt[4]{5^3} = \sqrt[4]{125}$

(b) $(6x)^{3/2} = \sqrt{(6x)^3} = \sqrt{216x^3}$

(c) $(-11y)^{2/3} = \sqrt[3]{(-11y)^2} = \sqrt[3]{121y^2}$

(d) $(8-r)^{2/5} = \sqrt[5]{(8-r)^2}$

Remember, Definition 3 is to be used *only* when $n > 0$ and *when m/n is reduced to lowest terms.* For example, we should know that it is incorrect to write $(-8)^{2/6} = (\sqrt[6]{-8})^2$, because $\frac{2}{6}$ is not reduced to lowest terms. First we must reduce $\frac{2}{6}$ to lowest terms:

$$(-8)^{2/6} = (-8)^{1/3}.$$

Then we can apply Definition 3:

$$(-8)^{1/3} = \sqrt[3]{-8} = -2.$$

EXAMPLE 4   Find the value of each expression.

(a) $27^{4/3}$ (b) $32^{3/5}$ (c) $(-4)^{3/2}$ (d) $(-64)^{8/12}$

SOLUTION   We use $a^{m/n} = (\sqrt[n]{a})^m$ of Definition 3:

(a) $27^{4/3} = (\sqrt[3]{27})^4 = 3^4 = 81$

(b) $32^{3/5} = (\sqrt[5]{32})^3 = 2^3 = 8$

(c) $(-4)^{3/2}$ is undefined, because $\sqrt{-4}$ does not exist (as a real number).

(d) $(-64)^{8/12} = (-64)^{2/3} = (\sqrt[3]{-64})^2 = (-4)^2 = 16$

Suppose that m and n are positive integers and that m/n is reduced to lowest terms. Then $-(m/n)$ represents a negative rational number. If Property (ii) (page 166) continues to hold true for negative rational exponents, we have

$$a^{-(m/n)} = (a^{1/n})^{-m}.$$

We use Definition 2 of a negative integer exponent (page 164). If $a \neq 0$, then

$$(a^{1/n})^{-m} = \frac{1}{(a^{1/n})^m}.$$

We have the following definition:

DEFINITION 4   **Negative Rational Exponents**

Let a be a nonzero real number. Suppose that m and n are positive integers, and that m/n is in lowest terms. Then, if $\sqrt[n]{a}$ exists,

$$a^{-m/n} = \frac{1}{a^{m/n}}.$$

EXAMPLE 5   Find the value of each expression.

(a) $16^{-3/4}$

(b) $(-8)^{-2/3}$

(c) $-8^{-2/3}$

(d) $64^{-5/6}$

(e) $(-125)^{-1/3}$

SOLUTION   We use Definition 4:

(a) $16^{-3/4} = \dfrac{1}{16^{3/4}} = \dfrac{1}{(\sqrt[4]{16})^3} = \dfrac{1}{2^3} = \dfrac{1}{8}$

(b) $(-8)^{-2/3} = \dfrac{1}{(-8)^{2/3}} = \dfrac{1}{(\sqrt[3]{-8})^2} = \dfrac{1}{(-2)^2} = \dfrac{1}{4}$

(c) $-8^{-2/3} = -\dfrac{1}{8^{2/3}} = -\dfrac{1}{(\sqrt[3]{8})^2} = -\dfrac{1}{2^2} = -\dfrac{1}{4}$

(d) $64^{-5/6} = \dfrac{1}{64^{5/6}} = \dfrac{1}{(\sqrt[6]{64})^5} = \dfrac{1}{2^5} = \dfrac{1}{32}$

(e) $(-125)^{-1/3} = \dfrac{1}{(-125)^{1/3}} = \dfrac{1}{\sqrt[3]{-125}} = \dfrac{1}{-5} = -\dfrac{1}{5}$

It is true that rational exponents (positive, negative, or zero) satisfy the properties of exponents listed in Section 5.1. We can therefore restate these properties to indicate that they hold for all rational exponents:

Properties of Rational Exponents

Let a and b be real numbers, and let p and q be rational numbers. Then, if all expressions are defined (as real numbers),

(i) $a^p a^q = a^{p+q}$ (ii) $(a^p)^q = a^{pq}$ (iii) $(ab)^p = a^p b^p$

(iv) $\left(\dfrac{a}{b}\right)^p = \dfrac{a^p}{b^p}$ (v) $\dfrac{a^p}{a^q} = a^{p-q}$

The following examples illustrate how these properties are used to simplify algebraic expressions involving rational exponents.

In Examples 6 and 7, rewrite each expression so that it contains only positive exponents, and simplify. You may assume that variables are restricted to values for which all expressions are defined.

EXAMPLE 6

(a) $7^{-1/2} \cdot 7^{5/2}$

(b) $(x^{-3/4})^{-8/3}$

(c) $(125x^{-18})^{-4/3}$

(d) $\left(\dfrac{32}{x^{-5}}\right)^{-2/5}$

(e) $\dfrac{x^{2/3}}{x^{-4/5}}$

SOLUTION

(a) $\begin{aligned}7^{-1/2} \cdot 7^{5/2} &= 7^{-1/2+5/2} \\ &= 7^{4/2} = 7^2 = 49\end{aligned}$ [by Property (i)]

(b) $(x^{-3/4})^{-8/3} = x^{(-3/4)(-8/3)} = x^2$ [by Property (ii)]

(c) $\begin{aligned}(125x^{-18})^{-4/3} &= 125^{-4/3}(x^{-18})^{-4/3} \\ &= (5^3)^{-4/3}(x^{-18})^{-4/3} \\ &= 5^{3(-4/3)}x^{-18(-4/3)} \\ &= 5^{-4}x^{24} = \dfrac{x^{24}}{5^4} = \dfrac{x^{24}}{625}\end{aligned}$ [by Property (iii)] [by Property (ii)]

(d) $\begin{aligned}\left(\dfrac{32}{x^{-5}}\right)^{-2/5} &= \dfrac{32^{-2/5}}{(x^{-5})^{-2/5}} \\[2mm] &= \dfrac{(2^5)^{-2/5}}{(x^{-5})^{-2/5}} = \dfrac{2^{5(-2/5)}}{x^{(-5)(-2/5)}} = \dfrac{2^{-2}}{x^2} \\[2mm] &= \dfrac{1}{2^2 x^2} = \dfrac{1}{4x^2}\end{aligned}$ [by Property (iv)]

(e) $\begin{aligned}\dfrac{x^{2/3}}{x^{-4/5}} &= x^{2/3-(-4/5)} \\[2mm] &= x^{10/15-(-12/15)} \\[2mm] &= x^{22/15}\end{aligned}$ [by Property (v)]

EXAMPLE 7

(a) $(2r^{1/3}t^{3/2})^6$ (b) $\left(\dfrac{2p^{2/3}}{q^{1/2}}\right)^2 \left(\dfrac{4p^{-5/3}}{q^{2/3}}\right)$

(c) $\left(\dfrac{125a^{-9}b^{-12}}{8c^{-15}}\right)^{2/3}$

SOLUTION (a) $(2r^{1/3}t^{3/2})^6 = 2^6(r^{1/3})^6(t^{3/2})^6$
$$= 64r^{(1/3)(6)}t^{(3/2)(6)}$$
$$= 64r^2t^9$$

(b) $\left(\dfrac{2p^{2/3}}{q^{1/2}}\right)^2\left(\dfrac{4p^{-5/3}}{q^{2/3}}\right) = \left(\dfrac{2^2(p^{2/3})^2}{(q^{1/2})^2}\right)\left(\dfrac{4p^{-5/3}}{q^{2/3}}\right)$

$$= \left(\frac{4p^{4/3}}{q}\right)\left(\frac{4p^{-5/3}}{q^{2/3}}\right) = \frac{16p^{4/3-5/3}}{q^{1+2/3}}$$

$$= \frac{16p^{-1/3}}{q^{5/3}} = \frac{16}{p^{1/3}q^{5/3}}$$

(c) $\left(\dfrac{125a^{-9}b^{-12}}{8c^{-15}}\right)^{2/3} = \dfrac{125^{2/3}(a^{-9})^{2/3}(b^{-12})^{2/3}}{8^{2/3}(c^{-15})^{2/3}}$

$$= \frac{(5^3)^{2/3}(a^{-9})^{2/3}(b^{-12})^{2/3}}{(2^3)^{2/3}(c^{-15})^{2/3}}$$

$$= \frac{5^2a^{-6}b^{-8}}{2^2c^{-10}} = \frac{25c^{10}}{4a^6b^8}$$

PROBLEM SET 5.2

In problems 1–8, find the principal root (if it is defined).

1. $\sqrt{16}$ **2.** $\sqrt{144}$ **3.** $\sqrt{441}$ **4.** $\sqrt[3]{-8}$

5. $\sqrt[3]{-27}$ **6.** $\sqrt[4]{-81}$ **7.** $\sqrt[6]{-64}$ **8.** $\sqrt[12]{-1}$

In problems 9–16, rewrite each expression in terms of rational exponents.

9. $\sqrt{13}$ **10.** $\sqrt[3]{-17}$ **11.** $\sqrt[5]{3x^2}$ **12.** $\sqrt[7]{(x+y)^3}$

13. $\sqrt[4]{a^3b^2}$ **14.** $\sqrt[9]{(3a+b)^2}$ **15.** $\sqrt[5]{-3a^2b^4}$ **16.** $\sqrt[11]{-2x^9}$

In problems 17–22, rewrite each expression in terms of radicals.

17. $3^{5/6}$ **18.** $11^{2/3}$ **19.** $(7y)^{3/2}$ **20.** $(3a+b)^{4/5}$

21. $(4xy)^{5/7}$ **22.** $(-7x^2y^3)^{5/9}$

In problems 23–40, find the value of each expression.

23. $8^{2/3}$ **24.** $-8^{4/3}$ **25.** $(-8)^{4/3}$ **26.** $(-27)^{2/3}$

27. $(-32)^{3/5}$ **28.** $128^{2/7}$ **29.** $81^{3/4}$ **30.** $(-64)^{4/3}$

31. $(-16)^{3/2}$ **32.** $(-36)^{5/2}$ **33.** $27^{8/12}$ **34.** $25^{7/14}$

35. $(-8)^{-4/3}$ **36.** $(-27)^{-2/3}$ **37.** $9^{-3/2}$ **38.** $(-32)^{-4/5}$

39. $(-128)^{-5/7}$ **40.** $(0.216)^{-4/6}$

In problems 41–68, rewrite each expression so that it contains only positive exponents, and simplify. You may assume that variables are restricted to values for which all expressions are defined.

41. $2^{1/3}2^{2/3}$

42. $5^{1/2}5^{-3/2}$

43. $x^{-2/3}x^{5/3}$

44. $y^{2/15}y^{-7/60}$

45. $(5^{1/7})^{14}$

46. $(8^{-2/3})^{-6}$

47. $(x^{-7/9})^{18/7}$

48. $(y^{-2})^{-15/2}$

49. $(8p^9)^{4/3}$

50. $(32u^{-5})^{-3/5}$

51. $(2^{-1/3}y^{-1/7})^{-21}$

52. $(81m^{12})^{-3/4}$

53. $\left(\dfrac{125}{y^3}\right)^{-1/3}$

54. $\left(\dfrac{x^{-5}}{32}\right)^{4/5}$

55. $\dfrac{5^{2/3}}{5^{-1/7}}$

56. $\dfrac{(x^{-3/4})^{-2}}{(x^2)^{-5/8}}$

57. $\dfrac{x^{1/3}}{x^{-1/6}}$

58. $\dfrac{y^{3/2}}{y^{-7/2}}$

59. $(4c^{2/3}d^{3/4})(2c^{-5/3}d^{1/4})$

60. $(5x^{1/3}y^{2/3})(3x^{-4/3}y^{4/3})$

61. $(6a^{7/2}b^{-3/2})^2(4a^{-1/3}b^{-2/3})^3$

62. $(u^{-3/4}v^{-5/6})^{-6}(2u^{1/4}v^{1/6})^6$

63. $\left(\dfrac{5^24^6}{5^{-4}4^7}\right)^{-1/2}$

64. $\left(\dfrac{2x^{3/2}y^{7/2}}{4x^2y^{-1}}\right)^{-4}$

65. $\dfrac{(p^{2/5}q^2)^5}{(p^{1/6}q^{1/3})^{12}}$

66. $\left[\dfrac{(x^{5/7}y^{3/2})(x^{2/7}y^{-5/2})}{x^2y^{-1/2}}\right]^{-2}$

67. $\left(\dfrac{81p^{-12}}{q^{16}}\right)^{-1/4}\cdot\left(-\dfrac{p^{-2/3}}{q^{1/3}}\right)^3$

68. $\left[\dfrac{(5x^2y^3)^{3/4}(5x^2y^3)^{1/4}}{(x^2y^4)^{-1/2}}\right]^{-4}$

© In problems 69–74, use a calculator with a y^x key to evaluate each expression. Round off your answer to four significant digits.

69. $4.18^{0.325}$

70. $0.6^{3.41}$

71. $7^{4/5}$

72. $5.61^{\sqrt{2}}$

73. $7.63^{\sqrt{3}}$

74. $2^{\sqrt{5}}$

© **75.** The formula $A = \sqrt{s(s-a)(s-b)(s-c)}$ gives the area of a triangle, where a, b, and c represent the lengths of the sides and

$$s = \frac{a+b+c}{2}.$$

Find the area of the triangle whose sides are 42.3, 28.7, and 37.1 meters.

© **76.** In statistics, the following formula is used,

$$\sigma = \sqrt{\frac{(x_1-\overline{x})^2 + (x_2-\overline{x})^2 + (x_3-\overline{x})^2}{3}},$$

where x_1, x_2, and x_3 represent the values of 3 observations, and

$$\overline{x} = \frac{x_1 + x_2 + x_3}{3}.$$

Find σ when $x_1 = 3.4$, $x_2 = 5.7$, and $x_3 = 6.8$.

5.3 **Radicals**

Radical expressions can often be simplified by using certain properties. Consider the following example:

$$\sqrt{9} \cdot \sqrt{25} = 3 \cdot 5 = 15.$$

We can also see that

$$\sqrt{9 \cdot 25} = \sqrt{225} = 15.$$

Thus, we conclude that

$$\sqrt{9 \cdot 25} = \sqrt{9} \cdot \sqrt{25}.$$

Now, notice that

$$\sqrt[3]{\frac{27}{8}} = \frac{3}{2}, \quad \text{and that} \quad \frac{\sqrt[3]{27}}{\sqrt[3]{8}} = \frac{3}{2}.$$

We conclude that

$$\sqrt[3]{\frac{27}{8}} = \frac{\sqrt[3]{27}}{\sqrt[3]{8}}.$$

Also observe that

$$\sqrt[3]{\sqrt{64}} = \sqrt[3]{8} = 2, \quad \sqrt{\sqrt[3]{64}} = \sqrt{4} = 2, \quad \text{and that} \quad \sqrt[6]{64} = 2.$$

This suggests that

$$\sqrt[3]{\sqrt{64}} = \sqrt{\sqrt[3]{64}} = \sqrt[6]{64}.$$

These examples imply the following properties of radicals.

Properties of Radicals

Let a and b be real numbers and let m and n be positive integers. Then, provided that all expressions are defined,

(i) $\sqrt[n]{ab} = \sqrt[n]{a} \cdot \sqrt[n]{b}$ (ii) $\sqrt[n]{\dfrac{a}{b}} = \dfrac{\sqrt[n]{a}}{\sqrt[n]{b}}$ (iii) $\sqrt[n]{\sqrt[m]{a}} = \sqrt[m]{\sqrt[n]{a}} = \sqrt[mn]{a}$

Because $a^{1/n} = \sqrt[n]{a}$, the properties of radicals can be proved by using the properties of rational exponents. For instance, we verify Property (i) as follows:

$$\sqrt[n]{a} \cdot \sqrt[n]{b} = a^{1/n} \cdot b^{1/n} = (ab)^{1/n} = \sqrt[n]{ab}.$$

(You may verify the other properties by using the properties of rational exponents on page 174.)

Note that, if

$$\sqrt{a} = b, \qquad \text{then} \qquad a = b^2, \qquad \text{that is,} \qquad (\sqrt{a})^2 = a.$$

Similarly, if

$$\sqrt[3]{a} = b, \qquad \text{then} \qquad a = b^3, \qquad \text{that is,} \qquad (\sqrt[3]{a})^3 = a.$$

In general: If n is a positive integer, and

$$\sqrt[n]{a} = b, \qquad \text{then} \qquad a = b^n.$$

That is, if $\sqrt[n]{a}$ exists, then we have the following property:

Property iv

$$(\sqrt[n]{a})^n = a.$$

For example,

$$(\sqrt{7})^2 = 7, \qquad (\sqrt[6]{3})^6 = 3, \qquad \text{and} \qquad (\sqrt[3]{-4})^3 = -4.$$

It also follows that if a is a nonzero real number and n is an *odd* positive integer, then we have the following property:

Property v

$$\sqrt[n]{a^n} = a.$$

For example,

$$\sqrt[5]{3^5} = 3, \qquad \sqrt[3]{(-5)^3} = -5, \qquad \text{and} \qquad \sqrt[7]{(-2)^7} = -2.$$

The symbol $\sqrt{a^2}$ represents the principal square root of a^2. Notice that if a is a nonnegative real number, then

$$\sqrt{a^2} = a.$$

On the other hand, if a is a negative real number, then

$$\sqrt{a^2} = -a.$$

In all cases,

$$\sqrt{a^2} = |a|$$

for any real number a. For example,

$$\sqrt{(-2)^2} = |-2| = 2,$$

which can also be obtained by writing

$$\sqrt{(-2)^2} = \sqrt{4} = 2.$$

In general, if n is an even positive integer, then we have the following property:

Property vi

$$\sqrt[n]{a^n} = |a|.$$

To *simplify* a radical expression, we use the properties above and we write the expression in a form that satisfies the following conditions:

(1) The power of any factor under the radical is less than the index of the radical, that is, in $\sqrt[n]{a^m}$, $m < n$.

(2) The exponents of factors under the radicals and the index of the radical have no common factors, that is, in $\sqrt[n]{a^m}$, m and n have no common factors.

(3) The radicand contains no fractions.

In Examples 1–4, use the properties of radicals to simplify each expression. Assume that variables are restricted to values for which all expressions are defined.

EXAMPLE 1 (a) $\sqrt{72}$

(b) $\sqrt[3]{-108}$

(c) $\sqrt[4]{\frac{7}{16}}$

SOLUTION (a) First, we look for the largest perfect square factor of 72. This factor is 36, so we have

$$\sqrt{72} = \sqrt{36 \cdot 2}$$
$$= \sqrt{36} \cdot \sqrt{2} = 6\sqrt{2}. \qquad \text{[by Property (i)]}$$

(b) $\sqrt[3]{-108} = \sqrt[3]{(-27)(4)}$
$$= \sqrt[3]{-27} \cdot \sqrt[3]{4} \qquad \text{[by Property (i)]}$$
$$= -3\sqrt[3]{4}$$

(c) $\sqrt[4]{\dfrac{7}{16}} = \dfrac{\sqrt[4]{7}}{\sqrt[4]{16}}$ [by Property (ii)]

$= \dfrac{\sqrt[4]{7}}{2}$

EXAMPLE 2   (a) $\sqrt[3]{x^5}$ (b) $\sqrt{9y^3}$ (c) $\sqrt[3]{\dfrac{4a}{27c^3}}$ (d) $\sqrt[5]{\dfrac{7}{-32w^5}}$

             (e) $\sqrt[3]{\sqrt{64}}$ (f) $\sqrt[3]{\sqrt[5]{-r^{15}}}$ (g) $\sqrt[7]{t^7}$ (h) $\sqrt[9]{-u^9}$

SOLUTION   (a) $\sqrt[3]{x^5} = \sqrt[3]{x^3 \cdot x^2} = \sqrt[3]{x^3} \cdot \sqrt[3]{x^2}$ [by Property (i)]

$= x\sqrt[3]{x^2}$

(b) $\sqrt{9y^3} = \sqrt{(9y^2)(y)} = \sqrt{9y^2} \cdot \sqrt{y}$ [by Property (i)]

$= 3y\sqrt{y}$

(c) $\sqrt[3]{\dfrac{4a}{27c^3}} = \dfrac{\sqrt[3]{4a}}{\sqrt[3]{27c^3}}$ [by Property (ii)]

$= \dfrac{\sqrt[3]{4a}}{3c}$

(d) $\sqrt[5]{\dfrac{7}{-32w^5}} = \dfrac{\sqrt[5]{7}}{\sqrt[5]{-32w^5}}$ [by Property (ii)]

$= \dfrac{\sqrt[5]{7}}{-2w}$

(e) $\sqrt[3]{\sqrt{64}} = \sqrt[6]{64}$ [by Property (iii)]

$= 2$

(f) $\sqrt[3]{\sqrt[5]{-r^{15}}} = \sqrt[15]{-r^{15}}$ [by Property (iii)]

$= \sqrt[15]{(-r)^{15}}$ [why?]

$= -r$

(g) $\sqrt[7]{t^7} = t$ [by Property (v)]

(h) $\sqrt[9]{-u^9} = -\sqrt[9]{u^9}$ [why?]

$= -u$

EXAMPLE 3   (a) $\sqrt[3]{-125x^8y^{10}}$

             (b) $\sqrt{3p^2q^3} \cdot \sqrt{6p^5q}$

             (c) $\dfrac{\sqrt{324a^7b} \cdot \sqrt{9a^5}}{\sqrt{36a^4b^3}}$

SOLUTION

(a) $\sqrt[3]{-125x^8y^{10}} = \sqrt[3]{(-125x^6y^9)(x^2y)}$
$= \sqrt[3]{-125x^6y^9} \cdot \sqrt[3]{x^2y}$
$= -5x^2y^3 \sqrt[3]{x^2y}$

(b) $\sqrt{3p^2q^3} \cdot \sqrt{6p^5q} = \sqrt{(3p^2q^3)(6p^5q)}$
$= \sqrt{18p^7q^4} = \sqrt{(9p^6q^4)(2p)}$
$= \sqrt{9p^6q^4} \cdot \sqrt{2p} = 3p^3q^2\sqrt{2p}$

(c) $\dfrac{\sqrt{324a^7b} \cdot \sqrt{9a^5}}{\sqrt{36a^4b^3}} = \dfrac{\sqrt{(324a^7b)(9a^5)}}{\sqrt{36a^4b^3}} = \sqrt{\dfrac{324(9)a^{12}b}{36a^4b^3}}$
$= \sqrt{\dfrac{81a^8}{b^2}} = \dfrac{9a^4}{b}$

In order to multiply or divide radicals with different indexes, we begin by building their indexes to a common index. The following property is a key to this operation.

Property (vii)

> If m and n are positive integers, and if each root exists, then
> $$\sqrt[cn]{a^{cm}} = \sqrt[n]{a^m}, \text{ where } c \text{ is a positive integer.}$$

EXAMPLE 4

Write $\sqrt[4]{x^3} \cdot \sqrt[3]{x^2}$ as a single radical, and then simplify the result. Assume that x is restricted to values for which all expressions are defined.

SOLUTION

We use the least common multiple (LCM) of the individual indexes as the common index. The LCM of the indexes 4 and 3 is 12, so that

$$\sqrt[4]{x^3} = \sqrt[4\cdot3]{x^{3\cdot3}} \qquad \text{[by Property (vii)]}$$
$$= \sqrt[12]{x^9},$$

and

$$\sqrt[3]{x^2} = \sqrt[3\cdot4]{x^{2\cdot4}} \qquad \text{[by Property (vii)]}$$
$$= \sqrt[12]{x^8}.$$

Hence,

$$\sqrt[4]{x^3} \cdot \sqrt[3]{x^2} = \sqrt[12]{x^9} \cdot \sqrt[12]{x^8} = \sqrt[12]{x^9 \cdot x^8}$$
$$= \sqrt[12]{x^{17}} = \sqrt[12]{x^{12} \cdot x^5} = \sqrt[12]{x^{12}} \cdot \sqrt[12]{x^5}$$
$$= x\sqrt[12]{x^5}.$$

PROBLEM SET 5.3

In problems 1–48, use the properties of radicals to simplify each expression. Assume
that variables are restricted to values for which all expressions are defined.

1. $\sqrt{27}$ **2.** $\sqrt{162}$ **3.** $\sqrt[3]{-54}$

4. $\sqrt[3]{-250}$ **5.** $\sqrt{288}$ **6.** $\sqrt[5]{64}$

7. $\sqrt[3]{8x^4}$ **8.** $\sqrt[5]{64x^6}$ **9.** $\sqrt[3]{-16y^{10}}$

10. $\sqrt[7]{-128p^{15}}$ **11.** $\sqrt{98p^3}$ **12.** $\sqrt[5]{-96x^{11}}$

13. $\sqrt{\dfrac{7}{4}}$ **14.** $\sqrt{\dfrac{3}{25}}$ **15.** $\sqrt[3]{\dfrac{-5}{8}}$

16. $\sqrt[3]{\dfrac{11}{-125}}$ **17.** $\sqrt[3]{\dfrac{-x^2}{64}}$ **18.** $\sqrt[5]{\dfrac{x}{y^{10}}}$

19. $\sqrt{\dfrac{3w^3}{4w^5}}$ **20.** $\sqrt[4]{\dfrac{m}{625n^8}}$ **21.** $\sqrt[5]{\dfrac{-4}{32b^{20}}}$

22. $\sqrt[7]{\dfrac{-3}{a^{14}}}$ **23.** $\dfrac{\sqrt[3]{54}}{\sqrt[3]{-3}}$ **24.** $\dfrac{\sqrt[3]{-x^4}}{\sqrt[3]{8x}}$

25. $\dfrac{\sqrt[5]{p^6}}{\sqrt[5]{-p}}$ **26.** $\dfrac{\sqrt[7]{a^{10}}}{\sqrt[7]{a^3}}$ **27.** $\sqrt[5]{\sqrt{1024}}$

28. $\sqrt[5]{\sqrt[3]{y^{15}}}$ **29.** $\sqrt{\sqrt[4]{p^{32}}}$ **30.** $\sqrt[3]{\sqrt[10]{m^{90}}}$

31. $(\sqrt[4]{5})^4$ **32.** $(\sqrt[3]{x})^3$ **33.** $(\sqrt[7]{-p})^7$

34. $\sqrt[5]{(-4)^5}$ **35.** $\sqrt[3]{w^3}$ **36.** $\sqrt[5]{-t^{15}}$

37. $\sqrt{18a^3b^7} \cdot \sqrt{2ab^3}$ **38.** $\sqrt[3]{25y^2} \cdot \sqrt[3]{5y^4}$ **39.** $\sqrt[3]{4xy^2} \cdot \sqrt[3]{2x^2y^3}$

40. $\sqrt[5]{-w^{15}t^4} \cdot \sqrt[5]{w^{10}t^6}$ **41.** $\sqrt[8]{x^{12}} \cdot \sqrt[8]{x^5y^{-8}} \cdot \sqrt[8]{x^2y^9}$ **42.** $\sqrt[5]{5p^2q^3} \cdot \sqrt[5]{5p^2q^3}$

43. $\dfrac{\sqrt[3]{u^{11}} \cdot \sqrt[3]{u^5}}{\sqrt[3]{u}}$ **44.** $\dfrac{\sqrt{m^2n} \cdot \sqrt{mn^4}}{\sqrt{mn^3}}$ **45.** $\sqrt[4]{\dfrac{a^4b^3}{a^3b}} \cdot \dfrac{\sqrt[4]{a^5b}}{\sqrt[4]{ab^{-1}}}$

46. $\dfrac{\sqrt{324x^5y} \cdot \sqrt{9x^2}}{\sqrt{25x^2y}}$ **47.** $\dfrac{\sqrt[3]{p^2q^3} \cdot \sqrt[3]{125p^3q^2}}{\sqrt[3]{8p^3q^4}}$ **48.** $\dfrac{\sqrt[3]{a^2b^4} \cdot \sqrt[3]{a^4b} \cdot \sqrt[3]{a^3b^4}}{\sqrt[3]{ab^2} \cdot \sqrt[3]{a^2b^7}}$

In problems 49–54, write each expression as a single radical, then simplify the result.
Assume that variables are restricted to values for which all expressions are defined.

49. $\sqrt[3]{x} \cdot \sqrt{x}$ **50.** $\sqrt[3]{p^2} \cdot \sqrt[4]{p^3}$ **51.** $\sqrt[3]{y^2} \cdot \sqrt[5]{y^4}$

52. $\sqrt[3]{p^2} \cdot \sqrt[7]{p^5}$ **53.** $\sqrt[5]{ab^2} \cdot \sqrt[4]{a^2b^3}$ **54.** $\sqrt[5]{c^2d^4} \cdot \sqrt[7]{c^3d^3}$

© In problems 55–58, use a calculator to verify each equation.

55. $\sqrt{7} \cdot \sqrt{5} = \sqrt{35}$ **56.** $\dfrac{\sqrt[5]{11}}{\sqrt[5]{3}} = \sqrt[5]{\dfrac{11}{3}}$ **57.** $\sqrt[3]{\sqrt[5]{22}} = \sqrt[15]{22}$

58. $\sqrt[5]{7^5} = 7$

5.4 Operations Involving Radicals

Expressions such as

$$\sqrt{2} + \sqrt{3}, \qquad \sqrt[3]{16} - \sqrt[3]{2}, \qquad \sqrt{2}(\sqrt{5} + \sqrt{3}), \qquad \text{and} \qquad \frac{\sqrt{3} - 2}{\sqrt{2}}$$

are **radical expressions.** In this section, we perform the operations of addition, subtraction, multiplication, and division using radical expressions.

Addition and Subtraction of Radicals

We add and subtract radical expressions in much the same way as we add or subtract polynomial expressions. To add or subtract radical expressions we apply the distributive property, and combine only *like* (or *similar*) terms. Two or more radical expressions are **like,** (or **similar**), if, after being simplified, they contain the same index and radicand. For instance,

$$3\sqrt{6}, \qquad 5\sqrt{6}, \qquad x\sqrt{6}, \qquad \text{and} \qquad \frac{\sqrt{6}}{3}$$

are like terms, whereas

$$5\sqrt{6} \qquad \text{and} \qquad 5\sqrt[3]{6}$$

are not like terms.

In Examples 1 and 2 below, combine like terms.

EXAMPLE 1 $2\sqrt{3} + 5\sqrt{3} - \sqrt{3}$

SOLUTION   All the terms are similar. Thus, we apply the distributive property and combine:
$$2\sqrt{3} + 5\sqrt{3} - \sqrt{3} = (2 + 5 - 1)\sqrt{3} = 6\sqrt{3}$$

EXAMPLE 2 $7\sqrt[3]{2} - 3\sqrt[3]{2} + 6\sqrt[3]{2}$

SOLUTION $7\sqrt[3]{2} - 3\sqrt[3]{2} + 6\sqrt[3]{2} = (7 - 3 + 6)\sqrt[3]{2} = 10\sqrt[3]{2}$

Each radical expression in Examples 1 and 2 is in simplified form. Occasionally two or more of the terms containing radical expressions do not appear to be similar but contain like terms when they are simplified. When this occurs, we write *each* expression in a simplified form, and then apply the distributive property to combine like terms, as shown in the following examples.

In Examples 3–6, combine like terms.

EXAMPLE 3 $\sqrt{8} + \sqrt{32}$

SOLUTION   First we write the terms in a simplified form:

$$\begin{aligned}
\sqrt{8} + \sqrt{32} &= \sqrt{4 \cdot 2} + \sqrt{16 \cdot 2} \\
&= \sqrt{4} \cdot \sqrt{2} + \sqrt{16} \cdot \sqrt{2} \quad \text{[by Property (i) of radicals on} \\
&\phantom{= \sqrt{4} \cdot \sqrt{2} + \sqrt{16} \cdot \sqrt{2} \quad} \text{page 178]} \\
&= 2\sqrt{2} + 4\sqrt{2} \\
&= 6\sqrt{2}.
\end{aligned}$$

EXAMPLE 4 $2\sqrt[3]{54} - 2\sqrt[3]{16}$

SOLUTION $$\begin{aligned}
2\sqrt[3]{54} - 2\sqrt[3]{16} &= 2\sqrt[3]{27 \cdot 2} - 2\sqrt[3]{8 \cdot 2} \\
&= 2\sqrt[3]{27} \cdot \sqrt[3]{2} - 2\sqrt[3]{8} \cdot \sqrt[3]{2} \\
&= 2(3)\sqrt[3]{2} - 2(2)\sqrt[3]{2} \\
&= 6\sqrt[3]{2} - 4\sqrt[3]{2} \\
&= 2\sqrt[3]{2}
\end{aligned}$$

EXAMPLE 5 $4\sqrt{12} + 5\sqrt{8} - \sqrt{50}$

SOLUTION $$\begin{aligned}
4\sqrt{12} + 5\sqrt{8} - \sqrt{50} &= 4\sqrt{4 \cdot 3} + 5\sqrt{4 \cdot 2} - \sqrt{25 \cdot 2} \\
&= 4\sqrt{4} \cdot \sqrt{3} + 5\sqrt{4} \cdot \sqrt{2} - \sqrt{25} \cdot \sqrt{2} \\
&= 4 \cdot 2\sqrt{3} + 5 \cdot 2\sqrt{2} - 5\sqrt{2} \\
&= 8\sqrt{3} + 10\sqrt{2} - 5\sqrt{2} \\
&= 8\sqrt{3} + (10 - 5)\sqrt{2} \\
&= 8\sqrt{3} + 5\sqrt{2}
\end{aligned}$$

EXAMPLE 6 $7\sqrt{4x} - 5\sqrt{9x}$, for $x > 0$.

SOLUTION $$\begin{aligned}
7\sqrt{4x} - 5\sqrt{9x} &= 7\sqrt{4} \cdot \sqrt{x} - 5\sqrt{9} \cdot \sqrt{x} \\
&= 7(2\sqrt{x}) - 5(3\sqrt{x}) \\
&= 14\sqrt{x} - 15\sqrt{x} \\
&= -\sqrt{x}
\end{aligned}$$

Multiplication of Radicals

Examples 7–13 illustrate how we can multiply radical expressions by using Property (i) of Section 5.3, written in the form:

$$\sqrt[n]{a} \cdot \sqrt[n]{b} = \sqrt[n]{ab}.$$

You will note that the multiplication of radical expressions is similar to the multiplication of polynomials.

In Examples 7–13, find each product, and simplify the result when possible.

EXAMPLE 7 $\sqrt{8} \cdot \sqrt{2}$

SOLUTION $\sqrt{8} \cdot \sqrt{2} = \sqrt{8 \cdot 2} = \sqrt{16} = 4$

EXAMPLE 8 $\sqrt{5}(3\sqrt{7} - 2\sqrt{5})$

SOLUTION $\sqrt{5}(3\sqrt{7} - 2\sqrt{5}) = \sqrt{5} \cdot 3\sqrt{7} - \sqrt{5} \cdot 2\sqrt{5}$ (by the distributive property)

$$= 3\sqrt{5 \cdot 7} - 2(\sqrt{5})^2$$
$$= 3\sqrt{35} - 2(5)$$
$$= 3\sqrt{35} - 10$$

EXAMPLE 9 $\sqrt{5}(\sqrt{15} + \sqrt{25})$

SOLUTION $\sqrt{5}(\sqrt{15} + \sqrt{25}) = \sqrt{5} \cdot \sqrt{15} + \sqrt{5} \cdot \sqrt{25}$

$$= \sqrt{75} + 5\sqrt{5}$$
$$= \sqrt{25 \cdot 3} + 5\sqrt{5} = \sqrt{25} \cdot \sqrt{3} + 5\sqrt{5}$$
$$= 5\sqrt{3} + 5\sqrt{5}$$

EXAMPLE 10 $(\sqrt{3} - \sqrt{2})(2\sqrt{3} + \sqrt{2})$

SOLUTION $(\sqrt{3} - \sqrt{2})(2\sqrt{3} + \sqrt{2})$

$$= 2\sqrt{3} \cdot \sqrt{3} + \sqrt{3} \cdot \sqrt{2} - 2\sqrt{3} \cdot \sqrt{2} - \sqrt{2} \cdot \sqrt{2}$$
$$= 2(3) + \sqrt{6} - 2\sqrt{6} - 2$$
$$= 6 - \sqrt{6} - 2 = 4 - \sqrt{6}$$

EXAMPLE 11 $(\sqrt{x} + 2\sqrt{y})^2$

SOLUTION   Using the special product $(A + B)^2 = A^2 + 2AB + B^2$, we have

$$(\sqrt{x} + 2\sqrt{y})^2 = (\sqrt{x})^2 + 2\sqrt{x}(2\sqrt{y}) + (2\sqrt{y})^2$$
$$= x + 4\sqrt{x} \cdot \sqrt{y} + 4y$$
$$= x + 4\sqrt{xy} + 4y.$$

EXAMPLE 12 $(\sqrt{10} + \sqrt{2})(\sqrt{10} - \sqrt{2})$

SOLUTION   Using the special product $(A + B)(A - B) = A^2 - B^2$, we have

$$(\sqrt{10} + \sqrt{2})(\sqrt{10} - \sqrt{2}) = (\sqrt{10})^2 - (\sqrt{2})^2$$
$$= 10 - 2 = 8.$$

EXAMPLE 13 $(3\sqrt{x} - 5\sqrt{y})(3\sqrt{x} + 5\sqrt{y})$

SOLUTION $(3\sqrt{x} - 5\sqrt{y})(3\sqrt{x} + 5\sqrt{y}) = (3\sqrt{x})^2 - (5\sqrt{y})^2$
$$= 9x - 25y$$

Division of Radicals

We can divide radical expressions by using Property (ii) of Section 5.3 written in the form:

$$\frac{\sqrt[n]{a}}{\sqrt[n]{b}} = \sqrt[n]{\frac{a}{b}}.$$

Thus,

$$\frac{\sqrt{18}}{\sqrt{2}} = \sqrt{\frac{18}{2}} = \sqrt{9} = 3,$$

$$\frac{\sqrt{108y^3}}{\sqrt{3y}} = \sqrt{\frac{108y^3}{3y}} = \sqrt{36y^2} = 6y \qquad \text{for } y > 0,$$

and

$$\frac{\sqrt{3}}{\sqrt{21}} = \sqrt{\frac{3}{21}} = \sqrt{\frac{1}{7}} = \frac{1}{\sqrt{7}}.$$

Radicals may appear in the denominator of a fraction, such as $1/\sqrt{7}$. It is sometimes easier to work with fractions if their denominators do not contain radicals. To rewrite a fraction so that there are no radicals in the denominator, we multiply the numerator and the denominator by a **rationalizing factor** for the denominator. Whenever the product of two radical expressions is free of radicals, we say that the two expressions are rationalizing factors of each other. For instance, $\sqrt{7}$ is a rationalizing factor of $\sqrt{7}$, because $\sqrt{7} \cdot \sqrt{7} = 7$. If we multiply the numerator and the denominator of $1/\sqrt{7}$ by the rationalizing factor $\sqrt{7}$, we have

$$\frac{1}{\sqrt{7}} = \frac{1 \cdot \sqrt{7}}{\sqrt{7} \cdot \sqrt{7}} = \frac{\sqrt{7}}{7}.$$

Thus, we have a fraction whose denominator is free of radicals. Note that when we multiply

$$(\sqrt{5} - \sqrt{3})(\sqrt{5} + \sqrt{3}) = (\sqrt{5})^2 - (\sqrt{3})^2 = 5 - 3 = 2,$$

the expressions $(\sqrt{5} - \sqrt{3})$ and $(\sqrt{5} + \sqrt{3})$ are rationalizing factors for each other. If we multiply the numerator and the denominator of

$$\frac{2 + \sqrt{3}}{\sqrt{5} - \sqrt{3}}$$

by the rationalizing factor $\sqrt{5} + \sqrt{3}$, we have

$$\frac{2 + \sqrt{3}}{\sqrt{5} - \sqrt{3}} = \frac{(2 + \sqrt{3})(\sqrt{5} + \sqrt{3})}{(\sqrt{5} - \sqrt{3})(\sqrt{5} + \sqrt{3})}$$

$$= \frac{2\sqrt{5} + 2\sqrt{3} + \sqrt{3} \cdot \sqrt{5} + \sqrt{3} \cdot \sqrt{3}}{(\sqrt{5})^2 - (\sqrt{3})^2}$$

$$= \frac{2\sqrt{5} + 2\sqrt{3} + \sqrt{15} + 3}{5 - 3} = \frac{2\sqrt{5} + 2\sqrt{3} + \sqrt{15} + 3}{2}.$$

The resulting denominator is free of radicals. The process of writing fractions so that there are no radicals in the denominator is called **rationalizing the denominator.**

In Examples 14–21, rationalize the denominator of each fraction and simplify the result. Assume that all variables are positive.

EXAMPLE 14 $\dfrac{7}{\sqrt{3}}$

SOLUTION   Here we use $\sqrt{3}$ as the rationalizing factor. We multiply the numerator and the denominator of the fraction by $\sqrt{3}$:

$$\frac{7}{\sqrt{3}} = \frac{7 \cdot \sqrt{3}}{\sqrt{3} \cdot \sqrt{3}} = \frac{7\sqrt{3}}{3}.$$

EXAMPLE 15 $\dfrac{\sqrt{5}}{\sqrt{6}}$

SOLUTION   The rationalizing factor is $\sqrt{6}$:

$$\frac{\sqrt{5}}{\sqrt{6}} = \frac{\sqrt{5} \cdot \sqrt{6}}{\sqrt{6} \cdot \sqrt{6}} = \frac{\sqrt{5 \cdot 6}}{6} = \frac{\sqrt{30}}{6}.$$

EXAMPLE 16 $\dfrac{3}{5\sqrt{2x}}$

SOLUTION   The rationalizing factor is $\sqrt{2x}$:

$$\frac{3}{5\sqrt{2x}} = \frac{3 \cdot \sqrt{2x}}{5\sqrt{2x} \cdot \sqrt{2x}} = \frac{3\sqrt{2x}}{5(2x)} = \frac{3\sqrt{2x}}{10x}.$$

EXAMPLE 17 $\dfrac{2}{\sqrt[3]{5}}$

SOLUTION   The rationalizing factor is $\sqrt[3]{5} \cdot \sqrt[3]{5}$:

$$\frac{2}{\sqrt[3]{5}} = \frac{2\sqrt[3]{5} \cdot \sqrt[3]{5}}{\sqrt[3]{5} \cdot \sqrt[3]{5} \cdot \sqrt[3]{5}} = \frac{2\sqrt[3]{5(5)}}{5} = \frac{2\sqrt[3]{25}}{5}.$$

EXAMPLE 18 $\dfrac{5}{\sqrt{3} - 1}$

SOLUTION   The rationalizing factor is $\sqrt{3} + 1$:

$$\frac{5}{\sqrt{3} - 1} = \frac{5(\sqrt{3} + 1)}{(\sqrt{3} - 1)(\sqrt{3} + 1)}$$

$$= \frac{5\sqrt{3} + 5}{(\sqrt{3})^2 - 1^2} = \frac{5\sqrt{3} + 5}{3 - 1} = \frac{5\sqrt{3} + 5}{2}.$$

EXAMPLE 19 $\dfrac{3\sqrt{5} + 7\sqrt{2}}{6\sqrt{5} - 3\sqrt{2}}$

SOLUTION   The rationalizing factor is $6\sqrt{5} + 3\sqrt{2}$:

$$\frac{3\sqrt{5} + 7\sqrt{2}}{6\sqrt{5} - 3\sqrt{2}} = \frac{(3\sqrt{5} + 7\sqrt{2})}{(6\sqrt{5} - 3\sqrt{2})} \cdot \frac{(6\sqrt{5} + 3\sqrt{2})}{(6\sqrt{5} + 3\sqrt{2})}$$

$$= \frac{3\sqrt{5} \cdot 6\sqrt{5} + 3\sqrt{5} \cdot 3\sqrt{2} + 7\sqrt{2} \cdot 6\sqrt{5} + 7\sqrt{2} \cdot 3\sqrt{2}}{(6\sqrt{5})^2 - (3\sqrt{2})^2}$$

$$= \frac{18(5) + 9\sqrt{10} + 42\sqrt{10} + 21(2)}{36(5) - 9(2)}$$

$$= \frac{90 + 51\sqrt{10} + 42}{180 - 18}$$

$$= \frac{132 + 51\sqrt{10}}{162} = \frac{\cancel{3}(44 + 17\sqrt{10})}{\cancel{3}(54)}$$

$$= \frac{44 + 17\sqrt{10}}{54}.$$

EXAMPLE **20** $\dfrac{3}{\sqrt{x} - 1}$

SOLUTION   The rationalizing factor is $\sqrt{x} + 1$:

$$\frac{3}{\sqrt{x} - 1} = \frac{3(\sqrt{x} + 1)}{(\sqrt{x} - 1)(\sqrt{x} + 1)}$$

$$= \frac{3\sqrt{x} + 3}{(\sqrt{x})^2 - 1^2}$$

$$= \frac{3\sqrt{x} + 3}{x - 1}.$$

EXAMPLE **21** $\dfrac{\sqrt{x} + \sqrt{y}}{\sqrt{x} - \sqrt{y}}$

SOLUTION   The rationalizing factor is $\sqrt{x} + \sqrt{y}$:

$$\frac{\sqrt{x} + \sqrt{y}}{\sqrt{x} - \sqrt{y}} = \frac{(\sqrt{x} + \sqrt{y})(\sqrt{x} + \sqrt{y})}{(\sqrt{x} - \sqrt{y})(\sqrt{x} + \sqrt{y})}$$

$$= \frac{(\sqrt{x})^2 + 2\sqrt{x} \cdot \sqrt{y} + (\sqrt{y})^2}{(\sqrt{x})^2 - (\sqrt{y})^2}$$

$$= \frac{x + 2\sqrt{xy} + y}{x - y}.$$

PROBLEM SET 5.4

In problems 1–20, simplify each expression by combining like terms. Assume that variables are restricted to values for which all radical expressions are defined.

1. $5\sqrt{7} + 3\sqrt{7}$ **2.** $8\sqrt{3} + 2\sqrt{3}$ **3.** $7\sqrt{5} + 3\sqrt{5} - 2\sqrt{5}$

4. $9\sqrt{11} + 8\sqrt{11} - 3\sqrt{11}$ **5.** $8\sqrt[3]{4} - 3\sqrt[3]{4} + 2\sqrt[3]{4}$ **6.** $8\sqrt[5]{2} - 4\sqrt[5]{2} + 3\sqrt[5]{2}$

7. $\sqrt{18} + \sqrt{8}$ **8.** $3\sqrt{18} + 4\sqrt{2}$ **9.** $\sqrt{72} - 2\sqrt{8} + \sqrt{2}$

10. $2\sqrt{50} - 3\sqrt{128} + 4\sqrt{2}$ **11.** $5\sqrt{3} + 2\sqrt{12} - 2\sqrt{27}$ **12.** $4\sqrt{12} + 2\sqrt{27} - \sqrt{48}$

13. $4\sqrt{20} - 2\sqrt{45} + \sqrt{80}$ **14.** $\sqrt[4]{162} + \sqrt[4]{32} - 2\sqrt[4]{2}$ **15.** $5\sqrt[3]{81} - 3\sqrt[3]{24} + \sqrt[3]{192}$

16. $3\sqrt[3]{192} + 4\sqrt[3]{24} - 2\sqrt[3]{3}$ **17.** $\sqrt{75x} - \sqrt{3x} - \sqrt{12x}$ **18.** $2\sqrt{108y} - \sqrt{27y} + \sqrt{363y}$

19. $\sqrt{p^3} + \sqrt{25p^3} + \sqrt{9p}$ **20.** $\sqrt{m^3} - 2m\sqrt{m^5} + 3m\sqrt{m^7}$

In problems 21–40, find each product and simplify. Assume that variables are restricted to values for which all radical expressions are defined.

21. $\sqrt{12} \cdot \sqrt{3}$ **22.** $\sqrt{15} \cdot \sqrt{135}$ **23.** $\sqrt{3}(\sqrt{18} - \sqrt{2})$

24. $\sqrt{5}(4\sqrt{5} - 3\sqrt{2})$

25. $\sqrt{8}(\sqrt{6} - \sqrt{18})$

26. $\sqrt{11}(\sqrt{3} - 4\sqrt{11})$

27. $(\sqrt{x} + 7)(2\sqrt{x} + 1)$

28. $(\sqrt{x} - 1)(3\sqrt{x} + 5)$

29. $(2\sqrt{2} + 5)^2$

30. $(\sqrt{3} + 5\sqrt{2})^2$

31. $(2\sqrt{x} - \sqrt{y})^2$

32. $(x + 3\sqrt{y})^2$

33. $(2\sqrt{3} - 1)(2\sqrt{3} + 1)$

34. $(5\sqrt{7} + \sqrt{2})(5\sqrt{7} - \sqrt{2})$

35. $(3\sqrt{5} - \sqrt{3})(3\sqrt{5} + \sqrt{3})$

36. $(2\sqrt{x} + 7)(2\sqrt{x} - 7)$

37. $(3\sqrt{x} - 11)(3\sqrt{x} + 11)$

38. $(4\sqrt{x} - \sqrt{y})(4\sqrt{x} + \sqrt{y})$

39. $(\sqrt{x} + \sqrt{3})(\sqrt{x} - \sqrt{3})$

40. $(\sqrt{7x} + \sqrt{2x})(\sqrt{7x} - \sqrt{2x})$

In problems 41–60, rationalize the denominator of each expression and simplify the result. Assume that all variables are restricted to values for which all radical expressions are defined.

41. $\dfrac{2}{\sqrt{3}}$

42. $\dfrac{9}{\sqrt{21}}$

43. $\dfrac{8}{7\sqrt{11x}}$

44. $\dfrac{10x}{3\sqrt{5x}}$

45. $\dfrac{5}{\sqrt[3]{7}}$

46. $\dfrac{8}{\sqrt[3]{36}}$

47. $\dfrac{10}{\sqrt{5} - 1}$

48. $\dfrac{36}{\sqrt{3} + 1}$

49. $\dfrac{\sqrt{2}}{1 + \sqrt{2}}$

50. $\dfrac{\sqrt{2} + 1}{\sqrt{3} + \sqrt{2}}$

51. $\dfrac{2 - 2\sqrt{3}}{\sqrt{7} - \sqrt{5}}$

52. $\dfrac{8}{6\sqrt{5} - 5\sqrt{3}}$

53. $\dfrac{\sqrt{y}}{3\sqrt{x} - 2\sqrt{y}}$

54. $\dfrac{3\sqrt{2} - \sqrt{3}}{2\sqrt{3} - 7\sqrt{2}}$

55. $\dfrac{x}{x + \sqrt{y}}$

56. $\dfrac{5}{y - \sqrt{y}}$

57. $\dfrac{4\sqrt{2} + 3\sqrt{5}}{7\sqrt{5} - 3\sqrt{2}}$

58. $\dfrac{1}{\sqrt{5} + \sqrt{3} + \sqrt{2}}$

59. $\dfrac{1}{\sqrt{7} - \sqrt{5} + 2}$

60. $\dfrac{1}{\sqrt[3]{5} - \sqrt[3]{2}}$

[*Hint*: In Problem 60, use the special product $(a - b)(a^2 + ab + b^2) = a^3 - b^3$.]

5.5 Complex Numbers

In many applications of electronics, engineering, and physical science, we must work with expressions involving radicals in the form $\sqrt{-x}$, where $x > 0$. The square of a number represented by such a symbol must be negative:

$$(\sqrt{-x})^2 = [(-x)^{1/2}]^2 = (-x)^1 = -x.$$

However, the square of any real number (positive, negative, or zero) must be a positive number or zero. Accordingly, we must consider a new set of numbers called *complex numbers*.

We define $\sqrt{-x}$, $x > 0$, to be the number such that

$$\sqrt{-x} \cdot \sqrt{-x} = -x.$$

In particular, if $x = 1$, we have

$$\sqrt{-1} \cdot \sqrt{-1} = -1.$$

We designate the number $\sqrt{-1}$ by the symbol i:

$$i = \sqrt{-1}.$$

We square both sides of this equation, and we have

$$i^2 = -1.$$

Remember, $\sqrt{-1} = i$, *not* $-i$, even though both i^2 and $(-i)^2 = -1$.

Positive integer exponents have the same meaning, in terms of repeated multiplication, for both complex numbers and real numbers. Therefore, we can extend the definition of positive integer exponents to include complex numbers:

$$i^1 = i$$
$$i^2 = -1$$
$$i^3 = i^2 \cdot i = -1(i) = -i$$
$$i^4 = i^2 \cdot i^2 = (-1)(-1) = 1.$$

If we continue the list, we repeat the same sequence of answers because we can replace every factor of i^4 by 1. Thus,

$$i^5 = i^4 \cdot i = 1(i) = i$$
$$i^6 = i^4 \cdot i^2 = 1(-1) = -1$$
$$i^7 = i^4 \cdot i^3 = 1(-i) = -i$$
$$i^8 = i^4 \cdot i^4 = 1(1) = 1$$
$$i^9 = i^8 \cdot i = 1(i) = i,$$

and so on.

EXAMPLE 1   Simplify each of the following.

(a) i^{18} (b) i^{27}

(c) i^{105} (d) i^{204}

SOLUTION   (a) $i^{18} = i^{16} \cdot i^2 = (i^4)^4 \cdot i^2 = 1^4(-1) = -1$

(b) $i^{27} = i^{24} \cdot i^3 = (i^4)^6 \cdot i^3 = 1^6(-i) = -i$

(c) $i^{105} = i^{104} \cdot i = (i^4)^{26} \cdot i = 1^{26}(i) = i$

(d) $i^{204} = (i^4)^{51} = 1^{51} = 1$

If $x > 0$, then

$$\sqrt{-x} = i\sqrt{x}.$$

For example,

$$\sqrt{-4} = \sqrt{4}i = 2i$$
$$\sqrt{-25} = \sqrt{25}i = 5i$$
$$\sqrt{-98} = \sqrt{98}i = 7i\sqrt{2}.$$

If we write an expression in the form

$$a + bi,$$

where a and b are real numbers and $i = \sqrt{-1}$, we obtain a number called a **complex number.**

For example

$$4 + 3i, \qquad 2 - 7i, \qquad 3 + i\sqrt{5}, \qquad \text{and} \qquad \frac{3}{4} - 2i\sqrt{7}$$

are complex numbers.

The numbers bi and a are complex numbers, because

$$bi = 0 + bi \qquad \text{and} \qquad a = a + 0i.$$

We add, subtract, and multiply complex numbers in the same way we add, subtract, and multiply polynomials. We can simplify our answers by combining similar terms, and by using the fact that $i^2 = -1$:

Addition, Subtraction, and Multiplication of Complex Numbers

Let a, b, c, and d be real numbers. Then

1. **Addition**

$$(a + bi) + (c + di) = a + bi + c + di = a + c + bi + di$$
$$= (a + c) + (b + d)i$$

2. **Subtraction**

$$(a + bi) - (c + di) = a + bi - c - di = a - c + bi - di$$
$$= (a - c) + (b - d)i$$

3. **Multiplication**

$$(a + bi)(c + di) = ac + adi + bic + bidi = ac + bdi^2 + adi + bci$$
$$= ac + bd(-1) + (ad + bc)i$$
$$= (ac - bd) + (ad + bc)i$$

In Examples 2 and 3, perform each operation.

EXAMPLE 2   (a) $(5 + 6i) + (9 + 3i)$ (b) $(4 - 2i) - (-3 + i)$

SOLUTION   (a) $(5 + 6i) + (9 + 3i) = 5 + 6i + 9 + 3i = 5 + 9 + 6i + 3i$
$$= (5 + 9) + (6 + 3)i = 14 + 9i$$

(b) $(4 - 2i) - (-3 + i) = 4 - 2i + 3 - i = 4 + 3 - 2i - i$
$$= (4 + 3) + (-2 - 1)i = 7 - 3i$$

EXAMPLE 3   (a) $4i(3 + 7i)$ (b) $(4 + 3i)(2 - 4i)$ (c) $(1 + 2i)(1 - 2i)$

SOLUTION   (a) $4i(3 + 7i) = 4i(3) + 4i(7i) = 12i + 28i^2$
$$= 12i + 28(-1) = -28 + 12i$$

(b) $(4 + 3i)(2 - 4i) = 4(2) + 4(-4i) + 3i(2) + 3i(-4i)$
$$= 8 - 16i + 6i - 12i^2$$
$$= 8 - 16i + 6i - 12(-1)$$
$$= 8 - 16i + 6i + 12$$
$$= 20 - 10i$$

(c) $(1 + 2i)(1 - 2i) = 1(1) + 1(-2i) + 2i(1) + 2i(-2i)$
$$= 1 - 2i + 2i - 4i^2$$
$$= 1 - 4(-1)$$
$$= 1 + 4 = 5$$

We divide complex numbers by applying the identical process we used to rationalize denominators. In this case, the rationalizing factor is called the **complex conjugate.**

If $a + bi$ is a complex number, then its complex conjugate is denoted by $\overline{a + bi}$ and $\overline{a + bi}$ is defined to be equal to $a - bi$.

$\overline{a + bi} = a - bi$ (read "the complex conjugate of $a + bi$ equals $a - bi$").

For example,

$$\overline{4 + 3i} = 4 - 3i$$

$$\overline{1 - 4i} = 1 + 4i$$

$$\overline{-5 - 7i} = -5 + 7i.$$

To perform the division

$$\frac{3i}{4 - 3i},$$

we multiply the numerator and the denominator by the complex conjugate $4 + 3i$ of the denominator:

$$\frac{3i}{4 - 3i} = \frac{3i(4 + 3i)}{(4 - 3i)(4 + 3i)} = \frac{12i + 9i^2}{16 - 9i^2} = \frac{12i - 9}{16 + 9} = \frac{-9 + 12i}{25}$$

$$= -\frac{9}{25} + \frac{12}{25}i.$$

Division of Complex Numbers

Let a, b, c, and d be real numbers. Then

$$\frac{a + bi}{c + di} = \frac{(a + bi)(c - di)}{(c + di)(c - di)} = \frac{ac - adi + bic - bidi}{c^2 - d^2i^2}$$

$$= \frac{(ac + bd) + (bc - ad)i}{c^2 + d^2}.$$

In Examples 4–6, perform each division and express the answer in the form $a + bi$.

EXAMPLE **4** $\dfrac{1}{3 - 2i}$

SOLUTION   We multiply the numerator and the denominator by the complex conjugate $3 + 2i$:

$$\frac{1}{3 - 2i} = \frac{1}{3 - 2i} \cdot \frac{3 + 2i}{3 + 2i} = \frac{3 + 2i}{9 - 4i^2} = \frac{3 + 2i}{9 + 4} = \frac{3 + 2i}{13} = \frac{3}{13} + \frac{2}{13}i.$$

EXAMPLE **5** $\dfrac{1 + i}{1 - i}$

SOLUTION   The complex conjugate of the denominator is $1 + i$. Therefore, we have

$$\frac{1 + i}{1 - i} = \frac{1 + i}{1 - i} \cdot \frac{1 + i}{1 + i} = \frac{1 + 2i + i^2}{1 - i^2} = \frac{2i}{2} = i.$$

EXAMPLE **6** $\dfrac{4 - 3i}{5 + 7i}$

SOLUTION   The complex conjugate of the denominator is $5 - 7i$:

$$\frac{4 - 3i}{5 + 7i} = \frac{4 - 3i}{5 + 7i} \cdot \frac{5 - 7i}{5 - 7i} = \frac{20 - 28i - 15i + 21i^2}{5^2 - 49i^2}$$

$$= \frac{(20 - 21) + (-28 - 15)i}{25 + 49}$$

$$= \frac{-1 - 43i}{74} = \frac{-1}{74} - \frac{43}{74}i.$$

PROBLEM SET 5.5

In problems 1–8, simplify each expression.

1. i^{29} **2.** i^{37} **3.** i^{54} **4.** i^{49}

5. i^{65} **6.** i^{74} **7.** i^{108} **8.** i^{119}

In problems 9–14, write each expression in the form $a + bi$, where a and b are real numbers. (If $a = 0$, write the number as bi.)

9. $3\sqrt{-25}$ **10.** $-5\sqrt{-8}$ **11.** $5 + \sqrt{-81}$ **12.** $-8 + \sqrt{-72}$

13. $2 - \sqrt{-4x^2}, x > 0$ **14.** $6 - \sqrt{-16y^2}, y > 0$

In problems 15–38, perform each operation.

15. $(-3 + 6i) + (2 + 3i)$ **16.** $(-2 + 5i) + (-5 + i)$ **17.** $(-5 + 3i) + (5i - 1)$

18. $(10 - 24i) + (3 + 7i)$ **19.** $(6 + 8i) + (6 - 8i)$ **20.** $(3 + 2i) + (-7 + 2i)$

21. $(-2 - 3i) - (-3 - 2i)$ **22.** $(7 + 24i) - (-3 - 4i)$ **23.** $(10 - 8i) - (10 + 8i)$

24. $(5 - 7i) - (5 - 13i)$ **25.** $(6 - 8i) - (5 + 3i)$ **26.** $(4 - 5i) - (2 - 6i)$

27. $3i(2 + 3i)$ **28.** $-7i(-2 + 4i)$ **29.** $(2 + 3i)(1 - 3i)$

30. $(1 - 6i)(2 + 5i)$ **31.** $(3 - 7i)(2 + 3i)$ **32.** $(-2 + 7i)(-2 - 7i)$

33. $(5 - 4i)(5 + 4i)$ **34.** $(3 - 2i)(3 + 2i)$ **35.** $(7 - 11i)(7 + 11i)$

36. $(5 - 12i)(5 + 12i)$ **37.** $(2 - 7i)(2 + 7i)$ **38.** $(2 + i^{11})(2 - i^{11})$

In problems 39–44, find the conjugate of each complex number.

39. $2 + 4i$ **40.** $3 - 4i$ **41.** $2i$ **42.** $-3i$

43. 5 **44.** -6

In problems 45–62, perform each division and express the answer in the form $a + bi$.

45. $\dfrac{1}{4 + 3i}$ **46.** $\dfrac{1}{3 - 5i}$ **47.** $\dfrac{7}{5 - 4i}$ **48.** $\dfrac{-3}{3 + 4i}$

49. $\dfrac{2 + i}{2 + 3i}$ **50.** $\dfrac{1 - i}{-1 + i}$ **51.** $\dfrac{3 + 4i}{-2 + 5i}$ **52.** $\dfrac{7 + 2i}{1 - 5i}$

53. $\dfrac{4 - i^2}{2 - 7i}$ **54.** $\dfrac{5 - i^2}{11 + 2i}$ **55.** $\dfrac{i}{-1 - i}$ **56.** $\dfrac{1 + i}{2 - 3i}$

57. $\dfrac{5 + 2i}{i}$ **58.** $\dfrac{3 - 2i}{-i}$ **59.** $\dfrac{3 - 2i}{5 - 2i}$ **60.** $\dfrac{7 + 4i}{1 - 2i}$

61. $\dfrac{3 - \sqrt{-9}}{2 + \sqrt{-25}}$ **62.** $\dfrac{2 + \sqrt{-36}}{1 - \sqrt{-4}}$

REVIEW PROBLEM SET

In problems 1–14, rewrite each expression so that it contains only positive exponents, and simplify.

1. 3^{-2} **2.** $(\tfrac{1}{5})^{-2}$ **3.** y^{-4} **4.** $(3u)^{-3}$

5. $t^{-4}s^{-2}$ **6.** $4x^{-3}y^{-7}$ **7.** $\left(\dfrac{x^2}{y^3}\right)^{-1}$ **8.** $\dfrac{u^{-7}}{v^{-5}}$

9. $(3x^4)^0$ **10.** $(7y^3 + 1)^0$ **11.** $\dfrac{1 + 3^{-2}}{2^{-2}}$ **12.** $\dfrac{1}{2^{-3} + 3^{-2}}$

13. $\dfrac{2}{x^{-3} + y^{-2}}$ **14.** $\dfrac{w^{-3}}{z^3} + \dfrac{z^{-3}}{w^3}$

In problems 15–32, use the properties of exponents to rewrite each expression so that it contains only positive exponents, and simplify.

15. $3^{-11} \cdot 3^{15}$ **16.** $19^{-3} \cdot 19^5$ **17.** $x^{-8} \cdot x^{15}$ **18.** $w^{12} \cdot w^{-7}$

19. $(2^{-3})^2$ **20.** $(z^4)^{-3}$ **21.** $(t^{-3})^{-4}$ **22.** $(3u^{-2})^{-2}$

23. $(4^{-1}x^{-3})^{-2}$ **24.** $(3x^{-3}y^{-4})^{-3}$ **25.** $\left(\dfrac{2}{u^{-2}}\right)^{-3}$ **26.** $\left(\dfrac{m^3}{n^{-4}}\right)^{-2}$

27. $\dfrac{5^{-4}}{5^{-6}}$ **28.** $\dfrac{11^{-3}}{11^{-2}}$ **29.** $\dfrac{r^{-5}}{r^{-2}}$ **30.** $\dfrac{(xy)^2}{(xy)^{-4}}$

31. $\dfrac{8y^2}{2y^{-3}}$ **32.** $\dfrac{(ab)^{-6}}{(ab)^{-11}}$

In problems 33–38, rewrite each expression without negative exponents, and simplify.

33. $\left(\dfrac{x^{-3}y^2}{x^{-2}y^{-4}}\right)^{-2}$ **34.** $\left(\dfrac{w^3z^{-4}}{w^{-4}z^2}\right)^{-1}$ **35.** $\left(\dfrac{u^4}{2v^{-3}}\right)^{-1}\left(\dfrac{u^{-1}v^2}{4^{-1}}\right)^{-2}$

36. $\left(\dfrac{m^{-4}n^{-3}}{m^{-7}n^{-1}}\right)^{-2}\left(\dfrac{m^4n}{m^{-3}n^4}\right)^{-1}$ **37.** $\left[\dfrac{r^3s^{-2}t^{-4}}{(rst)^{-1}}\right]^{-2}$ **38.** $\left[\dfrac{x^{-5}y^3z^{-4}}{(xy)^{-2}(yz)^{-3}}\right]^{-1}$

In problems 39–44, find the principal root (if it is defined as a real number).

39. $\sqrt{81}$ **40.** $\sqrt[3]{-64}$ **41.** $\sqrt[3]{-216}$
42. $\sqrt[4]{-16}$ **43.** $\sqrt{-9}$ **44.** $\sqrt{289}$

In problems 45–48, rewrite each expression in terms of rational exponents.

45. $\sqrt{15}$ **46.** $\sqrt[3]{-7}$ **47.** $\sqrt[4]{(a+b)^3}$ **48.** $\sqrt[6]{9x^5}$

In problems 49–52, rewrite each expression in terms of radicals.

49. $13^{2/3}$ **50.** $19^{4/5}$ **51.** $(2x+3y)^{2/7}$ **52.** $(-3u^2v^4)^{4/9}$

In problems 53–60, find the value of each expression.

53. $(-32)^{3/5}$ **54.** $(-27)^{-4/3}$ **55.** $16^{-3/4}$ **56.** $-(-8)^{1/3}$
57. $-4^{3/2}$ **58.** $-25^{-1/2}$ **59.** $(-8)^{-2/3}$ **60.** $36^{-5/2}$

In problems 61–84, rewrite each expression so that it contains only positive exponents, and simplify. Assume that variables are restricted to values for which all expressions are defined.

61. $7^{-2/3} \cdot 7^{5/3}$ **62.** $9^{-4/7} \cdot 9^{11/7}$ **63.** $x^{3/2} \cdot x^{-5/4}$

64. $y^{5/11} \cdot y^{-16/11}$ **65.** $(2^{-1/3})^9$ **66.** $(8^2)^{-5/12}$

67. $(x^{-11/3})^{3/22}$ **68.** $(y^{-51})^{3/17}$ **69.** $(-8t^6)^{1/3}$

70. $(16u^4)^{-1/2}$ **71.** $(-32m^5)^{2/5}$ **72.** $(4x^2y^8)^{3/2}$

73. $\left(\dfrac{8}{v^9}\right)^{-2/3}$ **74.** $\left(\dfrac{128}{p^7}\right)^{-2/7}$ **75.** $\left(\dfrac{c^{-5/2}}{d^{-2/5}}\right)^{-30}$

76. $\left(\dfrac{16z^4}{w^8}\right)^{-3/4}$ **77.** $\left(\dfrac{-x^5y^{10}}{32z^{15}}\right)^{-3/5}$ **78.** $\left(\dfrac{8m^{-6}n^{-30}}{27n^{-12}}\right)^{-1/3}$

79. $\dfrac{a^{-2/3}}{a^{-5/7}}$ **80.** $\dfrac{(y^2)^{-3/2}}{y^{7/2}}$ **81.** $\dfrac{(r^{-1}y^{-2/3})^{-3}}{(r^{-1}y^{-2/3})^5}$

82. $\dfrac{(w^{-7/3})^{2/7}}{w^{-3}}$ **83.** $\left(\dfrac{x^{3/2}y^{-1/3}}{w^{-2}}\right)^{-6}\left(\dfrac{x^{1/3}y^{-2/3}}{w^{-2}}\right)^6$ **84.** $\left[\dfrac{(u^{2/3}v^{-1/2})(u^{1/2}v^{-3})}{u^{5/6}v^{-2}}\right]^{-1}$

© In problems 85–88, use a calculator with y^x key to evaluate each expression. Round off your answer to four significant digits.

85. $3.12^{4.3}$ **86.** $0.91^{2.71}$ **87.** $9^{\sqrt{2}}$ **88.** $\sqrt{5}^{\sqrt{3}}$

In problems 89–110, use the properties of radicals to simplify each expression. Assume that variables are restricted to values for which all expressions are defined.

89. $\sqrt{125}$ **90.** $\sqrt[5]{-64}$ **91.** $\sqrt[4]{32t^5}$ **92.** $\sqrt[3]{54w^7}$

93. $\sqrt[3]{-24x^4y^{11}}$ **94.** $\sqrt{250z^3y^{10}}$ **95.** $\sqrt{\dfrac{4}{25}}$ **96.** $\sqrt[3]{-\dfrac{8}{27}}$

97. $\sqrt[5]{-\dfrac{t^{10}}{32}}$ **98.** $\sqrt[4]{\dfrac{16x^8y^{12}}{81}}$ **99.** $\sqrt[3]{\dfrac{-5}{c^6d^9}}$ **100.** $\sqrt{\dfrac{11}{w^4z^8}}$

101. $\dfrac{\sqrt{125u^7}}{\sqrt{5u}}$ **102.** $\dfrac{\sqrt[3]{-24m^{11}}}{\sqrt[3]{3m^2}}$ **103.** $\sqrt{\sqrt{16x^8}}$ **104.** $\sqrt[3]{\sqrt[3]{t^{18}}}$

105. $\sqrt[3]{\sqrt[5]{v^{15}}}$ **106.** $\sqrt[4]{\sqrt[6]{x^{24}}}$ **107.** $\sqrt{8u^3v^4} \cdot \sqrt{2uv^2}$ **108.** $\sqrt[3]{4m^2} \cdot \sqrt[3]{2m}$

109. $\sqrt[3]{\dfrac{x^3\sqrt{x^5}}{x^4}}$ **110.** $\sqrt[6]{y\sqrt[3]{z\sqrt{w}}}$

In problems 111–114, write each expression as a single radical, and then simplify the result. Assume that variables are restricted to values for which all expressions are defined.

111. $\sqrt[3]{z}\ \sqrt[4]{z}$

112. $\sqrt[5]{x^2}\ \sqrt[3]{x}$

113. $\sqrt{2u}\ \sqrt[3]{3u^2}$

114. $\sqrt[4]{a^2b^3}\ \sqrt[5]{a^4b^2}$

$\boxed{\text{c}}$ In problems 115–116, use a calculator to verify each equation.

115. $\sqrt{11}\ \sqrt{3} = \sqrt{33}$

116. $\sqrt[4]{19}\ \sqrt[4]{3} = \sqrt[4]{57}$

In problems 117–142, perform the indicated operations and simplify the result. Assume that variables are restricted to values for which all expressions are defined.

117. $7\sqrt{2} - 3\sqrt{2} + 4\sqrt{2}$

118. $\sqrt{5} - 6\sqrt{5} + 2\sqrt{5}$

119. $4\sqrt{x} + 7\sqrt{x} - 5\sqrt{x}$

120. $3\sqrt[3]{t} - \sqrt[3]{t} + 2\sqrt[3]{t}$

121. $\sqrt{128} + \sqrt{8}$

122. $\sqrt{48} - \sqrt{12}$

123. $\sqrt{108} - \sqrt{27}$

124. $\sqrt{45} + \sqrt{80}$

125. $\sqrt{32z} + \sqrt{72z}$

126. $\sqrt{27x} - \sqrt{48x}$

127. $\sqrt{63u} + 2\sqrt{112u} - \sqrt{252u}$

128. $\sqrt[3]{16y^2} - \sqrt[3]{54y^2} + \sqrt[3]{250y^2}$

129. $\sqrt{3}(\sqrt{2} + \sqrt{5})$

130. $\sqrt{5}(\sqrt{7} - \sqrt{2})$

131. $\sqrt{3}(\sqrt{6} - \sqrt{8})$

132. $\sqrt{5}(\sqrt{15} + \sqrt{10})$

133. $(\sqrt{10} + \sqrt{2})^2$

134. $(2\sqrt{6} - 3\sqrt{2})^2$

135. $(\sqrt{8} - \sqrt{2})(\sqrt{8} + \sqrt{2})$

136. $(\sqrt{a} - \sqrt{b})(\sqrt{a} + \sqrt{b})$

137. $(\sqrt{3t} + \sqrt{5})(\sqrt{3t} - \sqrt{5})$

138. $(\sqrt{xy} - \sqrt{z})(\sqrt{xy} + \sqrt{z})$

139. $(\sqrt{3} + \sqrt{6})(2\sqrt{3} - \sqrt{6})$

140. $(\sqrt{6} + 1)(\sqrt{8} + \sqrt{18})$

141. $(\sqrt{x} - \sqrt{y})(2\sqrt{x} + \sqrt{y})$

142. $(3\sqrt{u} - v)(5\sqrt{u} + 2v)$

In problems 143–156, rationalize the denominator of each expression and simplify the result. Assume that all variables are restricted to values for which all radical expressions are defined.

143. $\dfrac{4}{\sqrt{3}}$

144. $\dfrac{7}{\sqrt{14}}$

145. $\dfrac{5}{\sqrt{32}}$

146. $\dfrac{8}{\sqrt{18w}}$

147. $\dfrac{6t}{\sqrt{27ts}}$

148. $\dfrac{2}{\sqrt{7} - 1}$

149. $\dfrac{5}{\sqrt{11} + 5}$

150. $\dfrac{6}{\sqrt{2} + 3\sqrt{5}}$

151. $\dfrac{10}{\sqrt{x} - 2}$

152. $\dfrac{\sqrt{m}}{3 + \sqrt{m}}$

153. $\dfrac{\sqrt{7} - \sqrt{6}}{\sqrt{7} + \sqrt{6}}$

154. $\dfrac{2\sqrt{x} - \sqrt{y}}{\sqrt{x} + \sqrt{y}}$

155. $\dfrac{u - \sqrt{v}}{u - 2\sqrt{v}}$

156. $\dfrac{2}{1 + \sqrt{2} - \sqrt{3}}$

In problems 157–162, simplify.

157. i^{33}

158. i^{44}

159. i^{55}

160. i^{69}

161. i^{402}

162. i^{570}

In problems 163–178, write each expression in the form of $a + bi$, where a and b are real numbers. (If $a = 0$, write the number as bi.)

163. $4\sqrt{-9}$

164. $-2\sqrt{-25}$

165. $3 + \sqrt{-t^2}, t > 0$

166. $-7 - \sqrt{-4y^2}, y > 0$

167. $(-4 - i) + (3 + 17i)$

168. $(3 - 42i) + (17 + 4i)$

169. $(-7 + 6i) + (3 - 2i)$

170. $(12 + 6i) + (12 - 6i)$

171. $(-24 + i) - (3 - 9i)$

172. $(65 - 4i) - (43 - 2i)$

173. $(15 - 7i) - (-8 + 3i)$

174. $(-12 + 3i) - (-11 - 5i)$

175. $(6 - 2i)(1 + 3i)$

176. $(2 + 3i)(1 - 7i)$

177. $(3 + 8i)(2 - 5i)$

178. $(3 - 2i\sqrt{7})(3 + 2i\sqrt{7})$

In problems 179–182, find the conjugate of each complex number.

179. $3 + 5i$

180. $-6i$

181. $-4 - 7i$

182. $-11 + 3i$

In problems 183–188, perform each division and express the answer in the form $a + bi$.

183. $\dfrac{3 + 2i}{4 + i}$

184. $\dfrac{1 - i}{2 - 3i}$

185. $\dfrac{7 + 3i}{3 - 2i}$

186. $\dfrac{2 + 7i}{2 + 2i}$

187. $\dfrac{2 + i}{-5 - i}$

188. $\dfrac{3 + \sqrt{-16}}{2 - \sqrt{-9}}$

6 Nonlinear Equations and Inequalities

Nonlinear equations and inequalities are used frequently in the physical sciences and engineering. We examine some nonlinear equations and inequalities, including quadratic (second-degree) equations and inequalities, radical equations, and equations containing rational exponents. We then use these equations and inequalities to solve practical problems.

A **quadratic equation** in one variable can be written in the form

$$ax^2 + bx + c = 0,$$

in which a, b, and c are constants with $a \neq 0$. When a quadratic equation is written this way, it is said to be in **standard form.**

The easiest way to solve a quadratic equation in standard form is to factor the polynomial expression on the left side of the equation. (If it is impossible to factor the polynomial, we can use other methods—discussed in Sections 6.2 and 6.3—to solve the equation.)

6.1 Solving Quadratic Equations by Factoring

To solve quadratic equations of the form

$$ax^2 + bx + c = 0$$

by factoring, we use the methods of factoring developed in Sections 2.5 through 2.7, *and* the following special property:

Zero Property

> Assume that a and b are real numbers.
>
> $$\text{If } ab = 0, \text{ then } a = 0 \text{ or } b = 0.$$

For instance, to solve the equation

$$(2x - 3)(x + 2) = 0,$$

we set each factor equal to zero and solve the resulting first-degree equations:

$$
\begin{array}{c|c}
2x - 3 = 0 & x + 2 = 0 \\
2x = 3 & x = -2 \\
x = \dfrac{3}{2} &
\end{array}
$$

Thus, the solutions are $\frac{3}{2}$ and -2.

The following examples illustrate how to solve quadratic equations by factoring.

In Examples 1–4, use the factoring method to solve each equation.

EXAMPLE 1 $x^2 + 4x - 21 = 0$

SOLUTION   We factor the left side of the equation:

$$(x - 3)(x + 7) = 0.$$

Now we set each factor equal to zero and solve the resulting first-degree equations:

$$
\begin{array}{c|c}
x - 3 = 0 & x + 7 = 0 \\
x = 3 & x = -7
\end{array}
$$

Check   For $x = 3$,

$$3^2 + 4(3) - 21 \overset{?}{=} 0$$

$$3^2 + 4(3) - 21 = 9 + 12 - 21 = 0.$$

For $x = -7$,

$$(-7)^2 + 4(-7) - 21 \overset{?}{=} 0$$

$$(-7)^2 + 4(-7) - 21 = 49 - 28 - 21 = 0.$$

Therefore, the solutions are 3 and -7.

EXAMPLE 2 $2y^2 - y = 1$

SOLUTION   First we rewrite the equation in standard form: $2y^2 - y - 1 = 0$.

We factor the left side of the equation: $(2y + 1)(y - 1) = 0$.

We set each factor equal to zero and solve the resulting equations:

$$
\begin{array}{c|c}
2y + 1 = 0 & y - 1 = 0 \\
2y = -1 & y = 1 \\
y = -\dfrac{1}{2} &
\end{array}
$$

Therefore, the solutions are $-\frac{1}{2}$ and 1.

EXAMPLE 3 $5m^2 = -15m$

SOLUTION   First we rewrite the equation in standard form:

$$5m^2 + 15m = 0.$$

Next we factor the left side of the equation:

$$5m(m + 3) = 0.$$

We set each factor equal to zero and solve the resulting equations:

$$
\begin{array}{c|c}
5m = 0 & m + 3 = 0 \\
m = 0 & m = -3
\end{array}
$$

Therefore, the solutions are 0 and -3.

EXAMPLE 4 $9r^2 = 7(6r - 7)$

SOLUTION   First we rewrite the equation in standard form:

$$9r^2 - 42r + 49 = 0.$$

We factor the left side of the equation:

$$(3r - 7)^2 = 0.$$

We set each factor equal to zero:

$$
\begin{array}{c|c}
3r - 7 = 0 & 3r - 7 = 0 \\
3r = 7 & 3r = 7 \\
r = \dfrac{7}{3} & r = \dfrac{7}{3}
\end{array}
$$

Therefore, the only solution is $\frac{7}{3}$.

We often encounter equations that contain fractions. Sometimes, when this occurs, a quadratic equation may be obtained by multiplying each side of the given equation by the LCD of the fractions. This possibility is illustrated in Example 5 below:

EXAMPLE 5   Solve the equation

$$\frac{5}{t+4} - \frac{3}{t-2} = 4.$$

SOLUTION   First we multiply both sides of the equation by $(t+4)(t-2)$, the LCD of the fractions. This gives us an equation that contains no fractions:

$$(t+4)(t-2)\left(\frac{5}{t+4} - \frac{3}{t-2}\right) = 4(t+4)(t-2)$$

$$5(t-2) - 3(t+4) = 4(t^2 + 2t - 8)$$

$$5t - 10 - 3t - 12 = 4t^2 + 8t - 32$$

$$2t - 22 = 4t^2 + 8t - 32.$$

We write this equation in standard form:

$$4t^2 + 6t - 10 = 0$$

or

$$2t^2 + 3t - 5 = 0. \qquad \text{(We divided both sides by 2.)}$$

We factor the left side of the equation:

$$(2t + 5)(t - 1) = 0.$$

We set each factor equal to zero:

$$2t + 5 = 0 \qquad \bigg| \qquad t - 1 = 0$$
$$2t = -5 \qquad \bigg| \qquad t = 1$$
$$t = -\frac{5}{2} \qquad \bigg|$$

We must check the proposed solutions because the equation contains a variable in the denominator (see page 120).

Check For $t = 1$,

$$\frac{5}{1+4} - \frac{3}{1-2} \overset{?}{=} 4$$

$$\frac{5}{1+4} - \frac{3}{1-2} = \frac{5}{5} - \frac{3}{-1}$$

$$= 1 + 3 = 4.$$

For $t = -\frac{5}{2}$

$$\frac{5}{-\frac{5}{2} + 4} - \frac{3}{-\frac{5}{2} - 2} \stackrel{?}{=} 4$$

$$\frac{5}{-\frac{5}{2} + 4} - \frac{3}{-\frac{5}{2} - 2} = \frac{5}{\frac{3}{2}} - \frac{3}{-\frac{9}{2}}$$

$$= 5\left(\frac{2}{3}\right) - 3\left(-\frac{2}{9}\right) = \frac{10}{3} + \frac{2}{3} = 4.$$

Therefore, the solutions are 1 and $-\frac{5}{2}$.

PROBLEM SET 6.1

In problems 1–44, use the factoring method to solve each equation. Check the solutions in problems 39–44.

1. $x^2 - 3x + 2 = 0$

2. $y^2 - 7y - 8 = 0$

3. $c^2 - 6c + 8 = 0$

4. $u^2 + 5u - 66 = 0$

5. $t^2 - t = 20$

6. $z^2 + 3z = 10$

7. $y^2 + 2y = 35$

8. $b^2 + 9b = 10$

9. $u^2 - u = 12$

10. $y^2 + y = 12$

11. $3x^2 - 2x - 5 = 0$

12. $15x^2 - 19x + 6 = 0$

13. $10y^2 + y - 2 = 0$

14. $6z^2 + 17z - 3 = 0$

15. $9r^2 + 6r - 8 = 0$

16. $5w^2 + 34w - 7 = 0$

17. $-10t^2 + 11t - 3 = 0$

18. $3y^2 - y - 14 = 0$

19. $4z^2 + 20 = 21z$

20. $12c^2 + 5c = 2$

21. $6u^2 + 7u = 20$

22. $4x^2 = 27x + 7$

23. $10y^2 - 31y = 14$

24. $18w^2 + 61w = 7$

25. $x^2 = 7x$

26. $x^2 = -8x$

27. $y^2 + 7y = 0$

28. $3m^2 - 7m = 0$

29. $49z^2 - 14z + 1 = 0$

30. $25y^2 - 20y + 4 = 0$

31. $9y^2 - (y + 2)^2 = 0$

32. $4x^2 - (x + 1)^2 = 0$

33. $z(3z + 11) = 20$

34. $m(4m + 25) = -6$

35. $n(5n + 2) = 3$

36. $x(x - 2) = 9 - 2x$

37. $r(2r - 19) = 33$

38. $(x + 3)(2x + 3) = 11(x + 3)$

39. $\dfrac{12}{y} - 7 = \dfrac{12}{1 - y}$

40. $\dfrac{5}{4(t + 4)} - 1 = \dfrac{3}{4(t - 2)}$

41. $\dfrac{15}{(x - 2)^2} + \dfrac{2}{x - 2} = 1$

42. $\dfrac{3 - 2m}{4m} - 4 = \dfrac{3}{4m - 3}$

43. $\dfrac{2u - 5}{2u + 1} = \dfrac{7}{4} - \dfrac{6}{2u - 3}$

44. $\dfrac{6}{t^2 - 1} = \dfrac{1}{2} + \dfrac{1}{1 - t}$

In problems 45–52, solve each equation for the indicated unknown.

45. $x^2 - 2ax - 15a^2 = 0$, for x.

46. $12y^2 - 10my = 12m^2$, for y.

47. $w^2 + 2qw = p^2 - q^2$, for w.

48. $(at - bt)^2 = t(b - a)$, for t.

49. $6m^2 + mb = 2b^2$, for m.

50. $7y^2 - 11hy = 6h^2$, for y.

51. $z^2 + 4d^2 - z + 2d = 4dz$, for z.

52. $b^2t^2 - 4bt + 3 = 4t^2 - 4t$, for t.

6.2 Solving Quadratic Equations by Roots Extraction and Completing the Square

It is clear that factoring is an efficient method for solving quadratic equations in standard form. But it is not always possible to factor such equations. In this section, we develop other methods for solving quadratic equations.

Roots Extraction

In Section 6.1, we solved equations such as

$$x^2 - 4 = 0$$

by factoring to get $(x - 2)(x + 2) = 0$, and found that $x = 2$ and $x = -2$. We can also solve this equation by determining the numbers whose squares are 4. That is, we determine the two square roots of 4. Thus, if $x^2 = 4$, then

$$x = -\sqrt{4} = -2 \quad \text{or} \quad x = \sqrt{4} = 2,$$

since

$$(-2)^2 = 4 \quad \text{and} \quad 2^2 = 4.$$

This example suggests the following property:

> If $au^2 = p$, which is equivalent to $u^2 = \dfrac{p}{a}$, then
>
> $$u = -\sqrt{\frac{p}{a}} \quad \text{or} \quad u = \sqrt{\frac{p}{a}}.$$

This method is called **roots extraction.**

Solve each equation, in Examples 1–4, by the roots extraction method.

EXAMPLE 1 $x^2 = 81$

SOLUTION   We take the square roots of each side:

$$x = -\sqrt{81} \quad \bigg| \quad x = \sqrt{81}$$
$$x = -9 \quad \bigg| \quad x = 9$$

The solutions are -9 and 9.

EXAMPLE 2 $2t^2 = 24$

SOLUTION   We divide both sides by 2:

$$t^2 = 12.$$

We take the square roots of each side:

$$
\begin{array}{c|c}
t = -\sqrt{12} & t = \sqrt{12} \\
t = -\sqrt{4(3)} & t = \sqrt{4(3)} \\
t = -2\sqrt{3} & t = 2\sqrt{3}
\end{array}
$$

Therefore, the solutions are $-2\sqrt{3}$ and $2\sqrt{3}$.

EXAMPLE 3 $(2y + 3)^2 = 36$

SOLUTION   The equation is of the form

$$u^2 = 36 \qquad \text{with} \qquad u = 2y + 3.$$

We take the square roots of each side:

$$
\begin{array}{c|c}
2y + 3 = -\sqrt{36} & 2y + 3 = \sqrt{36} \\
2y + 3 = -6 & 2y + 3 = 6 \\
2y = -9 & 2y = 3 \\
y = -\dfrac{9}{2} & y = \dfrac{3}{2}
\end{array}
$$

Therefore, the solutions are $-\frac{9}{2}$ and $\frac{3}{2}$.

EXAMPLE 4 $w^2 + 25 = 0$

SOLUTION   First we rewrite the equation as

$$w^2 = -25.$$

Then we take the square roots of each side:

$$
\begin{array}{c|c}
w = -\sqrt{-25} & w = \sqrt{-25} \\
w = -i\sqrt{25} & w = i\sqrt{25} \\
w = -5i & w = 5i
\end{array}
$$

Therefore, the solutions are the complex numbers $-5i$ and $5i$.

Completing the Square

Consider the following quadratic equations:

$$x^2 - 4x - 8 = 0 \quad \text{and} \quad (x - 2)^2 = 12.$$

Although these two equations appear to be different, they are simply two forms of the same equation, as you will see when you put the second equation into standard form:

$$(x - 2)^2 = 12$$

$$x^2 - 4x + 4 = 12$$

$$x^2 - 4x - 8 = 0 \quad \text{(We subtracted 12 from both sides.)}$$

The first equation, $x^2 - 4x - 8 = 0$, cannot be solved by factoring. However, we can convert it into the second equation, $(x - 2)^2 = 12$, using the method known as **completing the square.** We then solve $(x - 2)^2 = 12$ by the roots extraction method.

The method of completing the square is based on the fact that

$$x^2 + 2kx + k^2 = (x + k)^2.$$

Thus, if we start with an expression of the form $x^2 + 2kx$, we see from the above equation that we can "complete the square" by adding k^2. We obtain the term k^2 by squaring one-half the coefficient of the first-degree term x. That is, we complete the square by adding

$$\left[\frac{1}{2}(2k)\right]^2 = k^2,$$

so that

$$x^2 + 2kx + \left[\frac{1}{2}(2k)\right]^2 = x^2 + 2kx + k^2 = (x + k)^2.$$

coefficient of x term

For example, given $x^2 + 8x$, we can complete the square as follows:

$$x^2 + 8x + \left[\frac{1}{2}(8)\right]^2 = x^2 + 8x + 16 = (x + 4)^2.$$

coefficient of x term

EXAMPLE 5   Complete the square of the given expression and write the result as the square of a binomial.

(a) $x^2 + 4x$ (b) $y^2 - 18y$

(c) $w^2 - 5w$ (d) $t^2 + 13t$

SOLUTION

In each case, we complete the square by adding the square of one-half the coefficient of the linear term.

(a) $x^2 + 4x + [\frac{1}{2}(4)]^2 = x^2 + 4x + 4 = (x + 2)^2$

(b) $y^2 - 18y + [\frac{1}{2}(-18)]^2 = y^2 - 18y + 81 = (y - 9)^2$

(c) $w^2 - 5w + [\frac{1}{2}(-5)]^2 = w^2 - 5w + \frac{25}{4} = (w - \frac{5}{2})^2$

(d) $t^2 + 13t + [\frac{1}{2}(13)]^2 = t^2 + 13t + \frac{169}{4} = (t + \frac{13}{2})^2$

Now we can combine the method of roots extraction and the procedure of completing the square to solve quadratic equations. For instance, to solve $x^2 - 4x - 8 = 0$, we proceed as follows:

$$x^2 - 4x - 8 = 0$$

$$x^2 - 4x = 8 \qquad \text{(We added 8 to both sides.)}$$

$$x^2 - 4x + \left[\frac{1}{2}(-4)\right]^2 = 8 + \left[\frac{1}{2}(-4)\right]^2 \qquad \text{(We completed the square on the left side and added the same quantity to the right side.)}$$

$$x^2 - 4x + 4 = 8 + 4$$

$$(x - 2)^2 = 12. \qquad \text{(We wrote the left side as the square of a binomial.)}$$

Then we use the method of roots extraction:

$$
\begin{array}{c|c}
x - 2 = -\sqrt{12} & x - 2 = \sqrt{12} \\
x - 2 = -2\sqrt{3} & x - 2 = 2\sqrt{3} \\
x = 2 - 2\sqrt{3} & x = 2 + 2\sqrt{3}
\end{array}
$$

Therefore, the solutions are $2 - 2\sqrt{3}$ and $2 + 2\sqrt{3}$.

The following procedure summarizes the process of solving a quadratic equation $ax^2 + bx + c = 0$ by completing the square:

Procedure for Solving a Quadratic Equation by Completing the Square

Step 1. Write the equation in the equivalent form $ax^2 + bx = -c$.

Step 2. Divide both sides of the equation by the coefficient a of the x^2 term. (If $a = 1$, then proceed to step 3.)

Step 3. Complete the square on the left side of the equation by adding the square of one-half the coefficient b/a of the x term to both sides.

Step 4. Write the left side of the equation as the square of a binomial expression and solve the resulting equation by roots extraction.

In Examples 6–9, solve each equation by completing the square.

EXAMPLE 6 $x^2 + 4x - 2 = 0$

SOLUTION   Step 1. We add 2 to both sides of the equation:

$$x^2 + 4x = 2.$$

Step 2. We proceed to the next step, because the coefficient of $x^2 = 1$.

Step 3. We complete the square on the left side of the equation and we add the same quantity to the right side of the equation:

$$x^2 + 4x + \left[\frac{1}{2}(4)\right]^2 = 2 + \left[\frac{1}{2}(4)\right]^2$$
$$x^2 + 4x + 4 = 2 + 4.$$

Step 4. We rewrite the equation as

$$(x + 2)^2 = 6,$$

and we solve the equation by taking the square roots of each side:

$$
\begin{array}{c|c}
x + 2 = -\sqrt{6} & x + 2 = \sqrt{6} \\
x = -2 - \sqrt{6} & x = -2 + \sqrt{6}
\end{array}
$$

Therefore, the solutions are $-2 - \sqrt{6}$ and $-2 + \sqrt{6}$.

EXAMPLE 7 $4y^2 - 2y - 3 = 0$

SOLUTION   Step 1. We add 3 to each side of the equation:

$$4y^2 - 2y = 3.$$

Step 2. We divide both sides of the equation by 4:

$$y^2 - \frac{2}{4}y = \frac{3}{4}$$

$$y^2 - \frac{1}{2}y = \frac{3}{4}.$$

Step 3. We complete the square on the left side of the equation and we add the same quantity to the right side of the equation:

$$y^2 - \frac{1}{2}y + \left[\frac{1}{2}\left(-\frac{1}{2}\right)\right]^2 = \frac{3}{4} + \left[\frac{1}{2}\left(-\frac{1}{2}\right)\right]^2$$

$$y^2 - \frac{1}{2}y + \frac{1}{16} = \frac{3}{4} + \frac{1}{16} = \frac{13}{16}.$$

Step 4. We rewrite the equation as

$$\left(y - \frac{1}{4}\right)^2 = \frac{13}{16},$$

and we solve the equation by taking the square roots of each side:

$$
\begin{array}{c|c}
y - \dfrac{1}{4} = -\sqrt{\dfrac{13}{16}} & y - \dfrac{1}{4} = \sqrt{\dfrac{13}{16}} \\[2mm]
y - \dfrac{1}{4} = \dfrac{-\sqrt{13}}{4} & y - \dfrac{1}{4} = \dfrac{\sqrt{13}}{4} \\[2mm]
y = \dfrac{1}{4} - \dfrac{\sqrt{13}}{4} & y = \dfrac{1}{4} + \dfrac{\sqrt{13}}{4} \\[2mm]
y = \dfrac{1 - \sqrt{13}}{4} & y = \dfrac{1 + \sqrt{13}}{4}
\end{array}
$$

Therefore, the solutions are

$$\frac{1 - \sqrt{13}}{4} \quad \text{and} \quad \frac{1 + \sqrt{13}}{4}.$$

EXAMPLE 8 $7w^2 - 4w - 1 = 0$

SOLUTION

Step 1. $7w^2 - 4w = 1$

Step 2. $w^2 - \frac{4}{7}w = \frac{1}{7}$

Step 3. $w^2 - \frac{4}{7}w + [\frac{1}{2}(-\frac{4}{7})]^2 = \frac{1}{7} + [\frac{1}{2}(-\frac{4}{7})]^2$

$\qquad\quad w^2 - \frac{4}{7}w + \frac{4}{49} = \frac{1}{7} + \frac{4}{49}$

$\qquad\quad w^2 - \frac{4}{7}w + \frac{4}{49} = \frac{11}{49}$

Step 4. $(w - \frac{2}{7})^2 = \frac{11}{49}$

$$
\begin{array}{c|c}
w - \dfrac{2}{7} = -\sqrt{\dfrac{11}{49}} & w - \dfrac{2}{7} = \sqrt{\dfrac{11}{49}} \\[2mm]
w - \dfrac{2}{7} = \dfrac{-\sqrt{11}}{7} & w - \dfrac{2}{7} = \dfrac{\sqrt{11}}{7} \\[2mm]
w = \dfrac{2}{7} - \dfrac{\sqrt{11}}{7} & w = \dfrac{2}{7} + \dfrac{\sqrt{11}}{7} \\[2mm]
w = \dfrac{2 - \sqrt{11}}{7} & w = \dfrac{2 + \sqrt{11}}{7}
\end{array}
$$

Therefore, the solutions are

$$\frac{2 - \sqrt{11}}{7} \quad \text{and} \quad \frac{2 + \sqrt{11}}{7}.$$

EXAMPLE 9 $2r^2 + 3r + 2 = 0$

SOLUTION

Step 1. $2r^2 + 3r = -2$

Step 2. $r^2 + \frac{3}{2}r = -1$

Step 3. $r^2 + \frac{3}{2}r + [\frac{1}{2}(\frac{3}{2})]^2 = -1 + [\frac{1}{2}(\frac{3}{2})]^2$

$\qquad r^2 + \frac{3}{2}r + \frac{9}{16} = -1 + \frac{9}{16} = -\frac{7}{16}$

Step 4. $(r + \frac{3}{4})^2 = -\frac{7}{16}$

$$r + \frac{3}{4} = -\sqrt{\frac{-7}{16}} \qquad\qquad r + \frac{3}{4} = \sqrt{\frac{-7}{16}}$$

$$r + \frac{3}{4} = -\frac{\sqrt{-7}}{4} \qquad\qquad r + \frac{3}{4} = \frac{\sqrt{-7}}{4}$$

$$r = -\frac{3}{4} - \frac{\sqrt{-7}}{4} \qquad\qquad r = -\frac{3}{4} + \frac{\sqrt{-7}}{4}$$

$$r = -\frac{3}{4} - \frac{\sqrt{7}}{4}i \qquad\qquad r = -\frac{3}{4} + \frac{\sqrt{7}}{4}i$$

Therefore, the solutions are

$$-\frac{3}{4} - \frac{\sqrt{7}}{4}i \qquad \text{and} \qquad -\frac{3}{4} + \frac{\sqrt{7}}{4}i.$$

PROBLEM SET 6.2

In problems 1–20, solve each equation by the roots extraction method.

1. $x^2 = 64$
2. $9x^2 = 4$
3. $2y^2 = 18$
4. $3t^2 = 75$
5. $4z^2 = 60$
6. $5u^2 = 75$
7. $(x - 1)^2 = 9$
8. $(y + 2)^2 = 25$
9. $(2m + 1)^2 = 36$
10. $(2b - 5)^2 = 11$
11. $(6x - 5)^2 - 4 = 0$
12. $(8x - 3)^2 - 49 = 0$
13. $(4y - 3)^2 - 16 = 0$
14. $(3x - 2)^2 - 81 = 0$
15. $w^2 + 81 = 0$
16. $4y^2 + 49 = 0$
17. $(u - 6)^2 + 25 = 0$
18. $(t - 3)^2 + 8 = 0$
19. $(3x - 2)^2 + 49 = 0$
20. $(3y - 7)^2 + 4 = 0$

In problems 21–30, complete the square for each expression and write the result as the square of a binomial.

21. $x^2 + 6x$
22. $t^2 + 8t$
23. $y^2 - 10y$
24. $z^2 + 12z$
25. $m^2 + 20m$
26. $b^2 + 18b$
27. $x^2 + 7x$
28. $t^2 + 11t$
29. $c^2 + 17c$
30. $y^2 + 9y$

In problems 31–50, solve each equation by completing the square.

31. $x^2 + 2x - 3 = 0$

32. $x^2 + 4x - 7 = 0$

33. $y^2 - 12y - 17 = 0$

34. $t^2 - 6t - 11 = 0$

35. $3m^2 - 12m - 3 = 0$

36. $3u^2 - 18u - 7 = 0$

37. $5x^2 - 10x - 1 = 0$

38. $4y^2 - 8y + 3 = 0$

39. $25y^2 - 25y - 14 = 0$

40. $5z^2 + 10z - 3 = 0$

41. $5m^2 - 8m + 17 = 0$

42. $9x^2 - 27x + 14 = 0$

43. $4y^2 + 7y + 5 = 0$

44. $2p^2 + 7p + 3 = 0$

45. $3t^2 - 8t + 4 = 0$

46. $7m^2 + 5m + 1 = 0$

47. $9x^2 - 6x - 1 = 0$

48. $25x^2 - 50x + 21 = 0$

49. $16y^2 - 24y = -5$

50. $9t^2 - 30t = -21$

In problems 51–54, solve each equation for the indicated unknown.

51. $ax^2 + bx + 3b = 0$, for x.

52. $a^2y^2 - 3by + c^2 = 0$, for y.

53. $-gt^2 + vt = s$, for t.

54. $b^2t^2 - 7bt - 3a^2 = 0$, for t.

6.3 Using the Quadratic Formula to Solve Quadratic Equations

The method of completing the square can be used to solve any quadratic equation. However, it is often more efficient to use what is called the **quadratic formula.** To derive this formula, we use the method of completing the square to solve, for x, the general quadratic equation

$$ax^2 + bx + c = 0$$

in which a, b, and c are constants and $a \neq 0$. We follow the procedure outlined in Section 6.2.

Step 1. We subtract c from each side of the equation:

$$ax^2 + bx = -c.$$

Step 2. We divide both sides by a:

$$x^2 + \frac{b}{a}x = -\frac{c}{a}.$$

Step 3. We complete the square of the left side by adding

$$\left[\frac{1}{2}\left(\frac{b}{a}\right)\right]^2 = \frac{b^2}{4a^2}$$

to both sides:

$$x^2 + \frac{b}{a}x + \frac{b^2}{4a^2} = \frac{b^2}{4a^2} - \frac{c}{a} = \frac{b^2 - 4ac}{4a^2}.$$

Step 4. We rewrite the equation as

$$\left(x + \frac{b}{2a}\right)^2 = \frac{b^2 - 4ac}{4a^2}.$$

We solve the equation by roots extraction:

$$x + \frac{b}{2a} = \sqrt{\frac{b^2 - 4ac}{4a^2}} \qquad\qquad x + \frac{b}{2a} = -\sqrt{\frac{b^2 - 4ac}{4a^2}}$$

$$x + \frac{b}{2a} = \frac{\sqrt{b^2 - 4ac}}{2a} \qquad\qquad x + \frac{b}{2a} = -\frac{\sqrt{b^2 - 4ac}}{2a}$$

$$x = -\frac{b}{2a} + \frac{\sqrt{b^2 - 4ac}}{2a} \qquad\qquad x = -\frac{b}{2a} - \frac{\sqrt{b^2 - 4ac}}{2a}$$

$$x = \frac{-b + \sqrt{b^2 - 4ac}}{2a} \qquad\qquad x = \frac{-b - \sqrt{b^2 - 4ac}}{2a}$$

These two solutions are usually written in the compact form shown in the following statement of the quadratic formula. The symbol $\pm$ means $+$ or $-$.

The Quadratic Formula

If $ax^2 + bx + c = 0$ and $a \neq 0$, then

$$x = \frac{-b \pm \sqrt{b^2 - 4ac}}{2a}.$$

Note that if $b^2 - 4ac$ is negative, the symbol $\sqrt{b^2 - 4ac}$ represents a complex number.

In Examples 1–4, use the quadratic formula to solve each equation.

EXAMPLE 1 $3x^2 + 4x - 4 = 0$

SOLUTION $a = 3$, $b = 4$, and $c = -4$. We substitute these values in the quadratic formula

$$x = \frac{-b \pm \sqrt{b^2 - 4ac}}{2a}$$

and we obtain

$$x = \frac{-4 \pm \sqrt{4^2 - 4(3)(-4)}}{2(3)} = \frac{-4 \pm \sqrt{64}}{6}.$$

Thus,

$$x = \frac{-4 + 8}{6} \qquad \qquad x = \frac{-4 - 8}{6}$$

$$x = \frac{4}{6} = \frac{2}{3} \qquad \qquad x = \frac{-12}{6} = -2$$

Therefore, the solutions are $\frac{2}{3}$ and -2.

EXAMPLE **2** $2y^2 - 6y + 3 = 0$

SOLUTION $a = 2$, $b = -6$, and $c = 3$. We substitute these values in the quadratic formula

$$y = \frac{-b \pm \sqrt{b^2 - 4ac}}{2a}$$

so that

$$y = \frac{-(-6) \pm \sqrt{(-6)^2 - 4(2)(3)}}{2(2)} = \frac{6 \pm \sqrt{12}}{4} = \frac{6 \pm 2\sqrt{3}}{4} = \frac{3 \pm \sqrt{3}}{2}.$$

Therefore, the solutions are

$$\frac{3 + \sqrt{3}}{2} \qquad \text{and} \qquad \frac{3 - \sqrt{3}}{2}.$$

EXAMPLE **3** $t^2 + 2t + 2 = 0$

SOLUTION $a = 1$, $b = 2$, and $c = 2$. We substitute these values in the quadratic formula

$$t = \frac{-b \pm \sqrt{b^2 - 4ac}}{2a}$$

and

$$t = \frac{-2 \pm \sqrt{2^2 - 4(1)(2)}}{2(1)}$$

$$= \frac{-2 \pm \sqrt{-4}}{2}$$

$$= \frac{-2 \pm 2i}{2} \qquad \qquad \text{(since } \sqrt{-4} = \sqrt{4}i = 2i\text{)}$$

$$= -1 \pm i.$$

Therefore, the solutions are $-1 + i$ and $-1 - i$.

EXAMPLE 4   ☐ Use the quadratic formula and a calculator to find approximate solutions of the equation $2.41x^2 - 12.3x - 8.39 = 0$. Round off the results to two decimal places.

SOLUTION $a = 2.41$, $b = -12.3$, and $c = -8.39$. We substitute these values in the quadratic formula:

$$x = \frac{-(-12.3) \pm \sqrt{(-12.3)^2 - 4(2.41)(-8.39)}}{2(2.41)}$$

$$= \frac{12.3 \pm \sqrt{232.1696}}{4.82}$$

$$= \frac{12.3 \pm 15.24}{4.82}.$$

Thus,

$$x = \frac{12.3 + 15.24}{4.82} \qquad x = \frac{12.3 - 15.24}{4.82}$$

$$x = 5.71 \qquad x = -0.61$$

Therefore, the approximate solutions are 5.71 and -0.61.

The Quadratic Discriminant

The expression $b^2 - 4ac$, which appears under the radical sign in the quadratic formula

$$x = \frac{-b \pm \sqrt{b^2 - 4ac}}{2a},$$

is called the **discriminant** of the quadratic equation $ax^2 + bx + c = 0$. We can use the algebraic sign of the discriminant to determine the number and the kind of solutions to a quadratic equation:

Case 1. If $b^2 - 4ac > 0$, the quadratic equation has two real and unequal solutions.

Case 2. If $b^2 - 4ac = 0$, the quadratic equation has only one real solution, a *double* solution.

Case 3. If $b^2 - 4ac < 0$, the quadratic equation has two complex solutions.

In Examples 5–7, use the discriminant to determine the number and the kind of solutions to each quadratic equation.

EXAMPLE 5 $2x^2 - 4x + 1 = 0$

SOLUTION $a = 2, b = -4$, and $c = 1$. Thus,

$$b^2 - 4ac = (-4)^2 - 4(2)(1) = 8 > 0.$$

Therefore, the equation has two unequal real solutions.

EXAMPLE 6 $4x^2 - 28x + 49 = 0$

SOLUTION $a = 4, b = -28$, and $c = 49$. Thus,

$$b^2 - 4ac = (-28)^2 - 4(4)(49) = 0.$$

Therefore, the equation has just one solution—a double solution—and this solution is a real number.

EXAMPLE 7 $2x^2 + 3x + 2 = 0$

SOLUTION $a = 2, b = 3$, and $c = 2$. Thus,

$$b^2 - 4ac = 3^2 - 4(2)(2) = 9 - 16 = -7 < 0.$$

Therefore, the equation has two complex solutions.

PROBLEM SET 6.3

In problems 1–32, use the quadratic formula to solve each equation. In problems 27–32, round off the answers to two decimal places.

1. $x^2 - 5x + 4 = 0$ **2.** $x^2 - 2x - 3 = 0$ **3.** $6y^2 - y - 1 = 0$

4. $4t^2 - 8t + 3 = 0$ **5.** $2u^2 - 5u - 3 = 0$ **6.** $4z^2 - 4z - 9 = 0$

7. $6y^2 - 7y - 5 = 0$ **8.** $6p^2 + 3p - 5 = 0$ **9.** $6t^2 - 8t - 3 = 0$

10. $6z^2 - 7z - 2 = 0$ **11.** $2x^2 - 5x + 1 = 0$ **12.** $6y^2 + 5y + 2 = 0$

13. $5y^2 - 2y + 7 = 0$ **14.** $9m^2 - 2m + 11 = 0$ **15.** $7x^2 - 8x + 3 = 0$

16. $12p^2 - 4p + 3 = 0$ **17.** $15y^2 + 2y - 8 = 0$ **18.** $4x^2 + 11x - 3 = 0$

19. $3x^2 - x + 7 = 0$ **20.** $6x^2 + 17x - 14 = 0$ **21.** $4y(y - 1) = 19$

22. $t(t + 8) + 8(t - 8) = 0$ **23.** $(t - 2)(t - 3) = 5$ **24.** $(3y - 1)(2y + 5) = 3$

25. $(y - 1)(y - 5) = 9$ **26.** $(t + 1)(2t - 1) = 4$ C **27.** $0.8w^2 - 0.16w - 1.91 = 0$

C **28.** $0.17u^2 - 0.55u - 3.87 = 0$ C **29.** $1.47y^2 - 3.82y - 5.71 = 0$ C **30.** $2.81t^2 - 7.14t - 31.72 = 0$

C **31.** $1.32x^2 + 2.78x - 9.31 = 0$ C **32.** $8.84x^2 - 71.41x - 94.03 = 0$

In problems 33–36, use the quadratic formula to solve each equation for the indicated unknown.

33. $nx^2 + mnx - m^2 = 0$, for x.

34. $2ay^2 - 7ay - 4ab^2 = 0$, for y.

35. $LI^2 + RI + \dfrac{1}{C} = 0$, for I.

36. $Mx^2 + 2Rx + K = 0$, for x.

In problems 37–46, use the discriminant to determine the number and the kind of solutions to each equation.

37. $x^2 + 6x - 7 = 0$

38. $6x^2 - x + 1 = 0$

39. $4y^2 + 12y + 9 = 0$

40. $16t^2 - 40t + 25 = 0$

41. $t^2 + 3t + 5 = 0$

42. $2u^2 + u + 5 = 0$

43. $3x^2 - 7x + 4 = 0$

44. $6y^2 + y - 2 = 0$

45. $9z^2 + 30z + 25 = 0$

46. $20x^2 - 11x + 3 = 0$

6.4 Applications of Quadratic Equations

In this section, we see how quadratic equations can be applied in a wide variety of fields. We use the procedure for solving linear equations given in Chapter 4 (page 131) to solve word problems that give rise to quadratic equations. When the solutions to "real-world" problems involve quadratic equations, it is important to check both solutions in terms of the original problem to determine if we must reject either solution.

EXAMPLE 1   Find two consecutive odd positive integers whose product is 195.

SOLUTION   Let

$$x = \text{the first odd integer.}$$

Then

$$x + 2 = \text{the next consecutive odd integer.}$$

The product of the two numbers is 195. Therefore, an equation is

$$x(x + 2) = 195$$

$$x^2 + 2x = 195 \qquad \text{or} \qquad x^2 + 2x - 195 = 0.$$

We solve this equation by factoring:

$$(x - 13)(x + 15) = 0.$$

$$
\begin{array}{c|c}
x - 13 = 0 & x + 15 = 0 \\
x = 13 & x = -15
\end{array}
$$

We reject the solution -15 because it is not a positive integer. Therefore, the numbers are 13 and $13 + 2 = 15$.

Check 13 and 15 are consecutive odd positive integers and $13 \cdot 15 = 195$.

EXAMPLE 2   A rectangular garden is 5 meters wide and 7 meters long. If both the width and the length of the garden are increased by the same amount, the area is increased by 28 square meters. By how much are the length and the width increased?

SOLUTION   Let

x = the number of meters each side of the garden is increased by (Figure 1).

Figure 1

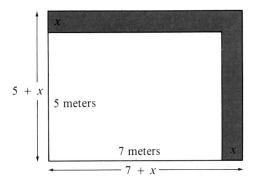

The new dimensions of the garden are $7 + x$ and $5 + x$. The new area is $(7 + x)(5 + x)$. Therefore, an equation is:

$$(7 + x)(5 + x) = 35 + 28 = 63$$

$$35 + 12x + x^2 = 63 \quad \text{or} \quad x^2 + 12x - 28 = 0.$$

We solve this equation by factoring:

$$(x - 2)(x + 14) = 0.$$

$$
\begin{array}{c|c}
x - 2 = 0 & x + 14 = 0 \\
x = 2 & x = -14
\end{array}
$$

Since an increase in size must be positive, 2 is the only solution we can use. Therefore, the width and the length of the garden each increase by 2 meters.

Check $5 \cdot 7 = 35$ and $7 \cdot 9 = 63$. The increase is $63 - 35 = 28$ square meters.

We learned in geometry that the Pythagorean theorem expresses the following property for a right triangle (Figure 2):

$$c^2 = a^2 + b^2.$$

Figure 2

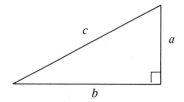

Use this theorem to solve the following examples.

EXAMPLE **3**   In order to support a solar collector at the correct angle, the roof trusses of a house are designed as right triangles. Rafters form the right angle, and the base of the truss is the hypotenuse (Figure 3). If the rafter on the same side as the solar collector is 10 feet shorter than the other rafter, and if the base of each truss is 50 feet long, how long is each of the rafters?

Figure 3

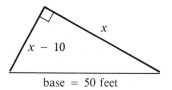

base = 50 feet

SOLUTION   Let

$$x = \text{the length of the longer rafter in feet.}$$

Then

$$x - 10 = \text{the length of the rafter on the side of the collector.}$$

We use the Pythagorean theorem:

$$x^2 + (x - 10)^2 = 50^2$$
$$x^2 + x^2 - 20x + 100 = 2{,}500$$
$$2x^2 - 20x - 2{,}400 = 0$$
$$x^2 - 10x - 1{,}200 = 0$$
$$(x + 30)(x - 40) = 0.$$

$$x + 30 = 0 \qquad \bigg| \qquad x - 40 = 0$$
$$x = -30 \qquad \bigg| \qquad x = 40$$

Because the length must be positive, the lengths of the rafters are 40 feet and $40 - 10 = 30$ feet.

EXAMPLE 4   A rocket is fired straight up from the ground at a time $t = 0$ with an initial speed of 560 feet per second. The rocket's distance of h feet above the ground t seconds later is given by the equation

$$h = 560t - 16t^2.$$

At what time will the rocket be 3,136 feet above the ground?

SOLUTION   When the rocket is 3,136 feet above the ground, $h = 3,136$, and the equation becomes

$$3,136 = 560t - 16t^2$$

or

$$16t^2 - 560t + 3,136 = 0$$

$$t^2 - 35t + 196 = 0.$$

We solve this equation by factoring:

$$(t - 28)(t - 7) = 0.$$

$$t - 28 = 0 \qquad \mid \qquad t - 7 = 0$$

$$t = 28 \qquad \mid \qquad t = 7$$

Both solutions make sense, because the rocket can be at a given height at two different times: once on its way up and once on its way down. Therefore, the rocket will be 3,136 feet above the ground after 7 seconds and after 28 seconds.

Check At $t = 7$,

$$560(7) - 16(7)^2 = 3,136.$$

At $t = 28$,

$$560(28) - 16(28)^2 = 3,136.$$

EXAMPLE 5   Carlos and Pedro both ran 26 miles in the Boston Marathon. Carlos averaged 2 miles per hour less than Pedro. If Carlos finished 26 minutes behind Pedro, what was Pedro's average speed?

SOLUTION   Let

$$x = \text{Pedro's average speed in miles per hour.}$$

Then

$$x - 2 = \text{Carlos's average speed in miles per hour.}$$

We use the formula $d = rt$ and solve for t:

$$t = \frac{d}{r}.$$

The following table summarizes the given information:

	Average Speed (in miles per hour)	Distance (in miles)	Time (in hours)
Pedro	x	26	$\dfrac{26}{x}$
Carlos	$x - 2$	26	$\dfrac{26}{x - 2}$

We are told that

$$\text{Carlos's time} - \text{Pedro's time} = 26 \text{ minutes} = \frac{26}{60} \text{ of an hour.}$$

The following equation expresses this relationship:

$$\frac{26}{x - 2} - \frac{26}{x} = \frac{26}{60}$$

or

$$\frac{1}{x - 2} - \frac{1}{x} = \frac{1}{60}.$$

We multiply both sides of the equation by $60x(x - 2)$, the LCD, and we solve the resulting equation by factoring:

$$60x(x - 2)\left(\frac{1}{x - 2} - \frac{1}{x}\right) = 60x(x - 2)\left(\frac{1}{60}\right)$$

$$60x - 60(x - 2) = x(x - 2)$$

$$x^2 - 2x - 120 = 0$$

$$(x - 12)(x + 10) = 0.$$

$$x - 12 = 0 \qquad \bigg| \qquad x + 10 = 0$$

$$x = 12 \qquad \bigg| \qquad x = -10$$

We reject $x = -10$ because speed cannot be negative. Therefore, Pedro's average speed is 12 miles per hour.

Check

$$\text{Pedro's time} = \frac{d}{r} = \frac{26}{12} = \frac{13}{6} \text{ hours}$$

$$\text{Carlos's time} = \frac{d}{r} = \frac{26}{10} = \frac{13}{5} \text{ hours}$$

Carlos finishes $\frac{13}{5} - \frac{13}{6} = \frac{13}{30} = \frac{26}{60}$ hour behind Pedro.

PROBLEM SET 6.4

In many of the following problems a calculator may be useful to speed up the arithmetic.

1. Find two consecutive positive integers whose product is 30.

2. Find two consecutive negative integers whose product is 56.

3. Two brothers were born in consecutive years. The product of their present ages is 156. How old are they now?

4. Maria throws two dice. The difference between the face values of the two dice is 2, and the product of their values is 24. What numbers did Maria throw?

5. If five times David's age plus 6 years is the same as the square of his age, how old is David?

6. Tom is 4 years older than Kim, and the product of their ages is three times what the product of their ages was 4 years ago. Find their present ages.

7. What are the dimensions of a rectangle if the area of the rectangle is 60 square feet and the perimeter is 32 feet?

8. A candy manufacturer makes rectangular chocolate bars in such a way that the length of each bar is four times the width. If the length of each bar is increased by 3 centimeters and the width is increased by 1 centimeter, the new area of each bar becomes 33 square centimeters. What were the original dimensions of the candy bar?

9. The area of a rectangular garden is 1,320 square feet. If the perimeter is 148 feet, what are the dimensions of the garden?

10. If increasing the lengths of the sides of a square by 8 feet results in a square whose area is 25 times the area of the original square, what was the length of a side of the original square?

11. A skating rink is 100 feet long and 70 feet wide. If we want to increase the area of the rink to 13,000 square feet by adding rectangular strips of equal width to one side and one end, maintaining the rectangular shape of the rink, how wide should these strips be?

12. A rectangular garden that is 4 meters wide and 6 meters long is surrounded by a path of uniform width. If the area of the path equals the area of the garden, what is the width of the path?

13. A lawn 30 feet long and 20 feet wide has a flower border of uniform width around it. If the area of the lawn and the area of the border are the same, find the width of the border.

14. A box without a top is to be constructed from a square piece of tin by removing a 3-inch square from each corner of the square, and bending up the sides. If the box is to have a volume of 300 cubic inches, what is the length of the side of the original tin square?

15. In order to support a solar collector at the correct angle, the roof trusses of a building are designed as right triangles. Rafters form the right angle, and the base of the truss is the hypotenuse (see Figure 3 on page 220). If the rafter on the same side as the solar collector is 7 meters shorter than the other rafter, and if the base of each truss is 13 meters long, how long is each of the rafters?

16. Two bicyclists, each moving at a uniform speed, leave the same point and travel at right angles to each other. One travels 5 kilometers per hour faster than the other. In 2 hours the bicyclists are 50 kilometers apart. How fast does each bicyclist travel?

17. Two joggers leave the same point and travel at right angles to each other, each jogging at a uniform speed. One jogs 7 miles per hour faster than the other. After 1 hour they are 13 miles apart. Find the speed of each jogger.

18. Two planes leave the same point and fly at right angles to each other. After 2 hours they are 1,000 miles apart. If one plane flies 100 miles per hour faster than the other, and each is moving at a uniform speed, how fast are the planes flying?

19. A ball thrown straight upward from ground level with an initial velocity of 128 feet per second will be at a height $h = 128t - 16t^2$, t seconds later (neglecting air resistance). How long will it take the ball to reach a height of 112 feet?

20. A rocket is fired upward from the ground at an initial velocity of 120 meters per second. The rocket's height h in meters above the ground t seconds later is given by the equation $h = 120t - 16t^2$. In how many seconds will the rocket reach a height of 176 meters?

21. A rocket is fired upward from the ground at an initial velocity of 480 feet per second. Its height h in feet above the ground t seconds later is $h = 480t - 16t^2$. In how many seconds will the rocket reach a height of 2,000 feet?

22. Howard leaves a ski lodge and accelerates as he skis downhill. After t seconds he has traveled a distance of h meters—h is given by the equation $h = 10t^2 + 10t$. How long does it take him to go 560 meters downhill?

23. In a 300-mile car race, Mario averages 20 miles per hour less than his competitor Andy. If Mario finishes 30 minutes behind Andy, what is Andy's average speed?

24. A carpenter, working alone, can finish a job in 2 hours less time than his assistant. They can do the job in 7 hours working together. Find the time that each requires to do the job alone.

25. Susan and June, working together, can type a chapter in a book in 8 hours. Typing alone, it would take June 12 hours longer than it would take Susan to type the chapter. How long would it take Susan to type the chapter by herself?

26. A motorist travels for 20 miles at one speed, and then increases her speed by 20 miles per hour for the next 30 miles. How fast was she traveling originally if the total trip took 1 hour?

27. A company determines that its total monthly revenue (in dollars) is given by the equation $R = 200x - \frac{1}{3}x^2$, in which x is the price (in dollars) of each unit. At what price per unit will the revenue be \$10,800, if the price per unit must be greater than \$60?

28. A cable television company plans to begin operations in a small town. The company foresees that 1,000 people will subscribe to the service if the price per subscriber is \$3 per month, but that for each 10-cent decrease in the monthly subscription price 50 more people will subscribe. The company begins operations, and its total revenue for the first month is \$3,125. How much did the company charge each subscriber?

6.5 Equations in Quadratic Form

An equation that is not quadratic may have the following characteristics: If you substitute a letter (for example, u) for an appropriate expression containing the unknown, the resulting equation is a quadratic equation in u. The original equation is called *an equation that is quadratic in form*. Examples of such equations are:

$$3x^4 + 2x^2 - 1 = 0 \quad (u = x^2),$$

$$(x^2 - 1)^2 + 2(x^2 - 1) - 3 = 0 \quad (u = x^2 - 1),$$

and

$$t^2 + t + \frac{12}{t^2 + t} = 8 \quad (u = t^2 + t).$$

To solve an equation that is quadratic in form, we make a substitution that transforms the given equation into a quadratic equation in another variable, such as u.

EXAMPLE 1   *In Examples 1–4, solve each equation.*

SOLUTION $x^4 - 5x^2 + 4 = 0$

First we write $x^4 - 5x^2 + 4 = 0$ as

$$(x^2)^2 - 5x^2 + 4 = 0.$$

If we let $u = x^2$, we have

$$u^2 - 5u + 4 = 0.$$

We solve this equation by factoring:

$$(u - 1)(u - 4) = 0.$$

$$\begin{array}{c|c} u - 1 = 0 & u - 4 = 0 \\ u = 1 & u = 4 \end{array}$$

If $u = 1$, then $x^2 = 1$; therefore,

$$\begin{array}{c|c} x = -\sqrt{1} & x = \sqrt{1} \\ x = -1 & x = 1 \end{array}$$

If $u = 4$, then $x^2 = 4$; therefore,

$$\begin{array}{c|c} x = -\sqrt{4} & x = \sqrt{4} \\ x = -2 & x = 2 \end{array}$$

Thus, -1, 1, -2, and 2 are the solutions.

EXAMPLE 2 $t^{-2} + t^{-1} - 6 = 0$

SOLUTION   First we write $t^{-2} + t^{-1} - 6 = 0$ as

$$(t^{-1})^2 + t^{-1} - 6 = 0.$$

If we let $u = t^{-1}$, we have

$$u^2 + u - 6 = 0.$$

We solve this equation by factoring:

$$(u + 3)(u - 2) = 0.$$

$$u + 3 = 0 \qquad\qquad u - 2 = 0$$
$$u = -3 \qquad\qquad u = 2$$

If $u = -3$, then $t^{-1} = -3$; therefore,

$$\frac{1}{t} = -3 \quad \text{or} \quad t = -\frac{1}{3}.$$

If $u = 2$, then $t^{-1} = 2$; therefore,

$$\frac{1}{t} = 2 \quad \text{or} \quad t = \frac{1}{2}.$$

Thus, $-\frac{1}{3}$ and $\frac{1}{2}$ are the solutions.

EXAMPLE 3 $3(y + 4)^2 - (y + 4) - 14 = 0$

SOLUTION   If we substitute u for $y + 4$, the equation becomes

$$3u^2 - u - 14 = 0.$$

We solve this equation by factoring:

$$(3u - 7)(u + 2) = 0.$$

$$3u - 7 = 0 \qquad\qquad u + 2 = 0$$
$$3u = 7 \qquad\qquad u = -2$$
$$u = \frac{7}{3}$$

If $u = \frac{7}{3}$, then $y + 4 = \frac{7}{3}$ and $y = -\frac{5}{3}$.
If $u = -2$, then $y + 4 = -2$ and $y = -6$.
Thus, $-\frac{5}{3}$ and -6 are the solutions.

EXAMPLE 4 $\left(r - \dfrac{8}{r}\right)^2 + \left(r - \dfrac{8}{r}\right) = 42$

SOLUTION   If we substitute u for $r - 8/r$, the equation becomes

$$u^2 + u = 42 \quad \text{or} \quad u^2 + u - 42 = 0.$$

We solve this equation by factoring:

$$(u + 7)(u - 6) = 0.$$

$$u + 7 = 0 \quad \bigg| \quad u - 6 = 0$$

$$u = -7 \quad \bigg| \quad u = 6$$

If $u = -7$, then

$$r - \frac{8}{r} = -7$$

$$r^2 - 8 = -7r$$

$$r^2 + 7r - 8 = 0.$$

We solve this equation by factoring:

$$(r + 8)(r - 1) = 0.$$

$$r + 8 = 0 \quad \bigg| \quad r - 1 = 0$$

$$r = -8 \quad \bigg| \quad r = 1$$

If $u = 6$, then

$$r - \frac{8}{r} = 6$$

$$r^2 - 6r - 8 = 0.$$

We solve this equation by using the quadratic formula:

$$r = \frac{6 \pm \sqrt{36 - 4(1)(-8)}}{2} = \frac{6 \pm \sqrt{68}}{2} = 3 \pm \sqrt{17}.$$

Thus, $-8, 1, 3 + \sqrt{17}$, and $3 - \sqrt{17}$ are the solutions.

Since the original equation contains a variable in a denominator, you should check the four proposed solutions.

PROBLEM SET 6.5

In problems 1–32, reduce each equation to a quadratic form by using an appropriate substitution, and solve the equation for the original variable.

1. $x^4 - 13x^2 + 36 = 0$, $u = x^2$

2. $y^4 - 17y^2 + 16 = 0$, $u = y^2$

3. $t^4 - 3t^2 - 4 = 0$

4. $z^4 - 10z^2 + 9 = 0$

5. $y^4 - 29y^2 + 100 = 0$

6. $9t^4 - 226t^2 + 25 = 0$

7. $t^{-2} - 2t^{-1} - 8 = 0, u = t^{-1}$

8. $6x^{-2} + 5x^{-1} - 4 = 0, u = x^{-1}$

9. $6y^{-2} + 13y^{-1} - 5 = 0$

10. $28p^{-2} - 17p^{-1} - 3 = 0$

11. $x^{-4} - 9x^{-2} + 20 = 0, u = x^{-2}$

12. $y^{-4} + 63y^{-2} - 64 = 0, u = y^{-2}$

13. $t^{-4} - 20t^{-2} + 64 = 0$

14. $w^{-4} - 34w^{-2} + 225 = 0$

15. $(x^2 + 1)^2 - 3(x^2 + 1) + 2 = 0, u = x^2 + 1$

16. $(2t^2 + 7t)^2 - 3(2t^2 + 7t) = 10, u = 2t^2 + 7t$

17. $(y^2 + 2y)^2 - 2(y^2 + 2y) = 3$

18. $(w^2 - w)^2 + 12 = 8(w^2 - w)$

19. $(w^2 + 2w)^2 - 14(w^2 + 2w) = 15$

20. $(2y^2 - y)^2 - 16(2y^2 - y) + 60 = 0$

21. $(m^2 + 2m)^2 + m^2 + 2m - 12 = 0$

22. $3(x + 1)^2 + 2(x + 1) = 2$

23. $(2p^2 + p + 4)^2 - 4(2p^2 + p + 4) = 5$

24. $(2w^2 - 3w)^2 - 2(2w^2 - 3w) = 3$

25. $\left(3x - \dfrac{2}{x}\right)^2 + 6\left(3x - \dfrac{2}{x}\right) + 5 = 0, u = 3x - \dfrac{2}{x}$

26. $\left(y - \dfrac{5}{y}\right)^2 - 2y + \dfrac{10}{y} = 8, u = y - \dfrac{5}{y}$

27. $t^2 + t + \dfrac{36}{t^2 + t} = 15$

28. $\dfrac{w^2 + 1}{w} + \dfrac{4w}{w^2 + 1} - 4 = 0$

29. $\dfrac{m + 1}{m} + 2 = \dfrac{3m}{m + 1}$

30. $\dfrac{p^2}{p + 1} + \dfrac{2(p + 1)}{p^2} = 3$

31. $\left(2x + \dfrac{1}{x}\right)^2 + 5\left(2x + \dfrac{1}{x}\right) + 6 = 0$

32. $\left(y + \dfrac{3}{y}\right)^2 - 2y - \dfrac{6}{y} = 15$

6.6 Equations Involving Radicals

The following equations are examples of **radical equations:**

$$\sqrt{x} = 3, \qquad \sqrt{5t + 1} = 4, \qquad \sqrt[3]{5y + 2} = 3, \qquad \text{and} \qquad u - \sqrt{u - 2} = 4.$$

To solve a radical equation, we use the following procedure:

Procedure for Solving a Radical Equation

> Step 1. Isolate one of the radical expressions containing the variable on one side of the equation.
>
> Step 2. Eliminate the radical by raising both sides of the equation to a power equal to the index of the radical. [Recall that $(\sqrt[n]{a})^n = a$.] You may have to repeat this technique in order to eliminate all radicals. When the equation is free of radicals, simplify and solve the equation.
>
> Step 3. Check all solutions in the original equation whenever a radical with an *even* index is involved.

In Examples 1–6, solve each equation.

EXAMPLE 1 $\sqrt{2x + 5} - 3 = 0$

SOLUTION

Step 1. We isolate the radical expression on one side of the equation by adding 3 to both sides:

$$\sqrt{2x + 5} = 3.$$

Step 2. We eliminate the radical by squaring both sides of the equation:

$$(\sqrt{2x + 5})^2 = 3^2$$
$$2x + 5 = 9$$
$$2x = 4$$
$$x = 2.$$

Step 3. *Check* For $x = 2$,

$$\sqrt{2(2) + 5} - 3 \overset{?}{=} 0$$
$$\sqrt{2(2) + 5} - 3 = \sqrt{4 + 5} - 3$$
$$= \sqrt{9} - 3 = 0.$$

Hence, 2 is the solution.

EXAMPLE 2 $\sqrt[3]{3t - 1} - 2 = 0$

SOLUTION

Step 1. We isolate the radical expression by adding 2 to both sides of the equation:

$$\sqrt[3]{3t - 1} = 2.$$

Step 2. We eliminate the radical by raising both sides of the equation to the third power:

$$(\sqrt[3]{3t - 1})^3 = 2^3$$
$$3t - 1 = 8$$
$$3t = 9$$
$$t = 3.$$

Step 3. We do not have to check the solution because the index is odd. Hence, 3 is the solution.

EXAMPLE **3** $\sqrt[4]{y^2 - 5y + 6} = \sqrt{y - 2}$

SOLUTION   We raise both sides of the equation to the power 4:

$$(\sqrt[4]{y^2 - 5y + 6})^4 = (\sqrt{y - 2})^4$$
$$y^2 - 5y + 6 = (y - 2)^2$$
$$y^2 - 5y + 6 = y^2 - 4y + 4$$
$$-5y + 4y = 4 - 6$$
$$-y = -2$$
$$y = 2.$$

Check For $y = 2$,

$$\sqrt[4]{4 - 10 + 6} \overset{?}{=} \sqrt{2 - 2}$$
$$\sqrt[4]{0} \overset{?}{=} \sqrt{0}$$
$$0 = 0.$$

Therefore, 2 is the solution.

EXAMPLE **4** $\sqrt{4w^2 - 3} = 2w + 1$

SOLUTION   We square both sides of the equation:

$$(\sqrt{4w^2 - 3})^2 = (2w + 1)^2$$
$$4w^2 - 3 = 4w^2 + 4w + 1$$
$$-4w = 1 + 3$$
$$-4w = 4$$
$$w = -1.$$

Check For $w = -1$,

$$\sqrt{4(-1)^2 - 3} \overset{?}{=} 2(-1) + 1$$
$$\sqrt{4 - 3} \overset{?}{=} -2 + 1$$
$$1 \neq -1.$$

The solution, $w = -1$, does not satisfy the original equation. Therefore, we say that -1 is an **extraneous** solution. The original equation has no solutions.

EXAMPLE **5** $\sqrt{z - 2} + 4 = z$

SOLUTION   We follow the procedure on page 228:

$$\sqrt{z - 2} = z - 4$$
$$(\sqrt{z - 2})^2 = (z - 4)^2$$
$$z - 2 = z^2 - 8z + 16$$
$$z^2 - 9z + 18 = 0$$
$$(z - 3)(z - 6) = 0.$$

$$z - 3 = 0 \quad \big| \quad z - 6 = 0$$
$$z = 3 \quad \big| \quad z = 6$$

Check For $z = 3$,

$$\sqrt{3 - 2} + 4 \overset{?}{=} 3$$
$$\sqrt{3 - 2} + 4 = 1 + 4 = 5 \neq 3.$$

For $z = 6$,

$$\sqrt{6 - 2} + 4 \overset{?}{=} 6$$
$$\sqrt{6 - 2} + 4 = \sqrt{4} + 4 = 2 + 4 = 6.$$

Therefore, 3 is an extraneous solution. The only solution that satisfies the original equation is 6.

EXAMPLE 6 $\sqrt{1 - 5x} + \sqrt{1 - x} = 2$

SOLUTION   We add $-\sqrt{1 - x}$ to both sides of the equation to isolate $\sqrt{1 - 5x}$:

$$\sqrt{1 - 5x} = 2 - \sqrt{1 - x}.$$

We square both sides of the equation:

$$(\sqrt{1 - 5x})^2 = (2 - \sqrt{1 - x})^2$$
$$1 - 5x = 4 - 4\sqrt{1 - x} + 1 - x.$$

The equation still contains a radical, so we simplify and isolate this radical:

$$-4 - 4x = -4\sqrt{1 - x}$$
$$1 + x = \sqrt{1 - x}. \qquad \text{(We divided both sides by } -4.)$$

Again, we square both sides of the equation:

$$(1 + x)^2 = (\sqrt{1 - x})^2$$
$$1 + 2x + x^2 = 1 - x$$
$$x^2 + 3x = 0$$
$$x(x + 3) = 0.$$

$$x = 0 \quad \big| \quad x + 3 = 0$$
$$\quad \big| \quad x = -3$$

Check For $x = -3$,

$$\sqrt{1 - 5(-3)} + \sqrt{1 - (-3)} \stackrel{?}{=} 2$$

$$\sqrt{1 - 5(-3)} + \sqrt{1 - (-3)} = \sqrt{16} + \sqrt{4}$$

$$= 4 + 2 = 6 \neq 2.$$

For $x = 0$.

$$\sqrt{1 - 0} + \sqrt{1 - 0} \stackrel{?}{=} 2$$

$$\sqrt{1 - 0} + \sqrt{1 - 0} = 1 + 1 = 2.$$

Therefore, -3 is an extraneous solution and 0 is the only solution.

PROBLEM SET 6.6

In problems 1–26, solve each equation (find only the real solutions) and check the solution whenever the equation has a radical with an even index.

1. $\sqrt{x} - 2 = 3$

2. $\sqrt{x} + 1 = 4$

3. $\sqrt{t + 1} - 2 = 0$

4. $\sqrt{y - 3} - 5 = 0$

5. $\sqrt{2w + 5} - 4 = 0$

6. $\sqrt{6t - 3} - 27 = 0$

7. $8 - \sqrt{y - 1} = 6$

8. $2 + \sqrt{7m - 5} = 6$

9. $\sqrt{11 - x} = \sqrt{x + 6}$

10. $\sqrt{3p + 1} = \sqrt{p + 1}$

11. $\sqrt{q + 14} = \sqrt{5q}$

12. $\sqrt{y + 5} - \sqrt{y} = 0$

13. $\sqrt{x^2 + 3x} = x + 1$

14. $\sqrt{9x^2 - 7} - 3x = 2$

15. $\sqrt[3]{y - 3} = 3$

16. $\sqrt[3]{t + 2} = -2$

17. $\sqrt[3]{3b - 4} = 2$

18. $\sqrt[4]{y - 1} = 2$

19. $\sqrt[4]{y^2 - 7y + 1} = \sqrt{y - 5}$

20. $\sqrt[4]{t^2 + 1} = \sqrt{t + 1}$

21. $5 + \sqrt[4]{x - 5} = 0$

22. $\sqrt[4]{2x - 1} + 3 = 0$

23. $\sqrt[4]{t + 8} = \sqrt{3t}$

24. $\sqrt[3]{x^2 + 2x - 6} = \sqrt[3]{x^2}$

25. $2 + \sqrt[4]{7x - 5} = 6$

26. $8 + \sqrt[4]{x - 1} = 3$

In problems 27–44, solve each equation and check the solution.

27. $\sqrt{t} = \sqrt{t + 16} - 2$

28. $\sqrt{2y^2 + 4} + 2 = 2y$

29. $\sqrt{p + 12} = 2 + \sqrt{p}$

30. $\sqrt{m^2 + 6m} = m + \sqrt{2m}$

31. $\sqrt{y + 5} = \sqrt{y} + 1$

32. $\sqrt{x + 7} = 5 + \sqrt{x - 2}$

33. $\sqrt{5t + 1} = 1 + \sqrt{3t}$

34. $\sqrt{2w + 1} = 1 + 2\sqrt{w}$

35. $\sqrt{2x - 5} = \sqrt{x - 2} + 2$

36. $\sqrt{3 - t} - \sqrt{2 + t} = 3$

37. $\sqrt{t + 2} + \sqrt{t - 3} = 5$

38. $\sqrt{3y + 1} - 1 = \sqrt{3y - 8}$

39. $\sqrt{m + 4} + 1 = \sqrt{m + 11}$

40. $\sqrt{6z + 7} = \sqrt{3z + 3} + 1$

41. $2\sqrt{1 - 3y} = 2 - \sqrt{2 - 4y}$

42. $\sqrt{7x - 6} = \sqrt{7x + 22} - 2$

43. $\sqrt{7 - 4t} - \sqrt{3 - 2t} = 1$

44. $\sqrt{2t} + \sqrt{2t + 12} = 2\sqrt{4t + 1}$

6.7 Equations Involving Rational Exponents

The following equations contain rational exponents:

$$x^{1/6} = 2, \qquad y^{2/3} = 4, \qquad (t - 3)^{2/5} = 1, \qquad \text{and} \qquad z^{2/3} - z^{1/3} - 12 = 0.$$

This type of equation can be solved using slight variations of the methods we have already discussed. Keep in mind that an equation containing rational exponents (in their lowest terms) with an even denominator is equivalent to a radical equation with even indexes. Therefore, you *must* check your solution.

In Examples 1–6, solve each equation.

EXAMPLE **1** $x^{1/6} = 2$

SOLUTION   The equation can be rewritten as

$$\sqrt[6]{x} = 2.$$

We eliminate the radical by raising both sides of the equation to the sixth power:

$$(\sqrt[6]{x})^6 = 2^6$$
$$x = 64.$$

Because the original equation contains a rational exponent (in its lowest terms) with an even denominator, we must check the solution.

Check For $x = 64$.
$$\sqrt[6]{64} = 2.$$

Hence, 64 is the solution.

EXAMPLE **2** $y^{2/3} = 4$

SOLUTION   The equation can be rewritten as

$$\sqrt[3]{y^2} = 4.$$

We raise both sides of the equation to the third power:

$$(\sqrt[3]{y^2})^3 = 4^3$$
$$y^2 = 64$$
$$y = \pm \sqrt{64}$$
$$y = \pm 8.$$

Therefore, -8 and 8 are the solutions.

EXAMPLE 3 $(t - 3)^{2/5} = 1$

SOLUTION   The equation can be rewritten as

$$\sqrt[5]{(t - 3)^2} = 1.$$

We raise both sides of the equation to the fifth power:

$$(\sqrt[5]{(t - 3)^2})^5 = 1^5$$

$$(t - 3)^2 = 1$$

$$t - 3 = \pm\sqrt{1}$$

$$t - 3 = \pm 1.$$

$$t - 3 = -1 \qquad \bigg| \qquad t - 3 = 1$$

$$t = 2 \qquad \bigg| \qquad t = 4$$

Hence, 2 and 4 are the solutions.

EXAMPLE 4 $2(w + 2)^{-2/7} = 1$

SOLUTION   We begin by rewriting the equation as

$$\frac{2}{(w + 2)^{2/7}} = 1 \qquad \text{or}$$

$$(w + 2)^{2/7} = 2.$$

The last equation can also be written as

$$\sqrt[7]{(w + 2)^2} = 2.$$

We raise both sides of the equation to the seventh power:

$$(w + 2)^2 = 2^7$$

$$(w + 2)^2 = 128$$

$$w + 2 = \pm\sqrt{128}.$$

$$w + 2 = -\sqrt{128} \qquad \bigg| \qquad w + 2 = \sqrt{128}$$

$$w + 2 = -8\sqrt{2} \qquad \bigg| \qquad w + 2 = 8\sqrt{2}$$

$$w = -2 - 8\sqrt{2} \qquad \bigg| \qquad w = -2 + 8\sqrt{2}$$

Hence, $-2 - 8\sqrt{2}$ and $-2 + 8\sqrt{2}$ are the solutions.

EXAMPLE 5 $z^{2/3} - z^{1/3} - 12 = 0$

SOLUTION   Let $u = z^{1/3}$ so that $u^2 = (z^{1/3})^2 = z^{2/3}$. Then the equation becomes

$$u^2 - u - 12 = 0$$

$$(u - 4)(u + 3) = 0.$$

$$\begin{array}{c|c} u - 4 = 0 & u + 3 = 0 \\ u = 4 & u = -3 \end{array}$$

Since $u = z^{1/3}$, we have

$$\begin{array}{c|c} z^{1/3} = 4 & z^{1/3} = -3 \\ \sqrt[3]{z} = 4 & \sqrt[3]{z} = -3 \\ z = 4^3 & z = (-3)^3 \\ z = 64 & z = -27 \end{array}$$

Hence, 64 and -27 are the solutions.

EXAMPLE 6 $x + 2 + (x + 2)^{1/2} - 2 = 0$

SOLUTION   Let $u = (x + 2)^{1/2}$, so that $u^2 = [(x + 2)^{1/2}]^2 = x + 2$. The equation becomes

$$u^2 + u - 2 = 0$$

$$(u + 2)(u - 1) = 0.$$

$$\begin{array}{c|c} u + 2 = 0 & u - 1 = 0 \\ u = -2 & u = 1 \end{array}$$

Since $u = (x + 2)^{1/2} = \sqrt{x + 2}$, we have

$$\begin{array}{c|c} \sqrt{x + 2} = -2 & \sqrt{x + 2} = 1 \\ x + 2 = (-2)^2 & x + 2 = 1^2 \\ x + 2 = 4 & x + 2 = 1 \\ x = 2 & x = -1 \end{array}$$

Here we must check the solutions. (Why?)

Check For $x = -1$,

$$[(-1) + 2] + \sqrt{(-1) + 2} - 2 \overset{?}{=} 0$$

$$[(-1) + 2] + \sqrt{(-1) + 2} - 2 = 1 + \sqrt{1} - 2 = 2 - 2 = 0.$$

Thus, $x = -1$, is a solution.

For $x = 2$,

$$(2 + 2) + \sqrt{2 + 2} - 2 \overset{?}{=} 0$$

$$(2 + 2) + \sqrt{2 + 2} - 2 = 4 + 2 - 2 \neq 0.$$

Thus, $x = 2$ is *not* a solution. Therefore, the only solution is -1.

PROBLEM SET 6.7

In problems 1–32, solve each equation (find only the real solutions). Be sure to check the solutions whenever the equation has a rational exponent (in its lowest terms) with an even denominator.

1. $y^{1/5} = -3$

2. $x^{1/6} = \frac{1}{2}$

3. $t^{1/7} = -2$

4. $y^{1/9} = -1$

5. $x^{2/3} = 16$

6. $m^{3/5} = 8$

7. $u^{4/7} = 16$

8. $t^{3/7} = -8$

9. $(x + 1)^{3/5} = 1$

10. $(t - 1)^{2/3} = 4$

11. $(z - 1)^{5/2} = 32$

12. $(x - 3)^{3/2} = 8$

13. $(3y - 7)^{4/3} = 1$

14. $(5y - 7)^{4/3} = 16$

15. $(2t - 1)^{2/7} = 4$

16. $(3m + 1)^{5/3} = -32$

17. $(2x - 1)^{-1/3} = 8$

18. $(3y - 7)^{-4/3} = 1$

19. $(3x + 1)^{-5/3} = \frac{1}{32}$

20. $(7t - 1)^{-3/5} = \frac{1}{27}$

21. $y^{2/3} + 2y^{1/3} = 8, u = y^{1/3}$

22. $x^{1/2} + 2x^{1/4} = 3, u = x^{1/4}$

23. $x^{1/3} - 1 - 12x^{-1/3} = 0$

24. $z^5 - 33z^{5/2} + 32 = 0$

25. $x^3 - 9x^{3/2} + 8 = 0$

26. $8t^{2/3} + 7t^{1/3} = 1$

27. $x + 7 - (x + 7)^{1/2} - 2 = 0$

28. $y^2 + 3y + (y^2 + 3y - 2)^{1/2} = 22$

29. $(m + 20)^{1/2} - 4(m + 20)^{1/4} + 3 = 0$

30. $2(1 - y)^{1/3} + 3(1 - y)^{1/6} = 2$

31. $2t^2 + t - 4(2t^2 + t + 4)^{1/2} = 1$

32. $x^2 + 6x - 6(x^2 + 6x - 2)^{1/2} + 3 = 0$

6.8 Nonlinear Inequalities

Inequalities occurring in many applications of mathematics contain quadratic and rational expressions. In this section, we solve quadratic and rational inequalities.

Quadratic Inequalities

Consider the quadratic inequality

$$x^2 - 5x + 6 < 0.$$

If we move x along the number line, the sign of the expression $x^2 - 5x + 6$ changes. For example, when we let $x = -1$, the expression has the value $1 + 5 + 6 = 12$, and $12 > 0$. When $x = \frac{5}{2}$, the value of the expression is $\frac{25}{4} - \frac{25}{2} + 6 = -\frac{1}{4}$, and $-\frac{1}{4} < 0$. When $x = 4$, the value of the expression becomes $16 - 20 + 6 = 2$, and $2 > 0$.

To solve the inequality, we must find the values of x for which $x^2 - 5x + 6$ is negative. Quadratic inequalities have a characteristic that will help us do this. As we move along the number line, substituting values for x, we find that the parts of the line where $x^2 - 5x + 6$ is positive are separated from the parts where $x^2 - 5x + 6$ is negative by the values of x for which $x^2 - 5x + 6$ is zero. If we solve the equation

$$x^2 - 5x + 6 = 0$$

by factoring, we obtain

$$(x - 2)(x - 3) = 0,$$

which has the solutions 2 and 3. Plotting the corresponding points on a number line, we see that the solutions 2 and 3 divide the number line into three parts, which we label A, B, and C (Figure 1). In part A, $x < 2$; in part B, $2 < x < 3$; and in part C, $x > 3$.

Figure 1

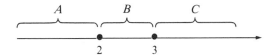

The quantity $x^2 - 5x + 6$ will have a constant algebraic sign over each of these parts. To find out whether $x^2 - 5x + 6$ is positive or negative over A (in which $x < 2$), we select any convenient test number in this part of the number line—say, $x = 0$—and substitute it into $x^2 - 5x + 6$:

$$0^2 - 5(0) + 6 = 6 > 0.$$

Because $x^2 - 5x + 6$ is positive at one number in part A, and cannot change its algebraic sign over this part, we conclude that $x^2 - 5x + 6 > 0$ for all values of x in A.

Similarly, to find the algebraic sign of $x^2 - 5x + 6$ over part B, we substitute a test number—say, $x = \frac{5}{2}$—in the expression to obtain

$$\left(\frac{5}{2}\right)^2 - 5\left(\frac{5}{2}\right) + 6 = \frac{25}{4} - \frac{25}{2} + 6$$

$$= -\frac{1}{4} < 0.$$

Thus we know that the expression is negative for all values of x in part B.

Finally, we choose $x = 4$ in part C:

$$4^2 - 5(4) + 6 = 16 - 20 + 6$$

$$= 2 > 0.$$

Therefore, the expression $x^2 - 5x + 6$ is positive for all values of x in part C.

The information we now have about the algebraic sign of $x^2 - 5x + 6$ in each part of the number line is summarized in Figure 2. Therefore, the solution set of $x^2 - 5x + 6 < 0$ consists of all real numbers x such that $2 < x < 3$.

Figure 2

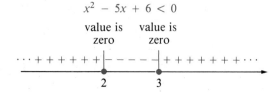

$$x^2 - 5x + 6 < 0$$

The method illustrated above can be used to solve any quadratic inequality in one unknown. This method is summarized in the following step-by-step procedure:

Procedure for Solving a Quadratic Inequality

Step 1. Set the quadratic expression equal to zero and find all real solutions of the resulting equation.

Step 2. Arrange the solutions obtained in step 1 in increasing order on a number line. These solutions will divide the number line into at most three parts (say, A, B, and C).

Step 3. Determine the algebraic sign for each part of the number line by selecting a test number from that part and substituting it for the unknown in the quadratic expression. The algebraic sign of the resulting value is the sign of the quadratic expression over the entire part.

Step 4. Using the information obtained in step 3, draw a figure showing the algebraic signs of the expression over the various parts of the number line. The solution set of the inequality can be read from this figure.

In Examples 1–4, solve each quadratic inequality.

EXAMPLE 1 $x^2 + x - 12 \leq 0$

SOLUTION   We carry out the steps in the procedure above.

Step 1. Set $x^2 + x - 12 = (x + 4)(x - 3) = 0$. The solutions of this equation are -4 and 3.

Step 2. The solutions -4 and 3 (of the equation in step 1) divide the number line into three parts, A, B, and C (Figure 3).

Figure 3

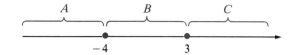

Step 3. We use test values to determine the algebraic sign of the expression $x^2 + x - 12$ in each part. The following table summarizes the results:

Part	Test Number	Test Value of $x^2 + x - 12$	Sign of $x^2 + x - 12$
A: $x < -4$	-5	$(-5)^2 + (-5) - 12 = 8$	$+$
B: $-4 < x < 3$	0	$0^2 + 0 - 12 = -12$	$-$
C: $x > 3$	4	$4^2 + 4 - 12 = 8$	$+$

Step 4. The information obtained in step 3 is illustrated in Figure 4a. We see that

$$x^2 + x - 12 \leq 0$$

if $-4 \leq x \leq 3$. Thus, the solution set consists of all real numbers x such that $-4 \leq x \leq 3$ (Figure 4b).

Figure 4

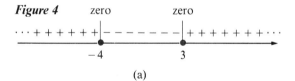

(a)

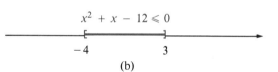

(b)

EXAMPLE 2 $x^2 + x - 2 > 0$

SOLUTION   Step 1. Set $x^2 + x - 2 = (x + 2)(x - 1) = 0$. The solutions of this equation are -2 and 1.

Step 2. The numbers -2 and 1 divide the number line into three parts, A, B, and C (Figure 5).

Figure 5

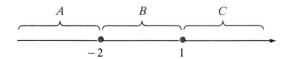

Step 3. We use test values to determine the algebraic sign of the expression $x^2 + x - 2$ in each part. The following table summarizes the results:

Part	Test Number	Test Value of $x^2 + x - 2$	Sign of $x^2 + x - 2$
A: $x < -2$	-3	$(-3)^2 + (-3) - 2 = 4$	$+$
B: $-2 < x < 1$	0	$0^2 + 0 - 2 = -2$	$-$
C: $x > 1$	2	$2^2 + 2 - 2 = 4$	$+$

Step 4. The information obtained in step 3 is illustrated in Figure 6a.

Figure 6a

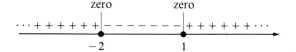

We see that $x^2 + x - 2 > 0$ if $x < -2$ or $x > 1$. Therefore, the solution set consists of all real numbers x such that $x < -2$ or $x > 1$ (Figure 6b).

Figure 6b

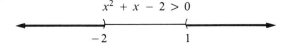

EXAMPLE 3 $9x^2 - 24x + 16 \geq 0$

SOLUTION   Step 1. Set $9x^2 - 24x + 16 = (3x - 4)^2 = 0$ so that $x = \frac{4}{3}$.

Step 2. The number $\frac{4}{3}$ divides the number line into two parts, A and B (Figure 7).

Figure 7

Step 3. We use test values to determine the sign of $9x^2 - 24x + 16$ in each part. The following table summarizes the results:

Part	Test Number	Test Value of $9x^2 - 24x + 16$	Sign of $9x^2 - 24x + 16$
$A: \quad x < \frac{4}{3}$	0	$9(0)^2 - 24(0) + 16 = 16$	$+$
$B: \quad x > \frac{4}{3}$	2	$9(2)^2 - 24(2) + 16 = 4$	$+$

Step 4. The information obtained in step 3 is illustrated in Figure 8a.

Figure 8a

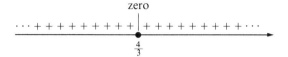

We see that $9x^2 - 24x + 16 \geq 0$ for all real numbers x. Therefore, the solution set consists of all real numbers (Figure 8b).

Figure 8b

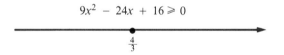

EXAMPLE 4 $4x^2 + 20x + 25 < 0$

SOLUTION   The inequality $4x^2 + 20x + 25 < 0$ is equivalent to $(2x + 5)^2 < 0$. The inequality is not true for any real values of x, because the square of the expression $2x + 5$ is *never* negative.

Rational Inequalities

The following are examples of **rational inequalities:**

$$\frac{2x + 5}{x + 1} < 0 \qquad \text{and} \qquad \frac{x + 3}{5x - 7} \geq 0.$$

Suppose that we want to solve the rational inequality

$$\frac{x - 1}{x + 2} > 0.$$

Note that the only possible values of x for which a fraction can change sign are values at which either the *numerator* or the *denominator* is zero. Thus the values of x for which a rational expression is positive are separated from the values of x for which the expression is negative by those values of x for which either the numerator or the denominator is zero. (Don't forget that a fraction is undefined when the denominator is zero.) The numerator of

$$\frac{x-1}{x+2}$$

is zero when $x = 1$, and the denominator is zero when $x = -2$. These values of x will divide the number line into three parts. We can use test values in each of the three parts as we did for the solution in the quadratic inequalities.

EXAMPLE 5   *In Examples 5 and 6, solve each rational inequality.*

SOLUTION $$\frac{3x-12}{x+2} < 0$$

Step 1. The numerator of

$$\frac{3x-12}{x+2}$$

equals zero if

$$3x - 12 = 0 \qquad \text{or}$$
$$x = 4.$$

The denominator is zero when

$$x + 2 = 0 \qquad \text{or}$$
$$x = -2.$$

Notice that -2 is not a solution of

$$\frac{3x-12}{x+2} < 0$$

since the expression is undefined for $x = -2$. Also, 4 is not a solution of

$$\frac{3x-12}{x+2} < 0 \qquad \text{since} \qquad \frac{3(4)-12}{4+2} = 0.$$

Step 2. The numbers -2 and 4 divide the number line into three parts, A, B, and C (Figure 9).

Figure 9

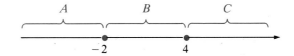

Step 3. We use test values to determine the algebraic sign of

$$\frac{3x - 12}{x + 2}$$

in each part:

Part	Test Number	Test Value of $\frac{3x - 12}{x + 2}$	Sign of $\frac{3x - 12}{x + 2}$
A: $x < -2$	-3	$\frac{-9 - 12}{-3 + 2} = 21$	$+$
B: $-2 < x < 4$	0	$\frac{0 - 12}{0 + 2} = -6$	$-$
C: $x > 4$	5	$\frac{15 - 12}{5 + 2} = \frac{3}{7}$	$+$

Step 4. The information obtained in step 3 is illustrated in Figure 10a.

Figure 10a

We see that

$$\frac{3x - 12}{x + 2} < 0$$

if $-2 < x < 4$. Hence, the solution set consists of all real numbers x such that $-2 < x < 4$ (Figure 10b).

Figure 10b

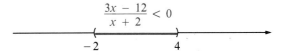

EXAMPLE **6** $\dfrac{2x + 1}{x - 2} \geq 1$

SOLUTION   The following inequalities are equivalent:

$$\frac{2x + 1}{x - 2} \geq 1$$

$$\frac{2x + 1}{x - 2} - 1 \geq 0$$

$$\frac{2x + 1 - x + 2}{x - 2} \geq 0$$

$$\frac{x + 3}{x - 2} \geq 0.$$

Step 1. The numerator of

$$\frac{x + 3}{x - 2}$$

equals zero if

$$x + 3 = 0 \qquad \text{or} \qquad x = -3.$$

The denominator is zero if

$$x - 2 = 0 \qquad \text{or} \qquad x = 2.$$

So the fraction

$$\frac{x + 3}{x - 2}$$

is undefined when $x = 2$.

Step 2. The numbers -3 and 2 divide the number line into three parts, A, B, and C (Figure 11).

Figure 11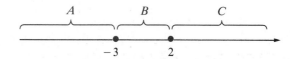

Step 3. We use test values to determine the algebraic sign of

$$\frac{x + 3}{x - 2}$$

in each part:

Part	Test Number	Test Value of $\dfrac{x+3}{x-2}$	Sign of $\dfrac{x+3}{x-2}$
$A:\quad x<-3$	-4	$\dfrac{-4+3}{-4-2}=\dfrac{1}{6}$	$+$
$B:\quad -3<x<2$	0	$\dfrac{0+3}{0-2}=-\dfrac{3}{2}$	$-$
$C:\quad x>2$	3	$\dfrac{3+3}{3-2}=6$	$+$

Therefore, $\dfrac{x+3}{x-2}>0$ for all values of x for which $x<-3$ or $x>2$.

Step 4. The information obtained in step 3 is illustrated in Figure 12a.

Figure 12a

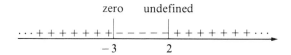

From this figure we see that

$$\frac{x+3}{x-2}\geq 0$$

if $x\leq -3$ or $x>2$. (Recall that 2 isn't included in the solution because the inequality is undefined at that point.) Therefore, the solution set consists of all real numbers x such that $x\leq -3$ or $x>2$ (Figure 12b).

Figure 12b

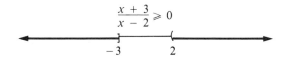

PROBLEM SET 6.8

In problems 1–22, solve each quadratic inequality and illustrate the solution set on a number line.

1. $x^2+x-2<0$

2. $x^2-x-6<0$

3. $x^2-3x+2\leq 0$

4. $x^2-6x+8\leq 0$

5. $x^2+2x-3>0$

6. $x^2+5x+6>0$

7. $x^2\geq 4x+12$

8. $x^2+3x\geq 10$

9. $x^2-x\leq 20$

10. $x^2 + 4x \leq 21$

11. $2x^2 + x - 1 \leq 0$

12. $2x^2 + 9x - 5 \geq 0$

13. $3x^2 - 2x - 5 > 0$

14. $40 - 3x - x^2 < 0$

15. $4x^2 \geq 27x + 7$

16. $10x^2 - 31x \leq 14$

17. $4x^2 - 20x + 25 \geq 0$

18. $16x^2 + 8x + 1 \geq 0$

19. $9x^2 + 6x + 1 > 0$

20. $25x^2 - 30x + 9 < 0$

21. $16x^2 - 24x + 9 < 0$

22. $49x^2 + 14x + 1 \leq 0$

In problems 23–36, solve each rational inequality and illustrate the solution set on a number line.

23. $\dfrac{x - 1}{x - 4} < 0$

24. $\dfrac{x + 3}{x - 3} > 0$

25. $\dfrac{x + 1}{x - 2} > 0$

26. $\dfrac{2x + 3}{4x + 8} < 0$

27. $\dfrac{x - 4}{5x + 2} \leq 0$

28. $\dfrac{x - 2}{3x - 1} \geq 0$

29. $\dfrac{5x - 1}{3x - 2} \geq 0$

30. $\dfrac{2x - 1}{5x + 1} \leq 0$

31. $\dfrac{1 - x}{3 - x} \geq 0$

32. $\dfrac{7x}{3x + 1} > 0$

33. $\dfrac{x + 1}{x - 3} \leq 1$

34. $\dfrac{x + 2}{x - 1} \geq 2$

35. $\dfrac{1 - x}{x} < 1$

36. $\dfrac{x}{x + 2} \leq 3$

REVIEW PROBLEM SET

In problems 1–14, use the factoring method to solve each equation.

1. $x^2 - 14x - 15 = 0$

2. $z^2 - 5z - 84 = 0$

3. $t^2 + t - 12 = 0$

4. $6y^2 - 15y = 0$

5. $5w^2 - 10w = 0$

6. $5 - 14x - 3x^2 = 0$

7. $8 - 2y - y^2 = 0$

8. $16t^2 - 24t + 9 = 0$

9. $4 + 5x = 9x^2$

10. $5x^2 = 2x + 24$

11. $z(6z - 7) = 20$

12. $y(4y - 31) = 8$

13. $\dfrac{3}{4x^2} + \dfrac{7}{8x} - \dfrac{5}{2} = 0$

14. $\dfrac{2}{w - 1} - \dfrac{3}{2w + 5} = \dfrac{5}{3}$

In problems 15–16, solve each equation for the indicated variable.

15. $x^2 + bx - 6b^2 = 0$, for x.

16. $6y^2 - 7cy - 3c^2 = 0$, for y.

In problems 17–24, solve each equation by the root extraction method.

17. $t^2 = 144$

18. $5u^2 - 80 = 0$

19. $3x^2 + 48 = 0$

20. $(v - 1)^2 + 4 = 0$

21. $(y - \frac{3}{5})^2 = \frac{16}{25}$

22. $(x + \frac{5}{2})^2 = \frac{7}{4}$

23. $(3m - 1)^2 - 256 = 0$

24. $(2z + 3)^2 + 25 = 0$

In problems 25–28, complete the square for each expression and write the result as the square of a binomial.

25. $u^2 - 8u$

26. $z^2 + 13z$

27. $x^2 + 5x$

28. $y^2 - 14y$

In problems 29–40, solve each equation by completing the square.

29. $x^2 - 6x + 7 = 0$
30. $y^2 + 4y - 21 = 0$
31. $t^2 - 8t + 9 = 0$
32. $u^2 - 4u + 5 = 0$
33. $9y^2 - 6y + 2 = 0$
34. $9t^2 - 12t + 1 = 0$
35. $4z^2 + 13 = 12z$
36. $16y^2 - 24y + 5 = 0$
37. $4x^2 + 4x - 3 = 0$
38. $2w^2 + 3 = 8w$
39. $y^2 + cy - 6c^2 = 0, c > 0$, solve for y
40. $3x^2 - dx + d^2 = 0, d < 0$, solve for x.

In problems 41–52, solve each equation by using the quadratic formula. In problems 51–52, round off your answers to two decimal places.

41. $x^2 + 10x - 13 = 0$
42. $5 - 8t + t^2 = 0$
43. $y^2 + y + 3 = 0$
44. $3 - z - z^2 = 0$
45. $4u^2 - u - 1 = 0$
46. $3x^2 - 8x + 1 = 0$
47. $2t^2 + 5t - 17 = 0$
48. $16 - 16y - 3y^2 = 0$
49. $(2x - 1)(x + 2) = 5$
50. $(w - 1)(3w + 5) = 2$
C 51. $0.9x^2 - 4.1x - 3.14 = 0$
C 52. $1.7y^2 + 3.8y - 1.56 = 0$

In problems 53–58, use the discriminant to determine the number and the kind of solutions to each equation.

53. $t^2 - 8t + 9 = 0$
54. $2y^2 - y - 1 = 0$
55. $5x^2 - 2x + 1 = 0$
56. $4m^2 - 12m + 13 = 0$
57. $4y^2 - 12y + 9 = 0$
58. $25z^2 - 20z + 4 = 0$

In problems 59–68, reduce each equation to the quadratic form by using an appropriate substitution and solve the equation for the original variable.

59. $x^4 - 5x^2 + 4 = 0$
60. $y^4 - 8y^2 + 16 = 0$
61. $16t^4 - 17t^2 + 1 = 0$
62. $v^8 - 2v^4 + 1 = 0$
63. $4z^{-4} - 11z^{-2} - 3 = 0$
64. $1 - 2x^{-2} - 3x^{-4} = 0$
65. $(x^2 + x)^2 - 8(x^2 + x) + 12 = 0$
66. $(y^2 + 4y)^2 - 17(y^2 + 4y) = 60$
67. $v^2 + 2v + \dfrac{3}{v^2 + 2v} = 4$
68. $2t^2 - t - \dfrac{6}{2t^2 - t} + 1 = 0$

In problems 69–84, find the real solutions of each equation. Check the solutions whenever the equation has a radical with even index.

69. $\sqrt{x - 1} = 4$
70. $\sqrt{2u + 5} - 7 = -4$
71. $\sqrt{t + 17} + 1 = 5$
72. $1 + \sqrt{x - 3} = 4$
73. $2\sqrt{2y - 3} + 4 = 1$
74. $\sqrt{z - 7} = 2\sqrt{z}$
75. $\sqrt[3]{3x - 2} = 4$
76. $1 - \sqrt[3]{t - 3} = 3$
77. $\sqrt[4]{5x - 7} = \sqrt[4]{x}$
78. $\sqrt[4]{4x - 3} = \sqrt[4]{5x - 7}$
79. $\sqrt{w - 3} - \sqrt{w} = -1$
80. $\sqrt{2x + 1} + \sqrt{8 - x} = 5$
81. $\sqrt{2t + 1} + \sqrt{t} = 1$
82. $\sqrt{5z - 4} - \sqrt{2z + 1} = 1$
83. $\sqrt{5x + 5} + \sqrt{x - 4} = 5$
84. $\sqrt{2x + \sqrt{3 + x}} = 2$

In problems 85–94, solve each equation. Check the solution whenever the equation has a rational exponent (in its lowest terms) with an even denominator.

85. $t^{1/3} = -2$
86. $x^{1/4} = 3$
87. $y^{3/4} = 8$
88. $z^{2/5} = 4$
89. $(x - 2)^{2/3} = 9$
90. $(2 - 3w)^{5/3} = 32$
91. $(3y + 2)^{-1/3} = 2$
92. $(2x - 1)^{-4/3} = 81$
93. $z^{2/3} + 2z^{1/3} - 3 = 0$
94. $x^{-3/2} - 28x^{-3/4} + 27 = 0$

In problems 95–102, solve each inequality and illustrate the solution set on a number line.

95. $x^2 + 2x - 15 < 0$ **96.** $u^2 - 2u - 3 \geq 0$ **97.** $t^2 - 3t \geq 10$

98. $3y^2 - 2y < 5$ **99.** $\dfrac{x - 2}{x + 3} > 0$ **100.** $\dfrac{3w + 1}{2w - 5} \leq 0$

101. $\dfrac{3y - 1}{5y - 7} \leq 0$ **102.** $\dfrac{7x - 3}{9x + 1} > 0$

103. Find three consecutive positive integers such that the sum of their squares is 149.

104. Find a number such that the sum of the number and its reciprocal is 4.

105. A boy has mowed a strip of uniform width around a rectangular lawn that is 80 feet by 60 feet. He still has half the lawn to mow. How wide is the strip?

106. A swimming pool 25 feet by 15 feet is bordered by a concrete walk of uniform width. The area of the walk is 329 square feet. How wide is it?

107. If the distance s in feet that a bomb falls in t seconds is given by the formula $s = 16t^2/(1 + 0.06t)$, how many seconds are required for a bomb released at 20,000 feet to reach its target?

108. A group of students chartered a bus for $60. When four students withdrew from the group, the share of each of the other students was increased by $2.50. How many students were in the group originally?

109. Two joggers leave the same point and travel at right angles to each other, each moving at a uniform speed. One jogs 1 mile per hour faster than the other. After 3 hours, they are 15 miles apart. Find the speed of each jogger.

110. A girl asked her father: "How old are you?" The father answered: "I was 30 years old when you were born and the product of our present ages is 736." How old is the father?

7 Graphing Linear Systems of Equations and Inequalities

We begin this chapter by extending the techniques of graphing to include points in a plane. This enables us to graph equations in two variables. We also examine equations and inequalities containing two or three variables. (Recall that in Chapter 4 we solved equations and inequalities in one variable. In practical applications of mathematics it is often difficult to define all unknown quantities in terms of a single variable. Therefore, it is useful to know how to solve systems of equations containing more than one variable.) Finally, we graph linear inequalities and systems of linear inequalities.

7.1 The Cartesian Coordinate System

Figure 1

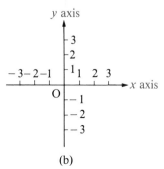

(a)

(b)

In Chapter 1, Section 1.1, we saw that a point P on a number line can be specified by a real number x, called the *coordinate* of that point. Similarly, if we use a **Cartesian coordinate system,** named in honor of the French mathematician René Descartes (1596–1650), a point P in a plane can be specified by two real numbers, also called *coordinates*.

A Cartesian coordinate system consists of two perpendicular number lines, called the **coordinate axes.** The point at which the two lines intersect is called the **origin** (Figure 1a). Ordinarily, one of the number lines is horizontal and is called the *x* **axis,** and the other is vertical and is called the *y* **axis**.

We normally use the same scale on the two axes, although in some figures space considerations make it convenient to use different scales. By convention, numerical coordinates increase to the right along the x axis and upward along the y axis. The positive portion of the x axis is to the right of the origin, and the negative portion is to the left of the origin. The positive portion of the y axis is above the origin, and the negative portion is below the origin (Figure 1b).

If P is a point in the Cartesian plane, the coordinates of P are the coordinates x and y of the points where the perpendiculars to the axes from P meet the two axes (Figure 2). The coordinates of P are traditionally written as (x, y). The notation (x, y) represents an **ordered pair,** a pair of numbers or symbols in which the order of listing is important. For example,

$$(5, 6) \neq (6, 5).$$

The first number, x, of an ordered pair is called the **abscissa** of P. The second number, y, of the ordered pair is called the **ordinate** of P. To **plot,** or **graph,** the point P with coordinates (x, y) means to draw Cartesian coordinate axes and to place a dot representing P at the point with abscissa x and ordinate y.

For example, to locate the point $(3, 2)$, start at the origin and move three units to the right along the x axis; then move two units up from the x axis. Similarly, to locate $(-4, 3)$, move four units to the left of 0 and three units up from the x axis; and to locate $(0, -\frac{5}{2})$, move no units along the x axis and $\frac{5}{2}$ units down from the x axis along the y axis (Figure 3).

Figure 2

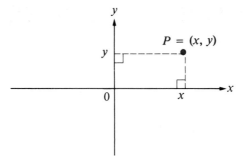

Figure 3

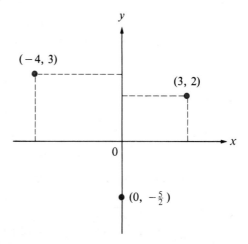

Figure 4

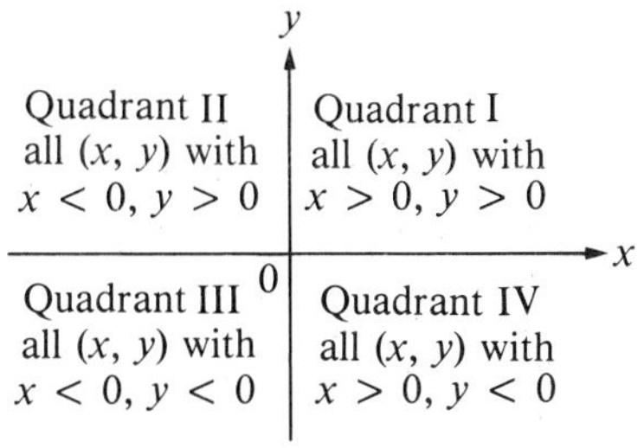

Quadrant II
all (x, y) with
$x < 0, y > 0$ | Quadrant I
all (x, y) with
$x > 0, y > 0$

Quadrant III
all (x, y) with
$x < 0, y < 0$ | Quadrant IV
all (x, y) with
$x > 0, y < 0$

You can think of the ordered pair (x, y) as the numerical "address" of P. We show the correspondence between P and (x, y) by identifying the point P with its "address" (x, y) and write $P = (x, y)$.

We relate each ordered pair of real numbers (x, y) to a **point,** and we refer to the set of all such ordered pairs as the **Cartesian plane** or the **xy plane.**

The x and y axes divide the plane into four regions called **quadrants** I, II, III, and IV (Figure 4).

EXAMPLE 1 Plot each point and indicate which quadrant or coordinate axis contains the point:

(a) $(4, 1)$ (b) $(-4, 2)$ (c) $(-2, -3)$ (d) $(2, -5)$

(e) $(2, 0)$ (f) $(-\frac{3}{2}, 0)$ (g) $(0, 5)$ (h) $(0, 0)$

SOLUTION The points are plotted in Figure 5.

Figure 5

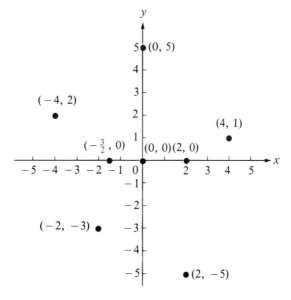

(a) $(4, 1)$ lies in quadrant I.   (b) $(-4, 2)$ lies in quadrant II.

(c) $(-2, -3)$ lies in quadrant III.   (d) $(2, -5)$ lies in quadrant IV.

(e) $(2, 0)$ lies on the positive x axis. (f) $(-\frac{3}{2}, 0)$ lies on the negative x axis.

(g) $(0, 5)$ lies on the positive y axis. (h) $(0, 0)$, the origin, lies on both axes.

EXAMPLE 2   Find the coordinates of each point on the Cartesian coordinate system in Figure 6.

Figure 6

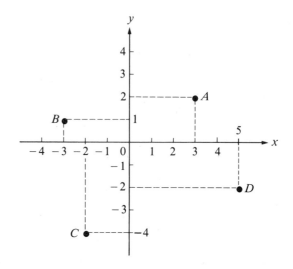

SOLUTION   Look at Figure 6. We see that the coordinates of each point are:

(a) Point A is three units to the right of the y axis and two units up from the x axis so that $A = (3, 2)$.

(b) Point B is three units to the left of the y axis and one unit up from the x axis so that $B = (-3, 1)$.

(c) Point C is two units to the left of the y axis and four units down from the x axis so that $C = (-2, -4)$.

(d) Point D is five units to the right of the y axis and two units down from the x axis so that $D = (5, -2)$.

The Distance Formula

One useful feature of the Cartesian coordinate system is that there is a formula that gives the distance between two points in the xy plane. We represent the distance between two points P_1 and P_2 by

$$d = |\overline{P_1 P_2}|.$$

1. If $P_1 = P_2$, then $|\overline{P_1 P_2}| = 0$.

2. If $P_1 = (x_1, y_1)$ and $P_2 = (x_2, y_2)$ lie on the same horizontal line, that is, if $y_1 = y_2$, then $d = |\overline{P_1 P_2}| = |x_2 - x_1|$ (Figure 7a).

Figure 7

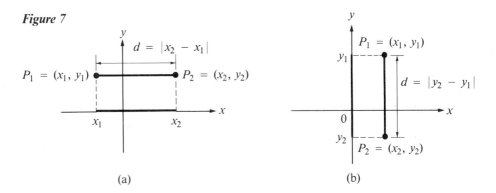

(a)   (b)

3. If $P_1 = (x_1, y_1)$ and $P_2 = (x_2, y_2)$ lie on the same vertical line, that is, if $x_1 = x_2$, then $d = |\overline{P_1 P_2}| = |y_2 - y_1|$ (Figure 7b).

Because $|x_2 - x_1| = |x_1 - x_2|$, the formula in (2) is true whether P_1 lies to the left of P_2 or to the right of P_2. Moreover, it does not matter in which quadrants the points lie.

Finally, let us consider the general case in which the points $P_1 = (x_1, y_1)$ and $P_2 = (x_2, y_2)$ do not lie on the same horizontal or vertical line, that is, the case in which the line segment $\overline{P_1 P_2}$ is neither horizontal nor vertical. To determine the length of $\overline{P_1 P_2}$, we can draw a right triangle with $\overline{P_1 P_2}$ as the hypotenuse and with one leg parallel to the x axis and the other leg parallel to the y axis (Figure 8). The two legs intersect at some point P_3. P_3 has the same y coordinate as P_1 and the same x coordinate as P_2. Therefore, we can represent P_3 by $P_3 = (x_2, y_1)$ (Figure 8). Then $|\overline{P_1 P_3}| = |x_2 - x_1|$ and $|\overline{P_3 P_2}| = |y_2 - y_1|$.

Figure 8

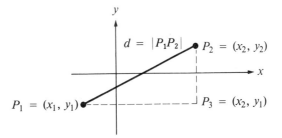

We can use the Pythagorean theorem to determine the length of the hypotenuse $\overline{P_1 P_2}$ of the right triangle $P_1 P_2 P_3$:

$$d^2 = |\overline{P_1 P_2}|^2 = |\overline{P_1 P_3}|^2 + |\overline{P_3 P_2}|^2,$$

or

$$d^2 = |\overline{P_1 P_2}|^2 = |x_2 - x_1|^2 + |y_2 - y_1|^2.$$

We use the fact that $|\overline{P_1P_2}|$ is nonnegative and $|a|^2 = a^2$, to write:

$$|x_2 - x_1|^2 = (x_2 - x_1)^2 \qquad \text{and}$$
$$|y_2 - y_1|^2 = (y_2 - y_1)^2.$$

We can now obtain the following formula:

The Distance Formula

If $P_1 = (x_1, y_1)$ and $P_2 = (x_2, y_2)$ are two points in the Cartesian plane, then the **distance** d between P_1 and P_2 is given by

$$d = |\overline{P_1P_2}| = \sqrt{(x_2 - x_1)^2 + (y_2 - y_1)^2}.$$

EXAMPLE 3   Plot the points $(2, -4)$ and $(-2, -1)$ and find the distance d between them.

SOLUTION   The points $P_1 = (2, -4)$ and $P_2 = (-2, -1)$ are plotted in Figure 9, and the distance d is given by

$$\begin{aligned}
d &= \sqrt{[-2 - 2]^2 + [-1 - (-4)]^2} \\
&= \sqrt{(-4)^2 + 3^2} \\
&= \sqrt{16 + 9} \\
&= \sqrt{25} \\
&= 5.
\end{aligned}$$

Figure 9

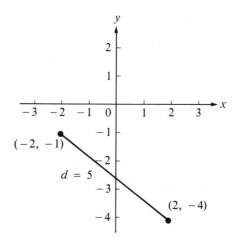

EXAMPLE 4   Plot the points $(-1, -2)$ and $(3, 4)$ and find the distance d between them.

SOLUTION   The points $P_1 = (-1, -2)$ and $P_2 = (3, 4)$ are plotted in Figure 10, and the distance d is given by

$$d = \sqrt{[3 - (-1)]^2 + [4 - (-2)]^2}$$
$$= \sqrt{4^2 + 6^2}$$
$$= \sqrt{16 + 36}$$
$$= \sqrt{52}$$
$$= 2\sqrt{13}.$$

Figure 10

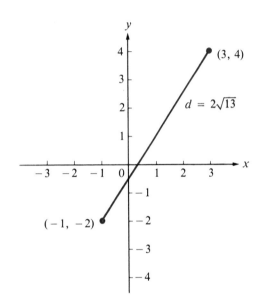

EXAMPLE 5   ⃞ Let $P_1 = (75.39, -14.06)$ and $P_2 = (18.64, 19.37)$. Find the distance d between P_1 and P_2. Round off the answer to four significant digits.

SOLUTION   The distance d between P_1 and P_2 is given by
$$d = \sqrt{(18.64 - 75.39)^2 + [19.37 - (-14.06)]^2}$$
$$= \sqrt{3220.5625 + 1117.5649}$$
$$= \sqrt{4338.1274} = 65.86.$$

EXAMPLE 6   Let $A = (-1, 3)$, $B = (2, 6)$, and $C = (3, 2)$.

(a) Plot the points A, B, and C, and draw the triangle ABC.

(b) Find the distances $|\overline{AB}|$, $|\overline{AC}|$, and $|\overline{BC}|$.

(c) Show that ABC is an isosceles triangle.

SOLUTION   (a) The points A, B, and C are plotted and the triangle ABC is drawn in Figure 11.

Figure 11

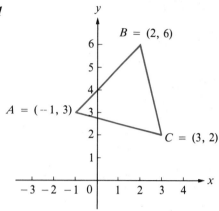

(b) $|\overline{AB}| = \sqrt{[2 - (-1)]^2 + (6 - 3)^2}$
$= \sqrt{9 + 9} = \sqrt{18} = 3\sqrt{2}$

$|\overline{AC}| = \sqrt{[3 - (-1)]^2 + (2 - 3)^2}$
$= \sqrt{16 + 1} = \sqrt{17}$

$|\overline{BC}| = \sqrt{(3 - 2)^2 + (2 - 6)^2}$
$= \sqrt{1 + 16} = \sqrt{17}.$

(c) Because $|\overline{AC}| = |\overline{BC}|$, we conclude that the triangle ABC is isosceles.

PROBLEM SET 7.1

1. Plot each point and indicate which quadrant or coordinate axis contains the point.
   (a) $(1, 2)$ (b) $(-1, 1)$ (c) $(-1, -2)$ (d) $(3, -2)$
   (e) $(0, 1)$ (f) $(-6, 0)$ (g) $(3, 0)$ (h) $(0, -4)$

2. Plot each point and indicate which quadrant or coordinate axis contains the point.
   (a) $(3, \sqrt{2})$ (b) $(-\sqrt{3}, 1)$ (c) $(-\sqrt{3}, -\sqrt{2})$ (d) $(0, -\frac{3}{4})$
   (e) $(\frac{2}{3}, 0)$ (f) $(-\pi, 0)$ (g) $(-\frac{1}{2}, \frac{1}{2})$ (h) $(\frac{2}{3}, -\frac{3}{4})$

3. Find the coordinates of each point on the Cartesian coordinate system in Figure 12.

Figure 12

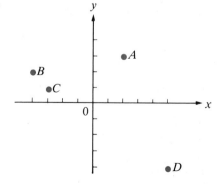

4. Find the coordinates of each point on the Cartesian coordinate system in Figure 13.

Figure 13

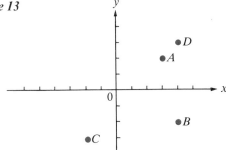

In problems 5–20, find the distance between the two points with the given coordinates.

5. (7, 10) and (1, 2)   **6.** $(-3, -4)$ and $(-5, -7)$

7. (1, 1) and $(-3, 2)$ **8.** $(-2, 5)$ and $(3, -1)$

9. $(4, -3)$ and (6, 2)   **10.** (1, 5) and (4, 9)

11. $(-5, 0)$ and $(-2, -4)$ **12.** $(-4, 7)$ and $(0, -8)$

13. $(7, -3)$ and $(3, -3)$ **14.** (6, 2) and $(6, -2)$

15. $(-4, -3)$ and (0, 0)   **16.** (0, 3) and $(-4, 0)$

17. $(7, -1)$ and (7, 3)   **18.** $(2, -b)$ and (4, b)

19. $(-\frac{1}{2}, \frac{1}{3})$ and (2, 3)   **20.** (t, 8) and $(t, -2)$

In problems 21–26, use the distance formula to determine whether or not the triangle ABC is isosceles.

21. $A = (-5, 1)$, $B = (-6, 5)$, $C = (-2, 4)$ **22.** $A = (3, 1)$, $B = (4, 3)$, $C = (6, 2)$

23. $A = (-2, -3)$, $B = (3, -1)$, $C = (1, 4)$ **24.** $A = (-4, 2)$, $B = (1, 4)$, $C = (3, -1)$

25. $A = (5, -3)$, $B = (2, 4)$, $C = (2, 5)$ **26.** $A = (-6, 3)$, $B = (2, 1)$, $C = (-2, -3)$

In problems 27–30, use the distance formula and the Pythagorean theorem to show that the triangle ABC is a right triangle.

27. $A = (1, 1)$, $B = (5, 1)$, $C = (5, 7)$ **28.** $A = (1, 2)$, $B = (5, 2)$, $C = (5, 5)$

29. $A = (-3, 1)$, $B = (3, 1)$, $C = (3, 10)$ **30.** $A = (-2, -2)$, $B = (0, 0)$, $C = (3, -3)$

31. Find all values of u so that the distance between the points $(-2, 3)$ and (u, u) is five units.

32. If P_1, P_2, and P_3 are points in a plane, then P_1, P_2, and P_3 are **collinear** if P_2 lies on the line segment $\overline{P_1P_3}$, that is, P_1, P_2, and P_3 are collinear if and only if $|\overline{P_1P_3}| = |\overline{P_1P_2}| + |\overline{P_2P_3}|$. Illustrate this fact for $P_1 = (-3, -2)$, $P_2 = (1, 2)$, and $P_3 = (3, 4)$.

33. Show that the points $A = (-2, 2)$, $B = (1, 4)$, $C = (3, 1)$, and $D = (0, -1)$ are the vertices of the square $ABCD$.

34. Show that the point

$$P_2 = \left(\frac{a+c}{2}, \frac{b+d}{2} \right)$$

is the **midpoint** of the line segment joining $P_1 = (a, b)$ and $P_3 = (c, d)$. (*Hint:* Use the distance formula to show that $|\overline{P_1 P_2}| = |\overline{P_2 P_3}|$, and that P_2 belongs to the line segment $\overline{P_1 P_3}$.)

In problems 35–38, find the coordinates of the midpoint P_2 of the line segment $\overline{P_1 P_3}$.

35. $P_1 = (5, 6)$ and $P_3 = (-7, 8)$ **36.** $P_1 = (-4, 7)$ and $P_3 = (-3, 0)$

37. $P_1 = (-2, 3)$ and $P_3 = (4, -2)$ **38.** $P_1 = (2, -5)$ and $P_3 = (-1, -3)$

© In problems 39–42, find the distance between the two points. Round off the answers to four significant digits.

39. $(81.31, 74.01)$ and $(93.72, 61.37)$ **40.** $(39.14, 61.78)$ and $(49.07, 32.14)$

41. $(17.81, 13.75)$ and $(45.03, 21.82)$ **42.** $(19.41, 27.72)$ and $(34.16, 12.14)$

7.2 Graphs of Linear Equations

Equations such as

$$2x + 3y = 6, \qquad y = 5 - \frac{1}{2}x, \qquad \text{and} \qquad \sqrt{3}x + 5y - 2 = 0$$

are called **linear equations** (or **first-degree equations**) in two variables, x and y. An ordered pair (a, b) is said to **satisfy** a linear equation if, when a is substituted for x and b is substituted for y, the resulting statement is true. A **solution** of a linear equation in two variables, x and y, is an ordered pair that satisfies the equation.

We can graphically display the solution of an equation in two variables by using the Cartesian coordinate system. The **graph** of an equation in the two variables x and y is defined to be the set of all points $P = (x, y)$ in the Cartesian plane such that (x, y) is a solution of the given equation. For example, to graph the equation

$$2x + 3y = 6$$

or

$$y = 2 - \frac{2}{3}x,$$

we choose some values of x, say -1, 0, 3, and 6, and then we determine the corresponding values for y. This procedure will yield some ordered pairs (x, y), which represent points on the graph. We show these points in the following table:

x	$y = 2 - \frac{2}{3}x$	Point (x, y)
-1	$y = 2 - \frac{2}{3}(-1) = \frac{8}{3}$	$(-1, \frac{8}{3})$
0	$y = 2 - \frac{2}{3}(0) \quad = 2$	$(0, 2)$
3	$y = 2 - \frac{2}{3}(3) \quad = 0$	$(3, 0)$
6	$y = 2 - \frac{2}{3}(6) \quad = -2$	$(6, -2)$

The points $(-1, \frac{8}{3})$, $(0, 2)$, $(3, 0)$, and $(6, -2)$ are plotted in Figure 1a. These points appear to lie on a straight line. It can be shown that every solution of the equation $2x + 3y = 6$ corresponds to a point on a single line. It can also be shown that the coordinates of each point on this line satisfy the equation $2x + 3y = 6$. Hence, the line is the graph of the equation $2x + 3y = 6$, and we can **sketch the graph** (Figure 1b). This sketch is understood to be a portion of the complete graph, because the line does not terminate, but continues on in both directions. In general, the graph of any first-degree equation in the two variables x and y of the form

$$Ax + By = C,$$

where A, B, and C are constants, and where A and B cannot both equal 0, is a **straight line.** For this reason, such equations are called **linear equations.**

Figure 1

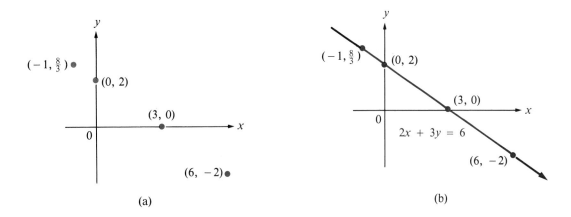

(a)   (b)

A straight line is completely determined by any two different points of the line. Therefore, we can graph a linear equation by finding two points whose coordinates are solutions of the equation, and drawing a straight line through these points. However, to avoid mistakes, we suggest that you perform the following steps:

Procedure for Graphing a Linear Equation

Step 1. Find three points on the graph of the given equation. (Choose three different values of x and calculate the corresponding values of y.)

Step 2. Plot the three points on a Cartesian coordinate system.

Step 3. Draw a straight line through the three points.

In Examples 1–7, sketch the graph of each equation.

EXAMPLE 1 $y = 3x$

SOLUTION   We choose three values of x and then determine the corresponding values for y. This will yield three points of the form (x, y). Let us choose the values -2, 0, and 1 for x, and find the corresponding values for y. We show the resulting points in the following table:

Figure 2

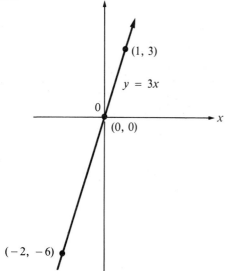

x	$y = 3x$	Point (x, y)
-2	$y = 3(-2) = -6$	$(-2, -6)$
0	$y = 3(0) = 0$	$(0, 0)$
1	$y = 3(1) = 3$	$(1, 3)$

We plot these points and draw a straight line through them (Figure 2).

EXAMPLE **2** $y = -2x$

SOLUTION   We choose three values of x and then determine the corresponding values for y to obtain three points of the form (x, y). Let us choose the values -2, 0, and 1 for x, and find the corresponding values for y. The resulting points are shown in the following table.

Figure 3

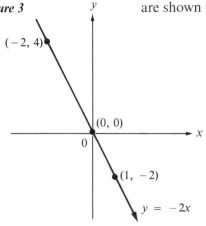

x	$y = -2x$	Point (x, y)
-2	$y = -2(-2) = 4$	$(-2, 4)$
0	$y = -2(0) = 0$	$(0, 0)$
1	$y = -2(1) = -2$	$(1, -2)$

We plot these points and draw a straight line through them (Figure 3).

EXAMPLE **3** $y = 2x + 1$

SOLUTION   We choose three values of x and then determine the corresponding values for y to obtain three points of the form (x, y). Let us choose the values -2, 0, and 1 for x, and find the corresponding values for y. The resulting points are shown in the following table.

Figure 4

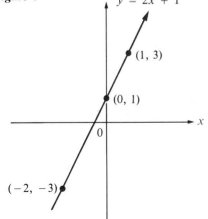

x	$y = 2x + 1$	Point (x, y)
-2	$y = 2(-2) + 1 = -3$	$(-2, -3)$
0	$y = 2(0) + 1 = 1$	$(0, 1)$
1	$y = 2(1) + 1 = 3$	$(1, 3)$

We plot these points and draw a straight line through them (Figure 4).

EXAMPLE 4 $y = -\frac{2}{3}x + 3$

SOLUTION   We choose the values -3, 0, and 3 for x, and find the corresponding values for y. The resulting points are shown in the following table.

Figure 5

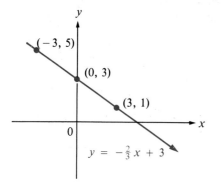

x	$y = -\frac{2}{3}x + 3$	Point (x, y)
-3	$y = -\frac{2}{3}(-3) + 3 = 5$	$(-3, 5)$
0	$y = -\frac{2}{3}(0) + 3 = 3$	$(0, 3)$
3	$y = -\frac{2}{3}(3) + 3 = 1$	$(3, 1)$

We plot these points and draw a straight line through them (Figure 5).

EXAMPLE 5 $3x - 4y - 12 = 0$

SOLUTION   We choose values for x and calculate the corresponding values for y. The calculations are more efficient if we solve the equation for y in terms of x:

$$3x - 4y - 12 = 0$$
$$-4y = -3x + 12$$
$$y = \frac{3}{4}x - 3.$$

We choose the values -4, 0, and 4 for x and calculate the y values:

Figure 6

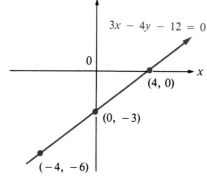

x	$y = \frac{3}{4}x - 3$	Point (x, y)
-4	$y = \frac{3}{4}(-4) - 3 = -6$	$(-4, -6)$
0	$y = \frac{3}{4}(0) - 3 = -3$	$(0, -3)$
4	$y = \frac{3}{4}(4) - 3 = 0$	$(4, 0)$

Now we plot these points and draw a straight line through them (Figure 6).

EXAMPLE 6 $2y = -4$

SOLUTION   The equation $2y = -4$ is equivalent to $2y = 0x - 4$ or $y = 0x - 2$. We choose the values -3, 0, and 1 for x and we obtain -2 in each case as a value for y. In fact, no matter what value is assigned to x, we obtain $y = -2$, since $0 \cdot x = 0$. The graph of this equation is a *horizontal line* crossing the y axis at -2 (Figure 7).

Figure 7

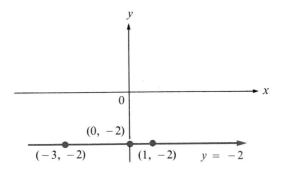

EXAMPLE 7 $x - 2 = 0$

SOLUTION   The equation $x - 2 = 0$ is equivalent to $x + 0y - 2 = 0$ or $x = 0y + 2$. If we choose the values -2, 0, and 3 for y, x will be 2. In fact, we find that $x = 2$ for any value of y. The graph of $x - 2 = 0$ is a *vertical line* crossing the x axis at 2 (Figure 8).

Figure 8

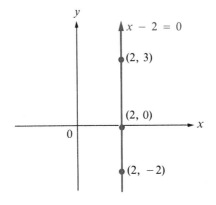

Many applications of mathematics involve the use of linear equations. The following example illustrates one of these applications.

EXAMPLE 8   The formula for changing temperature readings from Celsius C to Fahrenheit F is given by the linear equation

$$F = \frac{9}{5}C + 32.$$

(a) If the temperature is 15°C, what is the Fahrenheit temperature?

(b) If the temperature is 85°C, what is the Fahrenheit temperature?

(c) Sketch the graph of the equation by representing C on the horizontal axis and F on the vertical axis.

SOLUTION   (a) When $C = 15$,

Figure 9

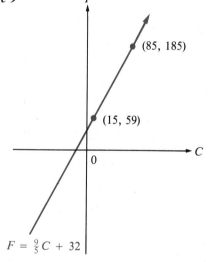

$$F = \frac{9}{5}(15) + 32$$
$$= 27 + 32$$
$$= 59.$$

(b) When $C = 85$,

$$F = \frac{9}{5}(85) + 32$$
$$= 153 + 32$$
$$= 185.$$

(c) We obtain the graph by drawing a straight line through the points (15, 59) and (85, 185) (Figure 9).

PROBLEM SET 7.2

In problems 1–36, sketch the graph of each equation.

1. $y = 2x$

2. $y = -5x$

3. $y = -\frac{2}{3}x$

4. $y = 7x$

5. $y = -4x$

6. $y = -3x$

7. $y = \frac{3}{7}x$

8. $y = \frac{4}{5}x$

9. $y = 3x + 1$

10. $y = 5x + 2$

11. $y = \frac{2}{3}x - 4$

12. $y = \frac{3}{4}x - 1$

13. $y = -2x + 5$

14. $y = -3x + 2$

15. $y = -\frac{2}{5}x + 1$

16. $y = -\frac{3}{4}x - 5$

17. $2x + 3y = -6$

18. $4x + 3y = 12$

19. $x - 3y = 5$

20. $4x - 3y - 12 = 0$

21. $-x + 5y = 10$

22. $-3x + 5y = 4$

23. $-2x - 4y + 1 = 0$

24. $-x - 2y + 4 = 0$

25. $-3x + 7y - 4 = 0$

26. $2x - 5y - 3 = 0$

27. $3y = 6$

28. $4y = 7$

29. $-2y = 5$

30. $-3y = 8$

31. $y - 6 = 0$

32. $2y - 3 = 0$

33. $3x - 2 = 0$

34. $5x + 1 = 0$

35. $-4x + 1 = 0$

36. $-7x + 2 = 0$

37. Chemists have found that the volume V (in cubic centimeters) of a liquid in a test tube changes as the temperature T (in Celsius) changes. The formula

$$V = \frac{1}{4}T + 32$$

describes this relationship. Sketch the graph of this equation by representing T on the horizontal axis and V on the vertical axis. Also find the volume when the temperature is 5°C.

38. A produce store makes a profit P (in cents) from selling x melons, and

$$P = 15x - 180.$$

(a) Sketch the graph of this equation by representing x on the horizontal axis and P on the vertical axis.
(b) How many melons must be sold for the store to break even?
(c) What is the store's profit if 100 melons are sold?

7.3 The Slope of a Line

Figure 1

$$\text{slope} = \frac{\text{rise}}{\text{run}}$$

We commonly use the word "slope" to refer to a steepness of an incline, or some deviation from the horizontal. For instance, we speak of a ski slope or the slope of a roof. In mathematics, the word slope has a similar meaning. Consider the line segment $\overline{AB}$ in Figure 1. The horizontal distance between A and B is called the **run.** The vertical distance between A and B is called the **rise.** The ratio of the rise to the run is called the **slope** of the line segment

A ski slope in the Swiss Alps.

Courtesy of the Swiss National Tourist Office

$\overline{AB}$, and is represented by the letter m:

$$m = \frac{\text{rise}}{\text{run}}.$$

If the line segment $\overline{AB}$ is horizontal, its rise is zero, and its slope $m = \text{rise}/\text{run} = 0$ (Figure 2a). If $\overline{AB}$ slants upward to the right, its rise is considered to be positive, and the slope $m = \text{rise}/\text{run}$ is positive (Figure 2b). If $\overline{AB}$ slants downward to the right, its rise is considered to be negative; hence, its slope $m = \text{rise}/\text{run}$ is negative (Figure 2c). The run is always considered to be nonnegative.

Figure 2

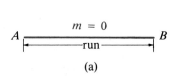

(a)

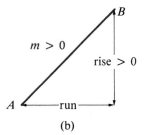

(b)

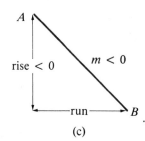

(c)

Now let's consider the line segment $\overline{AB}$ in which $A = (x_1, y_1)$ and $B = (x_2, y_2)$ (Figure 3). Note in Figure 3 that B is above and to the right of A. The line segment $\overline{AB}$ has rise $= y_2 - y_1$ and run $= x_2 - x_1$. Thus the slope is given by the formula

$$m = \frac{y_2 - y_1}{x_2 - x_1},$$

Figure 3

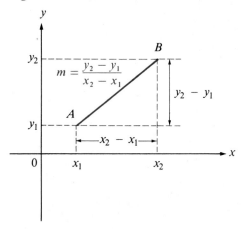

provided that $x_1 \neq x_2$. Notice that

$$\frac{y_2 - y_1}{x_2 - x_1} = \frac{y_1 - y_2}{x_1 - x_2}.$$

Therefore, the slope of a line segment is the same, no matter which endpoint is called (x_1, y_1) and which is called (x_2, y_2). Notice also that the slope m of a vertical line segment is undefined because the denominator (the run) is zero, that is,

$$m = \frac{y_2 - y_1}{0}$$

is undefined.

EXAMPLE **1**   Sketch each line segment $\overline{AB}$ and find its slope m.

(a) $A = (4, 1)$ and $B = (7, 6)$

(b) $A = (3, 9)$ and $B = (7, 4)$

(c) $A = (-1, 4)$ and $B = (3, 4)$

(d) $A = (2, -3)$ and $B = (2, 2)$

SOLUTION   The line segments are sketched in Figure 4.

Figure 4

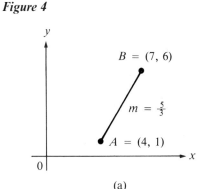

(a)

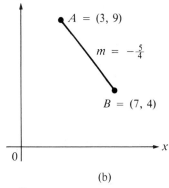

(b)

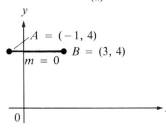

(c)

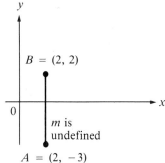

(a) $m = \dfrac{y_2 - y_1}{x_2 - x_1} = \dfrac{6 - 1}{7 - 4} = \dfrac{5}{3}$

(b) $m = \dfrac{y_2 - y_1}{x_2 - x_1} = \dfrac{4 - 9}{7 - 3} = -\dfrac{5}{4}$

(c) $m = \dfrac{y_2 - y_1}{x_2 - x_1} = \dfrac{4 - 4}{3 - (-1)} = \dfrac{0}{4} = 0$

(d) m is undefined because $x_2 - x_1 = 2 - 2 = 0$.

If two line segments $\overline{AB}$ and $\overline{CD}$ lie on the same line L, then they have the same slope (Figure 5). In fact, the slope of a line L is the same no matter which two points of the line are considered. The common slope of all line segments lying on a line L is called the **slope** of L.

Figure 5

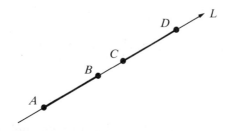

EXAMPLE 2 Sketch the line L that contains the point $P = (-2, 3)$ and that has slope
(a) $m = \frac{3}{5}$ and   (b) $m = -\frac{3}{5}$.

SOLUTION (a) The condition $m = \frac{3}{5}$ means that for every five units we move to the right from a point on L, we must move up three units to get back to L. If we start at the point $P = (-2, 3)$ on L and move five units to the right and three units up, we arrive at another point on line L, the point $Q = (-2 + 5, 3 + 3) = (3, 6)$. Because any two points on a line determine the line, we simply plot $P = (-2, 3)$ and $Q = (3, 6)$, and draw the line L (Figure 6a).

(b) The condition $m = -\frac{3}{5}$ means that for every five units we move to the right from a point on L, we must move down three units to get back to L. If we start at the point $P = (-2, 3)$ on L and move five units to the right and three units down, we arrive at the point $Q = (-2 + 5, 3 - 3) = (3, 0)$ on L. Thus, we plot $P = (-2, 3)$ and $Q = (3, 0)$ and draw the line L (Figure 6b).

Figure 6

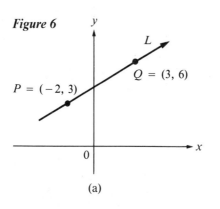

(a)

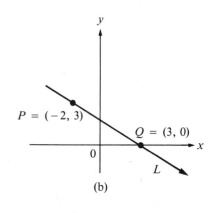

(b)

Figure 7

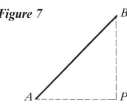

Parallel and Perpendicular Lines

Consider the similar triangles APB and CQD in Figure 7. The slopes of the two parallel line segments $\overline{AB}$ and $\overline{CD}$ are

$$\frac{|\overline{BP}|}{|\overline{AP}|} \quad \text{and} \quad \frac{|\overline{DQ}|}{|\overline{CQ}|},$$

respectively. These slopes are equal because they are the ratios of the corresponding sides of similar triangles:

$$\frac{|\overline{BP}|}{|\overline{AP}|} = \frac{|\overline{DQ}|}{|\overline{CQ}|}.$$

It follows that two parallel lines have the same slope. We can also use geometry to show that two different lines with the same slope are parallel. Therefore, we have the following condition:

The Parallelism Condition

> Two different nonvertical lines in the Cartesian plane with slopes m_1 and m_2 are **parallel** if and only if $m_1 = m_2$.

EXAMPLE 3 Determine whether or not the line containing the points A and B is parallel to the line containing the points C and D if $A = (-2, 5)$, $B = (2, -1)$, $C = (-4, 1)$, and $D = (0, -5)$. Sketch the lines.

SOLUTION The lines are sketched in Figure 8. We compute the slopes m_1 of $\overline{AB}$ and m_2 of $\overline{CD}$:

Figure 8

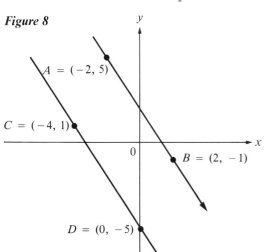

$$m_1 = \frac{y_2 - y_1}{x_2 - x_1}$$

$$= \frac{-1 - 5}{2 - (-2)} = -\frac{6}{4} = -\frac{3}{2}$$

$$m_2 = \frac{-5 - 1}{0 - (-4)} = -\frac{6}{4} = -\frac{3}{2}.$$

The two lines are parallel because the slopes are equal.

EXAMPLE 4
Find a constant real number k such that the line containing points A and B is parallel to the line containing points C and D, if $A = (-1, 1)$, $B = (1, k)$, $C = (3, 3)$, and $D = (5, 6)$.

SOLUTION
We compute the slopes m_1 of $\overline{AB}$ and m_2 of $\overline{CD}$:

$$m_1 = \frac{y_2 - y_1}{x_2 - x_1} = \frac{k - 1}{1 - (-1)} = \frac{k - 1}{2}$$

$$m_2 = \frac{6 - 3}{5 - 3} = \frac{3}{2}.$$

Because the two lines are to be parallel, we must have

$$m_1 = m_2.$$

We substitute the computed values for the slopes:

$$\frac{k - 1}{2} = \frac{3}{2},$$

or

$$k - 1 = 3$$

$$k = 4.$$

If we have two perpendicular lines with slopes m_1 and m_2, we can show that the product of their slopes $m_1 m_2$ is -1. Conversely, if we have two lines with slopes m_1 and m_2 such that the product of their slopes $m_1 m_2$ is -1, we can show that the lines are perpendicular.

The Perpendicularity Condition

> Two nonvertical lines in the Cartesian plane with slopes m_1 and m_2 are **perpendicular** if and only if $m_1 m_2 = -1$, or, equivalently, $m_2 = -1/m_1$.

EXAMPLE 5
Determine whether or not the line containing the points A and B is perpendicular to the line containing the points C and D, if $A = (-1, 1)$, $B = (1, 5)$, $C = (-2, -3)$, and $D = (2, -5)$.

SOLUTION
The lines are sketched in Figure 9. We compute the slopes m_1 of $\overline{AB}$ and m_2 of $\overline{CD}$:

$$m_1 = \frac{y_2 - y_1}{x_2 - x_1} = \frac{5 - 1}{1 - (-1)} = \frac{4}{2} = 2$$

$$m_2 = \frac{-5 - (-3)}{2 - (-2)} = \frac{-2}{4} = -\frac{1}{2}.$$

The two lines are perpendicular because $m_1m_2 = 2(-\frac{1}{2}) = -1$.

Figure 9

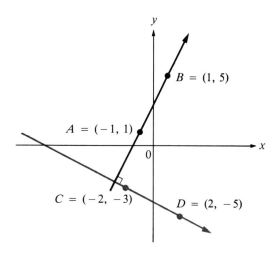

PROBLEM SET 7.3

In problems 1–10, sketch the line segment $\overline{AB}$ and find its slope m.

1. $A = (2, -1)$ and $B = (-3, 4)$ **2.** $A = (1, -4)$ and $B = (2, 3)$ **3.** $A = (1, 5)$ and $B = (-2, 3)$

4. $A = (4, 3)$ and $B = (-3, -4)$ **5.** $A = (6, -1)$ and $B = (0, 2)$ **6.** $A = (7, 1)$ and $B = (-8, 3)$

7. $A = (-2, 5)$ and $B = (4, 5)$ **8.** $A = (-1, 4)$ and $B = (3, 4)$ **9.** $A = (4, -1)$ and $B = (4, 3)$

10. $A = (3, 2)$ and $B = (3, -5)$

In problems 11–20, sketch the line L that contains the given point P and that has slope m.

11. $P = (1, 5)$ and $m = -3$ **12.** $P = (2, 5)$ and $m = 4$ **13.** $P = (2, -1)$ and $m = 2$

14. $P = (2, -3)$ and $m = -\frac{1}{2}$ **15.** $P = (7, -2)$ and $m = \frac{2}{3}$ **16.** $P = (-1, -6)$ and $m = \frac{3}{4}$

17. $P = (2, 3)$ and $m = -\frac{3}{7}$ **18.** $P = (1, 4)$ and $m = -\frac{4}{3}$ **19.** $P = (-1, -3)$ and $m = -\frac{4}{5}$

20. $P = (-1, -5)$ and $m = -\frac{3}{4}$

In problems 21–28, determine whether the line containing the points A and B is parallel or perpendicular (or neither) to the line containing the points C and D.

21. $A = (2, 4)$, $B = (3, 8)$, $C = (5, 1)$, and $D = (4, -3)$

22. $A = (2, -3)$, $B = (-4, 5)$, $C = (-1, 0)$, and $D = (-4, 4)$

23. $A = (1, 9)$, $B = (4, 0)$, $C = (0, 6)$, and $D = (5, 3)$

24. $A = (8, -1)$, $B = (2, 3)$, $C = (5, 1)$, and $D = (2, -7)$

25. $A = (2, 4)$, $B = (3, 8)$, $C = (8, -2)$, and $D = (-4, 1)$

26. $A = (-\frac{5}{3}, 0)$, $B = (0, 5)$, $C = (2, 1)$, and $D = (5, 0)$

27. $A = (8, -7)$, $B = (-7, 8)$, $C = (10, -7)$, and $D = (-4, 6)$

28. $A = (-2, 8)$, $B = (8, 2)$, $C = (-8, -2)$, and $D = (2, -8)$

In problems 29–34, find a constant real number k such that the lines containing the segments $\overline{AB}$ and $\overline{CD}$ are (a) parallel and (b) perpendicular.

29. $A = (2, 1)$, $B = (6, 3)$, $C = (4, k)$, and $D = (3, 1)$
30. $A = (k, 1)$, $B = (3, 2)$, $C = (7, 1)$, and $D = (6, 3)$
31. $A = (-10, k)$, $B = (-5, -5)$, $C = (4, 4)$, and $D = (10, 10)$
32. $A = (4, 9)$, $B = (10, k)$, $C = (-2, 11)$, and $D = (4, -2)$
33. $A = (2, 1)$, $B = (0, 4)$, $C = (2, k)$, and $D = (-3, 1)$
34. $A = (-9, -1)$, $B = (-6, 0)$, $C = (7, 1)$, and $D = (4, k)$
35. Show that the four points $A = (1, 9)$, $B = (4, 0)$, $C = (0, 6)$, and $D = (5, 3)$ are vertices of a parallelogram.
36. Show that the four points $A = (-4, -1)$, $B = (0, 2)$, $C = (-2, -1)$, and $D = (2, 2)$ are vertices of a parallelogram.

In problems 37–40, use the concept of slope to show that the triangle with vertices A, B, and C is a right triangle.

37. $A = (-4, 2)$, $B = (1, 4)$, and $C = (3, -1)$
38. $A = (2, 1)$, $B = (3, -1)$, and $C = (1, -2)$
39. $A = (8, 5)$, $B = (1, -2)$, and $C = (-3, 2)$
40. $A = (-6, 3)$, $B = (3, -5)$, and $C = (-1, 5)$

In problems 41–44, use the concept of slope to determine whether or not the points in each set are collinear. The points $P_1 = (x_1, y_1)$, $P_2 = (x_2, y_2)$, and $P_3 = (x_3, y_3)$ are **collinear** if the slope of the line between P_1 and P_2 is the same as the slope of the line between P_1 and P_3.

41. $(1, 1)$, $(2, 4)$, and $(3, 2)$
42. $(0, 3)$, $(1, 1)$, and $(2, -1)$
43. $(1, -3)$, $(-1, -11)$, and $(-2, -15)$
44. $(1, 5)$, $(-2, -1)$, and $(-3, -3)$

7.4 Equations of Lines

Suppose that we are given a point on a line and the slope of the line. We ask: Can we find an equation to describe all the points on the line? To answer this question, consider a nonvertical line L having slope m and containing the point $P_1 = (x_1, y_1)$ (Figure 1). If $P = (x, y)$ is any other point on L, then the slope m is given by

$$m = \frac{y - y_1}{x - x_1}.$$

We multiply both sides of this equation by $x - x_1$:

$$y - y_1 = m(x - x_1).$$

Figure 1

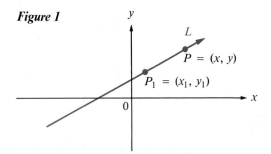

The latter equation is known as the **point-slope form** of an equation for the line L, which contains the point $P_1 = (x_1, y_1)$ and has the slope m.

In Examples 1 and 2, find an equation for the line L in point-slope form.

EXAMPLE **1** L contains the point $(-2, 3)$ and has slope $m = 4$.

SOLUTION   We substitute $x_1 = -2$, $y_1 = 3$, and $m = 4$ in $y - y_1 = m(x - x_1)$:

$$y - 3 = 4[x - (-2)]$$

or

$$y - 3 = 4(x + 2).$$

EXAMPLE **2** L contains the points $(6, 1)$ and $(8, 7)$.

SOLUTION   The slope m is given by

$$m = \frac{y_2 - y_1}{x_2 - x_1} = \frac{7 - 1}{8 - 6} = \frac{6}{2} = 3.$$

We use $P_1 = (x_1, y_1) = (6\ \ 1)$:

$$y - 1 = 3(x - 6).$$

Suppose that L is any nonvertical line with slope m. If $(a, 0)$ is the only point of intersection of the line L with the x axis, then a is called the **x intercept** of L. Similarly, if $(0, b)$ is the only point of intersection of L with the y axis, then b is called the **y intercept** of L. (Note that these intercepts are numbers rather than points) (Figure 2).

Figure 2

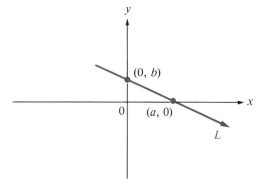

To determine the x intercept of line L, set $y = 0$ in any equation of L; to determine the y intercept, set $x = 0$.

Figure 3

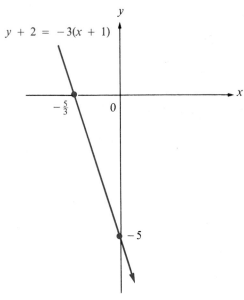

For example, to find the x intercept of the line $y + 2 = -3(x + 1)$ (Figure 3), set $y = 0$ and solve for x:

$$0 + 2 = -3(x + 1)$$
$$2 = -3x - 3$$
$$5 = -3x$$
$$x = -\frac{5}{3}.$$

To find the y intercept, set $x = 0$ and solve for y:

$$y + 2 = -3(0 + 1)$$
$$y + 2 = -3$$
$$y = -5.$$

We can find an equation of a line L in point-slope form if we know the slope m and a point on the line. If, for example, the point is $P_1 = (0, b)$, where b is the y intercept, then an equation of L is given by

$$y - b = m(x - 0).$$

We can simplify this equation:

$$\boxed{y = mx + b.}$$

The latter equation is called the **slope-intercept** form of an equation for L. In the equation $y = mx + b$ *the coefficient of x is the slope and the constant term b is the y intercept.*

EXAMPLE 3   Find the slope-intercept form of an equation of a line L having slope $m = -\frac{2}{3}$ and y intercept $b = 5$.

SOLUTION   We substitute $m = -\frac{2}{3}$ and $b = 5$ in the equation $y = mx + b$ to obtain

$$y = -\frac{2}{3}x + 5.$$

EXAMPLE 4   Write the equation $3x + 2y - 6 = 0$ in slope-intercept form. Find the slope m, the y intercept b, the x intercept, and sketch the graph.

SOLUTION   We begin by solving the equation for y in terms of x:

$$3x + 2y - 6 = 0$$

$$2y = -3x + 6$$

$$y = -\frac{3}{2}x + 3.$$

This last equation of the line is in slope-intercept form with $m = -\frac{3}{2}$ and y intercept $b = 3$.

To find the x intercept we substitute $y = 0$ in the original equation $3x + 2y - 6 = 0$:

$$3x + 2(0) - 6 = 0$$

$$3x - 6 = 0$$

$$3x = 6$$

$$x = 2.$$

Thus, we obtain the graph by drawing the line through the points $(2, 0)$ and $(0, 3)$ (Figure 4).

A **horizontal line** has slope $m = 0$. Hence, in slope-intercept form, an equation for a horizontal line is (Figure 5)

$$y = 0(x) + b$$

or

$$y = b.$$

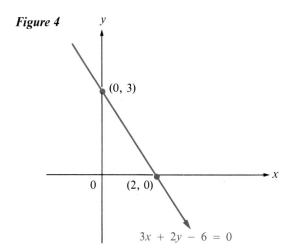

Figure 4

$3x + 2y - 6 = 0$

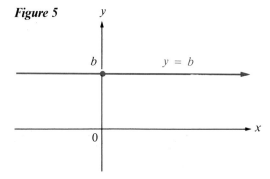

Figure 5

A **vertical line** has an undefined slope. Therefore, you cannot write an equation for a vertical line in slope-intercept form. However, because all points on a vertical line have the same abscissa or x coordinate—say, a—an equation of such a line is (Figure 6)

$$x = a.$$

Figure 6

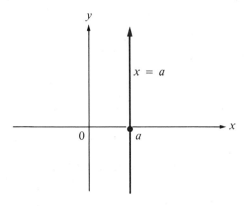

EXAMPLE 5

(a) Determine an equation for the horizontal line that contains the point $(1, 4)$.

(b) Determine an equation for the vertical line that contains the point $(-4, 3)$.

SOLUTION

(a) The slope of a horizontal line is 0. The y intercept is the y coordinate of the given point, that is, $b = 4$. Therefore, an equation is

$$y = 4.$$

(b) The slope is undefined because the line is vertical. The vertical line consists of all points with the x coordinate -4. Therefore, an equation is

$$x = -4.$$

If A, B, and C are constants, and if A and B are not both zero, any equation of the form

$$Ax + By + C = 0$$

represents a line. This equation is called the **general form** of an equation of the line.

In Examples 6 and 7, find an equation of the line L in (a) point-slope form, (b) slope-intercept form, and (c) general form.

EXAMPLE 6 *L* contains the point $(3, -5)$ and is parallel to the line L_1, whose equation is $3x - 4y + 9 = 0$.

SOLUTION We obtain the slope m_1 of L_1 by solving the equation $3x - 4y + 9 = 0$ for y in terms of x:

$$3x - 4y + 9 = 0$$
$$-4y = -3x - 9$$
$$y = \frac{3}{4}x + \frac{9}{4}.$$

Thus, the slope of L_1 is $m_1 = \frac{3}{4}$. We use the parallelism condition to find the slope m of L:

$$m = m_1 = \frac{3}{4}.$$

(a) Because L has the slope $m = \frac{3}{4}$ and contains the point $(3, -5)$, its equation in point-slope form is

$$y - (-5) = \frac{3}{4}(x - 3), \quad \text{or} \quad y + 5 = \frac{3}{4}(x - 3).$$

(b) To obtain an equation of L in slope-intercept form, we solve the equation in part (a) for y in terms of x.

$$y + 5 = \frac{3}{4}(x - 3)$$
$$y + 5 = \frac{3x}{4} - \frac{9}{4}$$
$$y = \frac{3}{4}x - \frac{9}{4} - 5$$
$$y = \frac{3}{4}x - \frac{29}{4}.$$

(c) To obtain an equation of L in general form, we multiply both sides of the equation in part (b) by 4 and rearrange terms.

$$y = \frac{3}{4}x - \frac{29}{4}$$
$$4y = 3x - 29$$
$$3x - 4y - 29 = 0.$$

EXAMPLE 7 L contains the point $(-4, 7)$ and is perpendicular to the line L_1, whose equation is $5x + 7y - 11 = 0$.

SOLUTION   We obtain the slope m_1 of L_1 by solving the equation $5x + 7y - 11 = 0$ for y in terms of x:

$$5x + 7y - 11 = 0$$

$$7y = -5x + 11$$

$$y = -\frac{5}{7}x + \frac{11}{7}.$$

Thus, the slope of L_1 is $m_1 = -\frac{5}{7}$. We use the perpendicularity condition to find the slope m of L:

$$m = -\frac{1}{m_1} = \frac{7}{5}.$$

(a) Because L has slope $m = \frac{7}{5}$ and contains the point $(-4, 7)$, its equation in point-slope form is

$$y - 7 = \frac{7}{5}[x - (-4)],$$

or

$$y - 7 = \frac{7}{5}(x + 4).$$

(b) We solve the equation in part (a) for y in terms of x and obtain an equation of L in slope-intercept form:

$$y - 7 = \frac{7}{5}(x + 4)$$

$$y - 7 = \frac{7}{5}x + \frac{28}{5}$$

$$y = \frac{7}{5}x + \frac{63}{5}.$$

(c) We multiply both sides of the equation in part (b) by 5 and rearrange terms to obtain an equation of L in general form:

$$y = \frac{7}{5}x + \frac{63}{5}$$

$$5y = 7x + 63$$

$$7x - 5y + 63 = 0.$$

EXAMPLE **8**   A book salesperson receives a monthly salary plus commission on sales. The relationship between income (salary plus commission) and the amount of sales is linear. The salesperson's income was $3,400 during a month when sales totaled $12,000 and was $2,900 in another month, when sales totaled $9,500.

(a) Write an equation that expresses the salesperson's monthly income I in terms of the total sales T for that month. What is the monthly salary?

(b) Find the salesperson's rate of commission.

(c) If sales were $14,000 in a particular month, what was the salesperson's income I for that month?

SOLUTION   (a) The information given can be expressed as the two ordered pairs

$$(T_1, I_1) = (9,500, 2,900) \quad \text{and} \quad (T_2, I_2) = (12,000, 3,400).$$

The slope of the line through these points is

$$m = \frac{I_2 - I_1}{T_2 - T_1} = \frac{3,400 - 2,900}{12,000 - 9,500}$$

$$= \frac{500}{2,500} = \frac{1}{5}.$$

Using the point (9,500, 2,900), we obtain the equation

$$I - 2,900 = \frac{1}{5}(T - 9,500).$$

This is equivalent to

$$I - 2,900 = \frac{1}{5}T - 1,900,$$

or

$$I = \frac{1}{5}T + 1,000.$$

This equation informs us that the monthly salary was $1,000, because if the total monthly sales T were zero, no commission would be earned and income I would equal salary, $1,000.

(b) The slope $m = \frac{1}{5}$ gives the salesperson's rate of commission. We can express this rate as a percentage:

$$\frac{1}{5} = 0.20$$

$$= 20\%.$$

(c) Here $T = 14{,}000$. Thus,

$$I = \frac{1}{5}(14{,}000) + 1{,}000$$

$$= 2{,}800 + 1{,}000$$

$$= 3{,}800.$$

Hence, the salesperson's income for a month in which sales were $14,000 was $3,800, made up of $2,800 in commission and $1,000 in salary.

PROBLEM SET 7.4

In problems 1–18, find an equation of the line L.

1. L contains $P = (-1, 2)$ and has slope $m = 5$.
2. L contains $P = (0, 3)$ and has slope $m = -7$.
3. L contains $P = (7, 3)$ and has slope $m = -3$.
4. L contains $P = (-1, 4)$ and has slope $m = \frac{2}{5}$.
5. L contains $P = (5, -1)$ and has slope $m = -\frac{3}{7}$.
6. L contains $P = (-4, 6)$ and has slope $m = -\frac{3}{4}$.
7. L contains $P = (0, 0)$ and has slope $m = \frac{3}{8}$.
8. L contains $P = (0, 0)$ and has slope $m = -\frac{4}{9}$.
9. L contains $P = (-1, -5)$ and has slope $m = 0$.
10. L contains $P = (-\frac{1}{2}, 4)$ and has slope $m = 0$.
11. L contains $P_1 = (-3, 2)$ and $P_2 = (3, 5)$.
12. L contains $P_1 = (-2, 4)$ and $P_2 = (0, 1)$.
13. L has slope $m = -3$ and y intercept $b = 5$.
14. L has slope $m = -7$ and y intercept $b = 2$.
15. L has slope $m = -\frac{3}{7}$ and y intercept $b = 0$.
16. L has slope $m = -\frac{5}{11}$ and y intercept $b = 0$.
17. L contains $P = (-3, 4)$ with undefined slope.
18. L contains $P = (2, -5)$ with undefined slope.

In problems 19–26, express each equation in slope-intercept form. Find the slope m, the x intercept, and the y intercept; and sketch the graph.

19. $2x - 3y - 1 = 0$
20. $2x + 3y + 12 = 0$
21. $y - 1 = -2(x - 2)$
22. $5x - 7y - 8 = 0$
23. $4x - y + 5 = 0$
24. $3x - 4y - 5 = 0$
25. $-2x + y = 0$
26. $y - 3 = -4(x - 3)$

In problems 27–40, find equations of the line L in (a) point-slope form, (b) slope-intercept form, and (c) general form.

27. L contains $P = (3, -1)$ and is parallel to the line with equation $x + 2y + 7 = 0$.
28. L contains $P = (-2, 1)$ and is parallel to the line with equation $5x - 7y - 8 = 0$.
29. L contains $P = (-\frac{1}{2}, 5)$ and is parallel to the line containing $P_1 = (4, -3)$ and $P_2 = (5, 7)$.
30. L contains $P = (6, -1)$ and is parallel to the line containing $P_1 = (-2, 3)$ and $P_2 = (3, -4)$.

31. L contains $P = (1, 5)$ and is perpendicular to the line whose equation is $2x + 3y - 1 = 0$.

32. L contains $P = (-3, 2)$ and is perpendicular to the line whose equation is $x + 2y - 5 = 0$.

33. L contains $P = (0, 0)$ and is perpendicular to the line containing $P_1 = (3, 0)$ and $P_2 = (-2, 3)$.

34. L contains $P = (-1, 2)$ and is perpendicular to the line containing $P_1 = (5, 1)$ and $P_2 = (-2, 3)$.

35. L has y intercept 3 and is parallel to the line $x - 2y = 5$.

36. L has y intercept 5 and is perpendicular to the line $y = 3x$.

37. L has x intercept -1 and is perpendicular to the line $3x - 2y = 7$.

38. L contains $P = (7, -3)$ and is parallel to the x axis.

39. L contains $P = (-1, 6)$ and is perpendicular to the y axis.

40. L is the perpendicular bisector of the line segment $\overline{AB}$, where $A = (4, -5)$ and $B = (8, 5)$.

41. Find a value of k so that each of the following conditions will hold.
   (a) The line $3x + ky + 2 = 0$ is parallel to the line $6x - 5y + 3 = 0$.
   (b) The line $y = (2 - k)x + 2$ is perpendicular to the line $y = 3x - 1$.

42. Suppose that the line L has nonzero x and y intercepts a and b, respectively. Show that an equation of L can be written in the **intercept form**

$$\frac{x}{a} + \frac{y}{b} = 1.$$

In problems 43–48, use the result of problem 42 to find the equation of the line in intercept form.

43. x intercept 5 and y intercept 6

44. x intercept -2 and y intercept 7

45. x intercept -3 and y intercept -1

46. x intercept -1 and y intercept -5

47. $5x - 2y = 5$

48. $9x + 5y = 13$

49. A jogger's heartbeat N (in beats per minute) is related to her speed V (in feet per second) by a linear equation. The jogger's heartbeat is 75 beats per minute when her speed is 10 feet per second, and her heartbeat is 80 beats per minute when her speed is 12 feet per second.
   (a) Write an equation that expresses the jogger's heartbeat N in terms of her speed V.
   (b) If the jogger's heartbeat is 90 beats per minute, what is her speed?

50. The annual simple interest earned I is related to the amount P invested in a bank by a linear equation. If you invest \$650, you earn \$45.50 in 1 year, and if you invest \$1,375, you earn \$96.25 in a year.
   (a) Write an equation that expresses the relationship between the annual interest earned I and the amount P invested.
   (b) How much money should you invest to earn \$515.55 in one year?

7.5 Systems of Linear Equations in Two Variables

Two linear equations in two variables may be associated in what is called a **system** of equations. The equations in such a system are usually written in a column with a brace on the left. For instance,

$$\begin{cases} 3x - y = 7 \\ 2x + y = 8 \end{cases}$$

is a system of two linear equations in two variables x and y. If each of the two equations in a system is a true statement when we substitute particular numbers for the variables, we say that the pair of numbers substituted is a **solution** to the system. For example, a solution to the system

$$\begin{cases} 2x + 3y = 12 \\ 6x - 2y = 14 \end{cases}$$

is $x = 3$ and $y = 2$. We usually write this solution as the ordered pair (3, 2). We assume that 3 is substituted for x and 2 for y.

The graph of each linear equation in the above system is a straight line. Therefore, the solution (3, 2) can be interpreted graphically as the point where the graph of $2x + 3y = 12$ *intersects* the graph of $6x - 2y = 14$ assuming, of course, that the graphs are drawn on the same coordinate system (Figure 1).

Figure 1

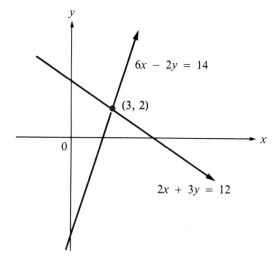

The graph of a system of linear equations yields considerable information about the solution to the system. If we graph the two lines of a system of linear equations on the same coordinate axes, one of the following cases will occur:

Case 1. The two lines intersect at exactly one point. Therefore, there is exactly one solution. In this case, we say that the system is **independent.**

Case 2. The two lines are parallel, and therefore do not intersect. In this case, there is no solution, and we say that the system is **inconsistent.**

Case 3. The two lines coincide. In this case, every point on the common line corresponds to a solution, and we say that the system is **dependent.**

EXAMPLE **1**   Use graphs to determine whether each system is independent, inconsistent, or dependent. Indicate the solutions (if any).

(a) $\begin{cases} 2x - 3y = -7 \\ x + 2y = 7 \end{cases}$ (b) $\begin{cases} 2x - 3y = -7 \\ 4x - 6y = 8 \end{cases}$ (c) $\begin{cases} 2x - 3y = -7 \\ 4x - 6y = -14 \end{cases}$

SOLUTION   The graphs of equations in systems (a), (b), and (c) are shown in Figure 2. Graphs of both equations in each system are drawn on the same coordinate system.

(a) In Figure 2a, the graphs intersect at one point. Therefore, there is one solution, and the system is independent. We see that the solution is $(1, 3)$.

(b) In Figure 2b, the two lines do not appear to intersect. In fact, they are parallel. Therefore, there is no solution, and the system is inconsistent.

(c) In Figure 2c, we see that the two equations have the same graph. Therefore, there are infinitely many solutions (one for each point on the graph of the line), and the system is dependent.

Figure 2

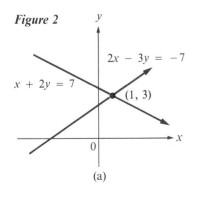

(a)

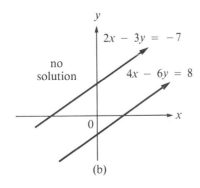

(b)

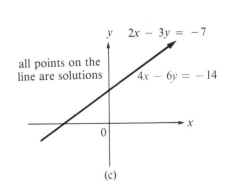

(c)

The Substitution Method

The graphic method for solving systems of linear equations is not always practical when we are interested in decimal-place accuracy. We can, however, use algebraic methods to obtain an exact solution of any independent system. The procedure for the algebraic technique known as the **substitution method** follows:

Procedure for Solving a Linear System by the Substitution Method

Step 1. Choose one of the equations and solve it for one of the variables in terms of the remaining variable.

Step 2. Substitute the resulting expression for the variable in the other equation to obtain an equation in one variable.

Step 3. Solve the equation resulting from step 2 for the variable.

Step 4. Substitute the value from step 3 in any of the equations involving both variables to obtain an equation in the remaining variable. Solve this equation.

In Examples 2–4, use the substitution method to solve each system.

EXAMPLE 2
$$\begin{cases} 3x + y = 7 \\ 2x - 3y = 1 \end{cases}$$

SOLUTION
We solve the equation $3x + y = 7$ for y in terms of x and obtain

$$y = -3x + 7.$$

We then substitute $-3x + 7$ for y in the equation $2x - 3y = 1$:

$$2x - 3(-3x + 7) = 1$$
$$2x + 9x - 21 = 1$$
$$11x = 22$$
$$x = 2.$$

Finally, we find y by substituting $x = 2$ in the equation $y = -3x + 7$:

$$y = -3(2) + 7$$
$$= -6 + 7$$
$$= 1.$$

Check To check the solution, we replace x by 2 and y by 1 in the original system:

$$\begin{cases} 3(2) + 1 = 7 \\ 2(2) - 3(1) = 1. \end{cases}$$

Therefore, the solution is $x = 2$ and $y = 1$, or $(2, 1)$.

EXAMPLE 3

$$\begin{cases} 7x + 5y = 62 \\ x - 9y = -30 \end{cases}$$

SOLUTION

We solve the equation $x - 9y = -30$ for x in terms of y and obtain

$$x = 9y - 30.$$

We then substitute $9y - 30$ for x in the equation $7x + 5y = 62$:

$$7(9y - 30) + 5y = 62$$

$$63y - 210 + 5y = 62$$

$$68y = 272$$

$$y = 4.$$

To obtain the corresponding value of x, we substitute $y = 4$ in the equation $x = 9y - 30$:

$$x = 9(4) - 30 = 36 - 30 = 6.$$

Therefore, the solution is $x = 6$ and $y = 4$, or $(6, 4)$.

EXAMPLE 4

$$\begin{cases} 2x - 3y = 16 \\ -5x + 7y = -39 \end{cases}$$

SOLUTION

We solve the equation $2x - 3y = 16$ for x:

$$x = \frac{3}{2}y + 8.$$

We then substitute $\frac{3}{2}y + 8$ for x in the equation $-5x + 7y = -39$:

$$-5\left(\frac{3}{2}y + 8\right) + 7y = -39$$

$$-\frac{15}{2}y - 40 + 7y = -39$$

$$-15y - 80 + 14y = -78$$

$$-y = 2$$

$$y = -2.$$

To obtain the corresponding value for x, we substitute $y = -2$ in the equation $x = \frac{3}{2}y + 8$:

$$x = \frac{3}{2}(-2) + 8 = -3 + 8 = 5.$$

Therefore, the solution is $x = 5$ and $y = -2$, or $(5, -2)$.

The Elimination Method

Another algebraic method can be used to obtain an exact solution of an independent system of linear equations in two variables. To use this second method we eliminate one of the variables by adding or subtracting the two equations in the system. This method, known as the **elimination method,** is also called the **addition or subtraction method.** The procedure for using the elimination method follows:

Procedure for Solving a Linear System by the Elimination Method

> Step 1. Multiply the terms of one or both equations by numbers that will make the corresponding coefficients of one of the variables the same or negatives of each other. (This step may not be necessary.)
>
> Step 2. Add or subtract the two equations resulting from step 1 to produce a single equation in a single variable.
>
> Step 3. Solve the equation resulting from step 2 for that variable.
>
> Step 4. Substitute the solution from step 3 in one of the equations in the original system, and solve the resulting equation to obtain the value of the other variable.

In Examples 5–7, use the elimination method to solve each system.

EXAMPLE 5
$$\begin{cases} x - y = 1 \\ 2x + y = 5 \end{cases}$$

SOLUTION
We notice that the y terms in the equations are negatives of each other. We add the two equations to eliminate the y term:

$$\begin{cases} x - y = 1 \\ \underline{2x + y = 5} \quad \text{add} \\ 3x \quad\quad = 6 \end{cases}$$

$$x = 2.$$

We substitute 2 for x in the equation $x - y = 1$:

$$2 - y = 1$$
$$-y = -1$$
$$y = 1.$$

Check We replace x by 2 and y by 1 in the original system:

$$\begin{cases} 2 - 1 = 1 \\ 2(2) + 1 = 5. \end{cases}$$

Therefore, the solution is $x = 2$ and $y = 1$, or (2, 1).

EXAMPLE 6

$$\begin{cases} 2r + 3s = 7 \\ 5r + s = -2 \end{cases}$$

SOLUTION

We eliminate s as follows:

$$\begin{cases} 2r + 3s = 7 \\ 5r + s = -2 \end{cases} \xrightarrow[\text{We multiply each side by 3.}]{} \begin{cases} 2r + 3s = 7 \\ 15r + 3s = -6 \end{cases}$$

We subtract the second equation from the first:

$$-13r = 13$$
$$r = -1.$$

We substitute $r = -1$ in the equation $5r + s = -2$:

$$5(-1) + s = -2$$
$$-5 + s = -2$$
$$s = 3.$$

Therefore, the solution is $r = -1$ and $s = 3$, or $(-1, 3)$.

EXAMPLE 7

$$\begin{cases} 3p + 2q = 8 \\ 2p - 3q = 14 \end{cases}$$

SOLUTION

We eliminate q as follows:

$$\begin{cases} 3p + 2q = 8 \\ 2p - 3q = 14 \end{cases} \begin{array}{c} \xrightarrow{\text{We multiply each side by 3.}} \\ \xrightarrow{\text{We multiply each side by 2.}} \end{array} \begin{cases} 9p + 6q = 24 \\ 4p - 6q = 28 \end{cases}$$

Adding the two equations, we have

$$13p = 52$$
$$p = 4.$$

We substitute $p = 4$ in the equation $3p + 2q = 8$:

$$3(4) + 2q = 8$$
$$12 + 2q = 8$$
$$2q = -4$$
$$q = -2.$$

Therefore, the solution is $p = 4$ and $q = -2$, or $(4, -2)$.

PROBLEM SET 7.5

In problems 1–6, sketch the graph of each system of linear equations. Use the graph to determine whether the system is dependent, inconsistent, or independent. If the system is independent, determine the coordinates of the solution graphically.

1. $\begin{cases} 3x - 2y = 1 \\ 6x - 8y = 2 \end{cases}$

2. $\begin{cases} y = 2x - 3 \\ 4x - 2y = 6 \end{cases}$

3. $\begin{cases} x + 3y = 6 \\ 2x + 6y = 8 \end{cases}$

4. $\begin{cases} 2x = y + 3 \\ 4x - 2y = 5 \end{cases}$

5. $\begin{cases} 2x - 3y = 1 \\ 5x + 2y = 12 \end{cases}$

6. $\begin{cases} 3x + 2y = 11 \\ -2x + y = 2 \end{cases}$

In problems 7–18, use the substitution method to solve each system.

7. $\begin{cases} 2x - y = 5 \\ x + 3y = 13 \end{cases}$

8. $\begin{cases} 3x + y = 4 \\ 7x - y = 6 \end{cases}$

9. $\begin{cases} 3x - y = -2 \\ x + y = 6 \end{cases}$

10. $\begin{cases} 5x + 7y = 4 \\ -x + 6y = 14 \end{cases}$

11. $\begin{cases} 5p - q = 13 \\ p + q = 1 \end{cases}$

12. $\begin{cases} 3r - 4s = -5 \\ 2r + 2s = 3 \end{cases}$

13. $\begin{cases} 13a + 11b = 21 \\ 7a + 6b = -3 \end{cases}$

14. $\begin{cases} 0.2m - 0.3n = 0.1 \\ 0.3m + 0.2n = 2.1 \end{cases}$

15. $\begin{cases} \frac{1}{2}x - \frac{3}{4}y = 1 \\ 3x + y = 1 \end{cases}$

16. $\begin{cases} \frac{1}{3}x + \frac{1}{2}y = 4 \\ \frac{1}{4}x - \frac{1}{3}y = -1 \end{cases}$

17. $\begin{cases} 0.5x - 1.2y = 0.3 \\ 0.7x + 1.5y = 3.6 \end{cases}$

18. $\begin{cases} 2u - v = 3 \\ 3u + v = 22 \end{cases}$

In problems 19–36, use the elimination method to solve each system.

19. $\begin{cases} 2x - y = 1 \\ x + y = 2 \end{cases}$

20. $\begin{cases} 2x + y = 10 \\ 3x - y = 5 \end{cases}$

21. $\begin{cases} u + 3v = 9 \\ u - v = 1 \end{cases}$

22. $\begin{cases} 5u + v = 14 \\ 2u + v = 5 \end{cases}$

23. $\begin{cases} 2r + 4s = 2 \\ -r + s = 8 \end{cases}$

24. $\begin{cases} 7p + q = 5 \\ -2p + q = 3 \end{cases}$

25. $\begin{cases} 3x + 2y = 4 \\ 5x + 3y = 7 \end{cases}$

26. $\begin{cases} 2x - 7y = 5 \\ x + y = -8 \end{cases}$

27. $\begin{cases} 7p + q = 3 \\ 5p + q = 6 \end{cases}$

28. $\begin{cases} 3y + 2z = 5 \\ 2y + 3z = 1 \end{cases}$

29. $\begin{cases} -3x + y = 3 \\ 4x + 2y = 10 \end{cases}$

30. $\begin{cases} 5x - 2y = 35 \\ x + 4y = 25 \end{cases}$

31. $\begin{cases} 4u - 3v = 1 \\ 3u + 4v = 6 \end{cases}$

32. $\begin{cases} 13y + 5z = 2 \\ 6y + 2z = 2 \end{cases}$

33. $\begin{cases} \frac{1}{2}x + \frac{1}{3}y = 13 \\ \frac{1}{5}x + \frac{1}{8}y = 5 \end{cases}$

34. $\begin{cases} \frac{1}{3}x - \frac{1}{4}y = 2 \\ \frac{1}{4}x - \frac{1}{2}y = 7 \end{cases}$

35. $\begin{cases} 3x + y = a \\ x - 3y = b \end{cases}$ a and b are constants

36. $\begin{cases} 3ax + 2by = 6 \\ 2ax - 5by = 7 \end{cases}$ a and b are constants

In problems 37–42, solve each system by using the substitution $u = 1/x$ and $v = 1/y$.

37. $\begin{cases} \dfrac{2}{x} - \dfrac{1}{y} = 9 \\[2mm] \dfrac{5}{x} - \dfrac{3}{y} = 14 \end{cases}$

38. $\begin{cases} \dfrac{5}{x} - \dfrac{2}{y} = 1 \\[2mm] \dfrac{8}{x} + 11 = \dfrac{5}{y} \end{cases}$

39. $\begin{cases} \dfrac{3}{x} + \dfrac{2}{y} = 2 \\[2mm] \dfrac{1}{x} - \dfrac{1}{y} = 9 \end{cases}$

40. $\begin{cases} \dfrac{4}{x} - \dfrac{3}{y} = 1 \\[2mm] \dfrac{3}{x} - \dfrac{4}{y} = 6 \end{cases}$

41. $\begin{cases} \dfrac{5}{x} + \dfrac{2}{y} = 1 \\[2mm] \dfrac{13}{x} + \dfrac{8}{y} = 11 \end{cases}$

42. $\begin{cases} \dfrac{4}{x} + \dfrac{1}{y} = 16 \\[2mm] \dfrac{3}{x} + \dfrac{1}{y} = 11 \end{cases}$

7.6 Systems of Linear Equations in Three Variables

An equation such as

$$3x + 5y + z = 8$$

is an example of a **linear equation in three variables,** because each of the three unknowns present is raised to the first power. One solution to this equation is $x = -1$, $y = 2$, and $z = 1$, because

$$3(-1) + 5(2) + 1 = -3 + 10 + 1$$

$$= 8.$$

This solution can be written as the **ordered triple** $(-1, 2, 1)$. Graphic methods for solving systems of linear equations in three unknowns are impractical because such graphs may only be drawn in a three-dimensional coordinate system. Therefore, we consider only the algebraic methods of substitution and elimination to solve systems of linear equations in three unknowns. To use the method of substitution, we choose one of the equations and solve it for one of the variables in terms of the remaining two. Then we substitute this solution into the remaining equations. This produces a system involving two equations and two unknowns.

In Examples 1 and 2, use the substitution method to solve each system of equations.

EXAMPLE 1

$$\begin{cases} x + y + z = 6 \\ 2x - y - z = 0 \\ x - y + 2z = 7 \end{cases}$$

SOLUTION

We begin by solving the first equation, $x + y + z = 6$, for z in terms of x and y:

$$z = 6 - x - y.$$

Now we substitute $6 - x - y$ for z in the second and third equations:

$$2x - y - (6 - x - y) = 0$$

and

$$x - y + 2(6 - x - y) = 7.$$

Simplifying these two equations, we have the system

$$\begin{cases} 3x = 6 \\ -x - 3y = -5. \end{cases}$$

We can now solve this latter system by using the substitution procedure again:

$$3x = 6$$

$$x = 2.$$

We substitute $x = 2$ in the equation $-x - 3y = -5$:

$$-2 - 3y = -5$$

$$-3y = -3$$

$$y = 1.$$

We have found that $x = 2$ and $y = 1$. Therefore, we need only substitute these values in the equation $z = 6 - x - y$ to find a value for z:

$$z = 6 - 2 - 1$$

$$= 3.$$

Check We replace x by 2, y by 1, and z by 3 in the original system:

$$\begin{cases} 2 + 1 + 3 = 6 \\ 2(2) - 1 - 3 = 0 \\ 2 - 1 + 2(3) = 7. \end{cases}$$

Hence, our solution is $x = 2$, $y = 1$, and $z = 3$, or $(2, 1, 3)$.

EXAMPLE 2

$$\begin{cases} r + 2s + 4t = 12 \\ 2r - 3s + t = 10 \\ 3r - s - 2t = 1 \end{cases}$$

SOLUTION We solve the first equation $r + 2s + 4t = 12$ for r in terms of s and t:

$$r = 12 - 2s - 4t.$$

Now we substitute $12 - 2s - 4t$ for r in the second and third equations:

$$2(12 - 2s - 4t) - 3s + t = 10$$

and

$$3(12 - 2s - 4t) - s - 2t = 1.$$

Simplifying these two equations gives us the system

$$\begin{cases} -7s - 7t = -14 \\ -7s - 14t = -35 \end{cases}$$

or

$$\begin{cases} s + t = 2 \\ s + 2t = 5. \end{cases}$$

We solve the equation $s + t = 2$ for t in terms of s:

$$t = 2 - s.$$

We substitute $2 - s$ for t in the equation $s + 2t = 5$:

$$s + 2(2 - s) = 5$$
$$s + 4 - 2s = 5$$
$$-s = 1$$
$$s = -1.$$

Now we substitute $s = -1$ in the equation $t = 2 - s$:

$$t = 2 - (-1) = 3.$$

Finally, we substitute $s = -1$ and $t = 3$ in the equation $r = 12 - 2s - 4t$:

$$r = 12 - 2(-1) - 4(3) = 2.$$

Therefore, the solution is $r = 2$, $s = -1$, and $t = 3$, or $(2, -1, 3)$.

The method of substitution can be quite efficient for solving a system of linear equations in three variables. However, it is sometimes more efficient to use the method of elimination, which has the additional advantage that it can be programmed on a computer.

In Examples 3 and 4, use the elimination method to solve each system.

EXAMPLE 3

$$\begin{cases} x + y + z = 2 \\ 2x + 3y - z = 3 \\ 3x + 5y + z = 8 \end{cases}$$

SOLUTION   First we eliminate z as follows:

$$\begin{cases} x + y + z = 2 \\ \underline{2x + 3y - z = 3} \quad \text{add} \\ 3x + 4y \quad\quad = 5 \end{cases}$$

and

$$\begin{cases} 2x + 3y - z = 3 \\ \underline{3x + 5y + z = 8} \quad \text{add} \\ 5x + 8y \quad\quad = 11. \end{cases}$$

Next we eliminate y as follows:

$$\begin{cases} 3x + 4y = 5 \\ 5x + 8y = 11 \end{cases} \xrightarrow{\text{We multiply each side by 2.}} \begin{cases} 6x + 8y = 10 \\ \underline{5x + 8y = 11} \quad \text{subtract} \\ x \quad\quad = -1. \end{cases}$$

We substitute $x = -1$ in the equation $3x + 4y = 5$:

$$3(-1) + 4y = 5$$
$$4y = 8$$
$$y = 2.$$

We substitute $x = -1$ and $y = 2$ in the equation $x + y + z = 2$:

$$-1 + 2 + z = 2$$
$$z = 1.$$

Therefore, the solution is $x = -1$, $y = 2$, and $z = 1$, or $(-1, 2, 1)$.

EXAMPLE 4

$$\begin{cases} p + 2q + 5r = 4 \\ 4p + q + 3r = 9 \\ 6p + 9q + r = 21 \end{cases}$$

SOLUTION First we eliminate p as follows:

We multiply each side by 4.

$$\begin{cases} p + 2q + 5r = 4 \\ 4p + q + 3r = 9 \\ 6p + 9q + r = 21 \end{cases} \longrightarrow \begin{cases} 4p + 8q + 20r = 16 \\ \underline{4p + q + 3r = 9} \quad \text{subtract} \\ 7q + 17r = 7 \end{cases}$$

We multiply each side by 6.

$$\begin{cases} p + 2q + 5r = 4 \\ 4p + q + 3r = 9 \\ 6p + 9q + r = 21 \end{cases} \longrightarrow \begin{cases} 6p + 12q + 30r = 24 \\ \underline{6p + 9q + r = 21} \quad \text{subtract} \\ 3q + 29r = 3. \end{cases}$$

We now have a system of two equations in two unknowns. We can eliminate q as follows:

We multiply each side by 3.

$$\begin{cases} 7q + 17r = 7 \\ 3q + 29r = 3 \end{cases} \xrightarrow{\text{We multiply each side by 7.}} \begin{cases} 21q + 51r = 21 \\ \underline{21q + 203r = 21} \quad \text{subtract} \\ -152r = 0 \end{cases}$$

or

$$r = 0.$$

We substitute $r = 0$ in the equation $3q + 29r = 3$:

$$3q + 29(0) = 3$$
$$3q = 3$$
$$q = 1.$$

Finally, we substitute $r = 0$ and $q = 1$ in the equation $p + 2q + 5r = 4$:

$$p + 2(1) + 5(0) = 4$$
$$p = 2.$$

Therefore, the solution is $p = 2$, $q = 1$, and $r = 0$, or $(2, 1, 0)$.

PROBLEM SET 7.6

In problems 1–6, use the substitution method to solve each system.

1. $\begin{cases} x + y = 5 \\ x + z = 1 \\ y + z = 2 \end{cases}$

2. $\begin{cases} 2x + 3y = 28 \\ 3y + 4z = 46 \\ 4z + 5x = 53 \end{cases}$

3. $\begin{cases} x + y + 2z = 11 \\ x - y + z = 3 \\ 2x + y + 3z = 17 \end{cases}$

4. $\begin{cases} x - 3y = -11 \\ 2y - 5z = 26 \\ 7x - 3z = -2 \end{cases}$

5. $\begin{cases} 2p - q + r = 8 \\ p + 2q + 3r = 9 \\ 4p + q - 2r = 1 \end{cases}$

6. $\begin{cases} s + 3t - u = 4 \\ 3s - 2t + 4u = 11 \\ 2s + t + 3u = 13 \end{cases}$

In problems 7–20, use the elimination method to solve each system.

7. $\begin{cases} x + y + 2z = 4 \\ x + y - 2z = 0 \\ x - y = 0 \end{cases}$

8. $\begin{cases} x + y + z = 2 \\ x + 2y - z = 4 \\ 2x - y + z = 0 \end{cases}$

9. $\begin{cases} x + y + z = 6 \\ x - y + 2z = 12 \\ 2x + y + z = 1 \end{cases}$

10. $\begin{cases} x + y + 2z = 4 \\ x - 5y + z = 5 \\ 3x - 4y + 7z = 24 \end{cases}$

11. $\begin{cases} x + y = 4 \\ 3x - y + 3z = 7 \\ 5x - 7y + 2z = -2 \end{cases}$

12. $\begin{cases} 7x + y + 3z = -6 \\ 4x - 5y + 6z = -27 \\ x + 15y - 9z = 64 \end{cases}$

13. $\begin{cases} 2p + q - 3r = 9 \\ p - 2q + 4r = 5 \\ 3p + q - 2r = 15 \end{cases}$

14. $\begin{cases} u + 3v - w = -2 \\ 7u - 5v + 4w = 11 \\ 2u + v + 3w = 21 \end{cases}$

15. $\begin{cases} 2r + 3s + t = 6 \\ r - 2s + 3t = 3 \\ 3r + s - t = 8 \end{cases}$

16. $\begin{cases} 8s + 3t - 18u = -76 \\ 10s + 6t - 6u = -50 \\ 4s + 9t + 12u = 10 \end{cases}$

17. $\begin{cases} a - 5b + 4c = 8 \\ 3a + b - 2c = 4 \\ 9a - 3b + 6c = 6 \end{cases}$

18. $\begin{cases} 2p + 3q - 2r = 3 \\ 8p + q + r = 2 \\ 2p + 2q + r = 1 \end{cases}$

19. $\begin{cases} \dfrac{p}{2} + \dfrac{q}{3} - \dfrac{r}{4} = -1 \\[2mm] \dfrac{p}{3} + \dfrac{r}{2} = 8 \\[2mm] \dfrac{2p}{3} + \dfrac{q}{3} - \dfrac{3r}{4} = -6 \end{cases}$

20. $\begin{cases} 0.5u + 1.5v - 0.5w = 2 \\ -1.5u - 2.5v + 0.5w = -4 \\ -0.5v + 1.5w = 7 \end{cases}$

In problems 21–22, solve each system by using the substitutions $u = 1/x$, $v = 1/y$, and $w = 1/z$.

21. $\begin{cases} \dfrac{3}{x} - \dfrac{4}{y} + \dfrac{6}{z} = 1 \\[2mm] \dfrac{9}{x} + \dfrac{8}{y} - \dfrac{12}{z} = 3 \\[2mm] \dfrac{9}{x} - \dfrac{4}{y} + \dfrac{12}{z} = 4 \end{cases}$

22. $\begin{cases} \dfrac{3}{x} + \dfrac{1}{y} - \dfrac{1}{z} = 5 \\[2mm] \dfrac{4}{x} - \dfrac{1}{y} + \dfrac{2}{z} = 13 \\[2mm] \dfrac{2}{x} + \dfrac{2}{y} + \dfrac{3}{z} = 22 \end{cases}$

7.7 Applications Involving Linear Systems

In Chapter 4, Section 4.4, we worked applied problems that gave rise to linear equations in one variable. Problems in applied mathematics often contain two or more variables, rather than one. We can solve word problems involving linear systems with two or more variables by using a slight variation of the procedure in Chapter 4 (page 131).

EXAMPLE 1   The difference between two numbers is 12. Also, the sum of the larger number and twice the smaller number is 75. Find the numbers.

SOLUTION   Let
$$x = \text{the larger number,}$$
and let
$$y = \text{the smaller number.}$$
We have the system of linear equations
$$\begin{cases} x + 2y = 75 \\ x - y = 12. \end{cases}$$
To solve this system, we subtract the second equation from the first equation:
$$\begin{cases} x + 2y = 75 \\ \underline{x - y = 12} \\ 3y = 63 \end{cases}$$
$$y = 21$$
and
$$x - 21 = 12 \quad \text{or} \quad x = 33.$$
Therefore, the two numbers are 33 and 21.

Check $33 + 2(21) = 75$ and $33 - 21 = 12.$

EXAMPLE 2   A cash register contains 58 bills with a total value of $178. If the money is all in $1 and $5 bills, how many bills of each denomination are there?

SOLUTION   Let
$$x = \text{the number of \$1 bills}$$
$$y = \text{the number of \$5 bills.}$$
We have the following system of linear equations:
$$\begin{cases} x + 5y = 178 \\ x + y = 58. \end{cases}$$
We solve this system by subtracting:
$$\begin{cases} x + 5y = 178 \\ \underline{x + y = 58} \\ 4y = 120 \end{cases}$$
$$y = 30.$$

Substituting $y = 30$ in the equation $x + y = 58$, we obtain

$$x + 30 = 58 \qquad \text{or}$$

$$x = 28.$$

Therefore, the cash register contains 28 $1 bills and 30 $5 bills.

Check $\qquad\qquad\qquad$ $28(1) + 30(5) = \$28 + \$150$

$$= \$178.$$

EXAMPLE 3   The proprietor of a television repair shop charges a fixed charge plus an hourly rate to repair a set. If he charged $50 to repair a set that was worked on for 1 hour, and $90 to repair a set that was worked on for 3 hours, find the proprietor's fixed charge and hourly rate.

SOLUTION   Let

$$x = \text{the proprietor's fixed charge}$$

$$y = \text{the proprietor's hourly rate.}$$

The fixed charge plus the hourly rate for 1 hour equals $50, so

$$x + y = 50.$$

Also, the fixed charge plus the hourly rate for 3 hours equals $90, so

$$x + 3y = 90.$$

Thus, the system of equations is

$$\begin{cases} x + y = 50 \\ x + 3y = 90. \end{cases}$$

We solve this system by subtracting:

$$\begin{cases} x + y = 50 \\ \underline{x + 3y = 90} \\ -2y = -40 \end{cases}$$

$$y = 20.$$

We substitute $y = 20$ in the equation $x + y = 50$:

$$x + 20 = 50$$

$$x = 30.$$

Therefore, the fixed charge is $30 and the hourly rate is $20.

Check $\qquad\qquad$ $30 + 20 = 50 \qquad \text{and} \qquad 30 + 3(20) = 90.$

EXAMPLE 4   A bank customer bought two commercial papers for $40,000. One commercial paper paid 17% simple annual interest and the other paid 18% simple annual interest.

If the total interest from both papers, at the end of 1 year, is $7,050, how much did the customer pay for each paper?

SOLUTION   Let

$$x = \text{the amount (in dollars) paid for the 17\% paper}$$

$$y = \text{the amount (in dollars) paid for the 18\% paper.}$$

Then

$$x + y = 40{,}000,$$

and the total interest is

$$0.17x + 0.18y = 7{,}050.$$

The system is

$$\begin{cases} x + y = 40{,}000 \\ 0.17x + 0.18y = 7{,}050 \end{cases}$$

or

$$\begin{cases} x + y = 40{,}000 \\ 17x + 18y = 705{,}000. \end{cases}$$

We solve this system as follows:

We multiply each side by -17.

$$\begin{cases} x + y = 40{,}000 \\ 17x + 18y = 705{,}000 \end{cases} \longrightarrow \begin{cases} -17x - 17y = -680{,}000 \\ \underline{17x + 18y = 705{,}000} \\ \quad\quad\quad y = 25{,}000. \end{cases} \text{add}$$

We substitute $y = 25{,}000$ in the equation $x + y = 40{,}000$:

$$x + 25{,}000 = 40{,}000$$

$$x = 15{,}000.$$

Therefore, the bank customer paid $15,000 for the commercial paper carrying 17% interest and $25,000 for the commercial paper carrying 18% interest.

Check

$$0.17(15{,}000) + 0.18(25{,}000) = 2{,}550 + 4{,}500 = 7{,}050$$

$$15{,}000 + 25{,}000 = 40{,}000.$$

EXAMPLE 5   A concert was held in a sports arena that can seat 20,000 people. Tickets were available for $20, $15, and $10. There were twice as many $15 seats as $20 seats. If the concert was sold out and grossed $260,000, how many tickets of each kind were sold?

SOLUTION   Let

$$x = \text{the number of \$10 tickets}$$

$$y = \text{the number of \$15 tickets}$$

$$z = \text{the number of \$20 tickets.}$$

Then we have the following system of equations:

$$\begin{cases} x + y + z = 20{,}000 \\ y = 2z \\ 10x + 15y + 20z = 260{,}000. \end{cases}$$

We solve this system as follows:

We multiply each side by -10.

$$\begin{cases} x + y + z = 20{,}000 \\ 10x + 15y + 20z = 260{,}000 \end{cases} \longrightarrow \begin{cases} -10x - 10y - 10z = -200{,}000 \\ \underline{10x + 15y + 20z = 260{,}000} \quad \text{add} \\ 5y + 10z = 60{,}000. \end{cases}$$

so that we now have the system

$$\begin{cases} 5y + 10z = 60{,}000 \\ y - 2z = 0 \end{cases} \xrightarrow[\text{We multiply each side by 5.}]{} \begin{cases} 5y + 10z = 60{,}000 \\ \underline{5y - 10z = 0} \qquad \text{add} \\ 10y = 60{,}000 \\ y = 6{,}000. \end{cases}$$

We substitute $y = 6{,}000$ in the equation $y = 2z$:

$$6{,}000 = 2z$$

$$3{,}000 = z.$$

Next we substitute $y = 6{,}000$ and $z = 3{,}000$ in $x + y + z = 20{,}000$:

$$x + 6{,}000 + 3{,}000 = 20{,}000$$

$$x = 11{,}000.$$

Thus, there were 11,000 tickets sold at $10, 6,000 tickets sold at $15, and 3,000 tickets sold at $20.

Check

$$11,000 + 6,000 + 3,000 = 20,000$$

$$6,000 = 2(3,000)$$

$$10(11,000) + 15(6,000) + 20(3,000) = 260,000.$$

PROBLEM SET 7.7

© In many of the following problems, a calculator may be useful to speed up the arithmetic.

1. The sum of two numbers is 12. If one of the numbers is multiplied by 5, and the other is multiplied by 8, the sum of the products is 75. Find the numbers.

2. Find two numbers such that twice the first plus five times the second is 20, and four times the first less three times the second is 14.

3. The sum of the ages of Pete and Sal is 15. If Pete is four times as old as Sal, how old is each boy?

4. A bus has 31 passengers. If the number of adults exceeds the number of children by 5, how many adults are in the bus?

5. In a certain habitat, the number of prey exceeds the number of predators by 4,200, while together they total 5,650. How many prey and how many predators are in the habitat?

6. The sum of the digits of a two-place number is 13, and the number is increased by 9 when the order of the digits is reversed. Find the number.

7. Joe and Dawn are arguing over the difference in their ages. Joe says that he is 3 years older than Dawn. Dawn claims that 4 years ago the sum of their ages was 23. If they are both right, how old is each now?

8. Joan is 3 years older than Steve; 8 years ago she was four times as old as Steve. How old is each now?

9. Raul has 18 coins worth $3.60 in his pocket, and the coins are all quarters and dimes. How many quarters and how many dimes does he have?

10. Nadia has 11 coins, consisting of nickels and dimes. How many of each coin does she have if the total amount is 90 cents?

11. A cashier has 45 coins in dimes and quarters. If she has $8.70, how many coins of each type does she have?

12. After counting his cash, a bookstore cashier finds that he has three times as many dimes as nickels. If the value of the dimes and nickels is $4.20, how many coins of each type does he have?

13. A bankteller has 78 bills in $5 and $10 denominations. The total value is $465. How many bills of each denomination does the teller have?

14. A theater sells tickets at $2.50 for children and $4 for adults. If 375 tickets were sold to one show and $1,398 was collected, how many tickets of each kind were sold?

15. A plumber charges a fixed charge plus an hourly rate for service on a house call. The plumber charged $70 to repair a water tank that required 2 hours of labor and $100 to repair a water tank that took 3.5 hours. Find the plumber's fixed charge and hourly rate.

16. An agency providing temporary help charges $8 per hour for a bookkeeper and $5 per hour for a typist. A school called the agency to obtain a bookkeeper and a typist. The bill from the agency for their services came to $88. If the typist worked 2 more hours than the bookkeeper, what were the total number of hours worked by each employee, and how much did each earn?

17. If Tom and John work together, they can complete a job in 4 hours. When Tom works by himself, it takes him twice the time that it takes John, working alone, to finish the same job. How long does it take each man, working alone, to complete the job?

18. A 100-gallon vat in a chemical plant has two intake pipes, a large one carrying water and a small one carrying acid. If both pipes are turned on together, the vat can be filled in 40 minutes. Ordinarily, however, the water pipe is allowed to run for 45 minutes and is shut off before the acid pipe is turned on. If it then takes the acid pipe 20 minutes to fill the tank, what are the delivery rates of the water pipe and the acid pipe?

19. An electrician works for 8 hours and his assistant works for 6 hours on a job that pays them a total of $440. If the electrician is paid twice as much per hour as his assistant, what is the hourly rate for each?

20. A produce person bought a number of cases of vegetables for $221. He obtained some of the vegetables at the rate of three cases for $17 and the rest at the rate of four cases for $17. When he sold all the vegetables at $10 per case, he cleared $209. How many cases of each kind did the produce person buy?

21. A businesswoman has invested a total of $40,000 in two certificates. The first certificate pays 10.5% simple annual interest, and the second pays 13.5%. At the end of 1 year, her combined interest on the two certificates is $4,650. How much did she originally invest in each certificate?

22. A person invests part of $35,000 in a certificate that yields 11.5% simple annual interest, and the rest in a certificate that yields 13.2%. At the end of 1 year, the combined interest on the two certificates is $4,382. How much was invested in each certificate?

23. A man invested a total of $4,000 in securities. Part was invested at 8.5% simple annual interest and the rest was invested at 7.9%. His annual income from both investments was $329.50. What amount did he invest at each rate?

24. A chemist has in her laboratory the same acid in two strengths. Six parts of the first mixed with four parts of the second gives a mixture that is 86% pure, and four parts of the first mixed with six parts of the second gives a mixture that is 84% pure. What is the percentage of purity of each of the original solutions of acid?

25. Victor has loaned part of $30,000 at 18% simple annual interest and the rest at 19.5%. If after 1 year the income from the money loaned at 18% is $1,650 more than the income from the money loaned at 19.5%, what was the amount of each loan?

26. An 80% acid solution is mixed with an 18% acid solution. The result is 3 gallons of a solution that is one-third acid. How much of each solution was used?

27. A cashier has $48 consisting of 294 coins in half-dollars, quarters, and dimes. There are $3\frac{1}{2}$ times as many dimes as quarters. How many coins of each kind are there?

28. Judy bought three different bonds for $20,000, one paying a 6% annual dividend, one paying a 7% dividend, and the other paying an 8% dividend. If the sum of the dividends from the 6% and the 7% bonds amounts to $940 in a year, and the sum of the dividends from the 6% and the 8% bonds is $720, how much did she invest in each bond?

29. A department store has sold 80 men's suits of three different types at a discount. If the suits had been sold at their original prices—type I suits for $80, type II suits for $90, and type III suits for $95—the total receipts would have been $6,825. However, the suits were sold for $75, $80, and $85, respectively, and the total receipts amounted to $6,250. Determine the number of suits of each type sold during the sale.

30. A watch, a chain, and a ring together cost $225. The watch cost $50 more than the chain, and the ring cost $25 more than the watch and the chain together. What was the cost of each item?

7.8 Systems of Linear Inequalities

Linear inequalities play an important part in a number of practical applications of mathematics, especially those that involve the allocation of limited resources. In this section, we graph linear inequalities in two unknowns, and then we graph systems of inequalities on the same coordinate system.

Graphs of Linear Inequalities

Recall from Section 7.2 that the graph of a linear equation in two unknowns is a straight line. Now we graph linear inequalities such as

$$y < 3x + 6, \qquad 5x - 7 > y, \qquad \text{and}$$
$$y \geq 5x + 3.$$

The **graph** of an inequality in two unknowns is defined to be the set of all points (x, y) in the plane whose coordinates satisfy the inequality. To study the graph of a linear inequality in x and y, we begin by considering the graph of the associated linear equation. This latter graph is obtained by (temporarily) replacing the inequality sign with an equals sign. The graph is a straight line that divides the xy plane into two regions called **half-planes,** one above the line and one below it (assuming the line is not vertical). In Figure 1, the graph of the line $y = 3x + 6$ divides the xy plane into an upper half-plane A and a lower half-plane C. The set of all points on the line is represented by B. The solution of the inequality $y \leq 3x + 6$ consists of all points in the half-plane C and includes all points on the line B (Figure 1).

Figure 1

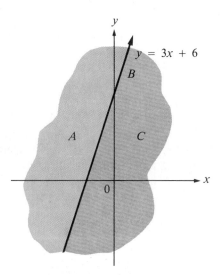

The procedure for sketching the graph of a linear inequality follows:

Procedure for Sketching Graphs of Linear Inequalities

Step 1. Sketch the graph of the linear equation obtained by replacing the inequality sign with an equals sign. If the inequality has the symbols $<$ or $>$, draw a dashed line; if the inequality has the symbols $\leq$ or $\geq$, draw a solid line.

Step 2. Determine which half-plane corresponds to the inequality. To do this, select any convenient test point (x, y) *not* on the line.
  (i) If the point satisfies the original inequality, shade the half-plane containing the test point.
  (ii) If the point does not satisfy the inequality, shade the half-plane not containing the test point.

In Examples 1–3, sketch the graph of each inequality.

EXAMPLE 1 $y < x + 2$

SOLUTION   Step 1. Because the inequality contains the symbol $<$, we draw the graph of $y = x + 2$ as a dashed line (Figure 2).

Step 2. We select the test point $(1, 2)$ not on the line. We see that $2 < 1 + 2$ is true.
        Therefore, we shade the lower half-plane, which contains the point $(1, 2)$ (Figure 2).

Figure 2

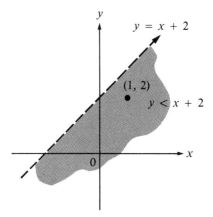

EXAMPLE 2 $2x - y \leq 1$

SOLUTION   Step 1. Because the inequality contains the symbol $\leq$, we draw the graph of $2x - y = 1$ as a solid line (Figure 3).

Figure 3

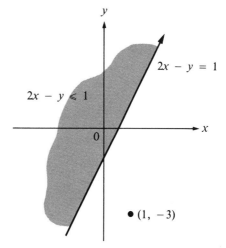

Step 2. We test the inequality at the point $(1, -3)$ and find that

$$2(1) - (-3) \leq 1$$

or

$$5 \leq 1$$

is *false*.

Therefore, we shade the upper half-plane, which does not contain the point $(1, -3)$ (Figure 3).

EXAMPLE **3** $y \geq 3$

SOLUTION

Step 1. We draw the graph of $y = 3$ as a solid line (Figure 4).

Step 2. We test the inequality at the point $(-2, 1)$ and find that

$$1 \geq 3$$

is *false*.

Therefore, we shade the upper half-plane, which does not contain the point $(-2, 1)$ (Figure 4).

Figure 4

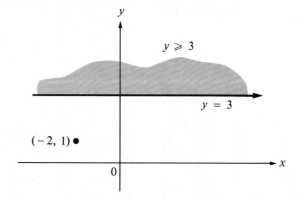

Graphs of Systems of Linear Inequalities

We define the **graph** of a system of linear inequalities as the set of all points (x, y) in the xy plane whose coordinates satisfy every inequality in the system. Such a graph is obtained by sketching the graphs of all inequalities on the same coordinate axes. The region where the half-planes intersect is the **solution** of the system.

In Examples 4–6, sketch the graph of each system of inequalities.

EXAMPLE 4 $\begin{cases} x + y > 3 \\ 3x - y \geq 6 \end{cases}$

SOLUTION First we sketch the graphs of $x + y > 3$ and $3x - y \geq 6$ on the same coordinate axes (Figure 5). The graph of $x + y > 3$ is the half-plane above the line $x + y = 3$, and the graph of $3x - y \geq 6$ is the half-plane below the line $3x - y = 6$. The two half-planes overlap in the region shaded in Figure 5. Hence, this shaded region is the graph of the system of inequalities.

EXAMPLE 5 $\begin{cases} 2x - y \geq 4 \\ 2x + 3y \leq 6 \end{cases}$

SOLUTION First we sketch the graphs of $2x - y \geq 4$ and $2x + 3y \leq 6$ on the same coordinate system (Figure 6). The graph of $2x - y \geq 4$ is the half-plane below the line $2x - y = 4$, and the graph of $2x + 3y \leq 6$ is the half-plane below the line $2x + 3y = 6$ (including all points on the lines). The two half-planes overlap in the region shaded in Figure 6. Hence, this shaded region is the graph of the system of inequalities.

Figure 5

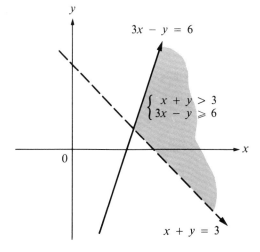

Figure 6

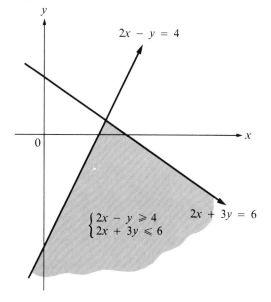

EXAMPLE 6 $\begin{cases} -2x + y \geq 0 \\ y - 2 > 0 \end{cases}$

SOLUTION First we sketch the graphs of $-2x + y \geq 0$ and $y - 2 > 0$ on the same coordinate system (Figure 7). The graph of $-2x + y \geq 0$ is the half-plane above and including the line $-2x + y = 0$ and the graph of $y - 2 > 0$ is the half-plane above the line $y - 2 = 0$. The two half-planes overlap in the region shaded in Figure 7. Hence, this shaded region is the graph of the system of inequalities.

Figure 7

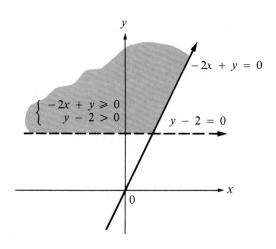

PROBLEM SET 7.8

In problems 1–12, sketch the graph of each inequality.

1. $y \leq 2x + 5$ **2.** $y \geq 3x + 4$ **3.** $y > 3x$ **4.** $y \leq -4x$

5. $y > -2x + 3$ **6.** $2y > 3x + 7$ **7.** $2x + y \leq 3$ **8.** $3x + 2y > 4$

9. $3y \leq -4$ **10.** $5y \geq 7$ **11.** $y \geq 2$ **12.** $y \leq -1$

In problems 13–24, sketch the graph of each system of linear inequalities.

13. $\begin{cases} x + y < 2 \\ 2x - y < -1 \end{cases}$ **14.** $\begin{cases} -x + y < -1 \\ -x + 3y \leq 5 \end{cases}$ **15.** $\begin{cases} x + y \leq 3 \\ x - y \geq 3 \end{cases}$ **16.** $\begin{cases} x + y > 5 \\ x - y < 9 \end{cases}$

17. $\begin{cases} x + 2y \leq 12 \\ x - y > 6 \end{cases}$ **18.** $\begin{cases} x + 2y > 12 \\ 3x - y < 1 \end{cases}$ **19.** $\begin{cases} 3x + y > 6 \\ x - y \geq 1 \end{cases}$ **20.** $\begin{cases} 2x - y \leq -2 \\ x - 2y \geq -2 \end{cases}$

21. $\begin{cases} 2x - 3y \leq -3 \\ 5x - 2y > 9 \end{cases}$ **22.** $\begin{cases} y \leq 2x + 4 \\ y \geq 3 - x \end{cases}$ **23.** $\begin{cases} x + y \leq 2 \\ -x + 3y \geq 4 \end{cases}$ **24.** $\begin{cases} 2x + y \geq 2 \\ x - 2y \leq 3 \end{cases}$

REVIEW PROBLEM SET

In problems 1–8, plot each point and indicate which quadrant or coordinate axis contains the point.

1. $(3, 2)$ **2.** $(-2, -3)$ **3.** $(-4, 3)$ **4.** $(1, -1)$

5. $(0, 2)$ **6.** $(3, 0)$ **7.** $(2, -3)$ **8.** $(-2, 2)$

In problems 9–12, find the distance between the two points.

9. $(3, 1)$ and $(15, 6)$ **10.** $(4, -3)$ and $(-1, 7)$ **11.** $(-2, 1)$ and $(-5, 5)$ **12.** $(a, -b)$ and $(-a, b)$

In problems 13–14, use the distance formula to show that the triangle ABC is isosceles.

13. $A = (2, 1)$, $B = (9, 3)$, and $C = (4, -6)$ **14.** $A = (0, 2)$, $B = (-1, 4)$, and $C = (-3, 3)$

In problems 15–20, sketch the graph of each equation.

15. $y = \frac{3}{4}x$ **16.** $y = -\frac{5}{3}x$ **17.** $y = 2x - 3$ **18.** $y = -2x + 1$

19. $2x - 5y = 10$ **20.** $-3x + 6y + 5 = 0$

In problems 21–24, sketch the line segment $\overline{AB}$ and find its slope m.

21. $A = (7, 3)$ and $B = (2, -2)$ **22.** $A = (11, 12)$ and $B = (7, 4)$

23. $A = (-3, 5)$ and $B = (1, 3)$ **24.** $A = (3, 0)$ and $B = (0, -2)$

In problems 25–28, sketch the line L that contains the given point P and has slope m.

25. $P = (1, 2)$ and $m = \frac{3}{4}$ **26.** $P = (4, -5)$ and $m = -\frac{1}{3}$

27. $P = (-8, 6)$ and $m = -\frac{5}{2}$ **28.** $P = (-3, -2)$ and $m = \frac{4}{7}$

In problems 29–32, determine whether the line containing the points A and B is parallel or perpendicular (or neither) to the line containing the points C and D.

29. $A = (1, 3)$, $B = (-1, -1)$, $C = (2, 1)$, and $D = (-3, -9)$

30. $A = (4, 6)$, $B = (-2, -3)$, $C = (3, 0)$, and $D = (6, -2)$

31. $A = (8, -6)$, $B = (-4, 3)$, $C = (3, 3)$, and $D = (9, 11)$

32. $A = (5, 3)$, $B = (0, 1)$, $C = (2, 3)$, and $D = (-4, -12)$

In problems 33–40, find an equation of the line L.

33. L contains $P = (1, 1)$ and has slope $m = 3$.   **34.** L contains $P = (2, -3)$ and has slope $m = -2$.

35. L contains $P = (-3, -2)$ and has slope $m = 0$.   **36.** L contains $P_1 = (4, 1)$ and $P_2 = (2, 3)$.

37. L has slope $m = 2$ and y intercept $b = 4$.   **38.** L has slope $m = -3$ and y intercept $b = -2$.

39. L contains $P = (3, 5)$ with undefined slope.   **40.** L contains $P = (-4, 7)$ with undefined slope.

In problems 41–44, express each equation in slope-intercept form. Find the slope m, the x intercept, the y intercept, and sketch the graph.

41. $2x + y = 4$ **42.** $x + y - 2 = 0$ **43.** $-3x + 4y - 12 = 0$ **44.** $\dfrac{x}{2} + \dfrac{y}{3} = 1$

In problems 45–48, find equations of the line L in: (a) point-slope form, (b) slope-intercept form, and (c) general form.

45. L contains $P = (2, 3)$ and is parallel to the line $3x - 2y + 5 = 0$.

46. L contains $P = (-3, 4)$ and is parallel to the line whose equation is $2x - 5y = 7$.

47. L contains $P = (1, -2)$ and is perpendicular to the line containing $P_1 = (3, 4)$ and $P_2 = (-5, 6)$.

48. L contains $P = (-7, 5)$ and is perpendicular to the line whose equation is $y = -\frac{2}{3}x + 6$.

In problems 49–52, use graphs to determine whether the system is dependent, inconsistent, or independent.

49. $\begin{cases} y = -2x + 2 \\ y = x - 4 \end{cases}$ **50.** $\begin{cases} y = 5x + 2 \\ 10x - 2y + 4 = 0 \end{cases}$ **51.** $\begin{cases} 3u + 2v = 1 \\ 3u + 2v = 3 \end{cases}$ **52.** $\begin{cases} r + s = 4 \\ 2r - s = 8 \end{cases}$

In problems 53–56, use the substitution method to solve each system.

53. $\begin{cases} 2x - y = 5 \\ x + 3y = 6 \end{cases}$ **54.** $\begin{cases} 3p + q = -1 \\ 2p - 2q = -14 \end{cases}$ **55.** $\begin{cases} 2x + y + z = 2 \\ x + 3y - z = 9 \\ 3x - y - 2z = 5 \end{cases}$ **56.** $\begin{cases} r + s + 2t = 0 \\ 2r - 3s - t = 5 \\ 3r + 7s - 2t = 8 \end{cases}$

In problems 57–62, use the elimination method to solve each system.

57. $\begin{cases} x - y = 3 \\ 2x + y = 3 \end{cases}$ **58.** $\begin{cases} 5m + 2p = 3 \\ 2m - 3p = 5 \end{cases}$ **59.** $\begin{cases} u - v + 2w = 0 \\ 3u + v + w = 2 \\ 2u - v + 5w = 5 \end{cases}$

60. $\begin{cases} 3x + 2y - z = -4 \\ x - y + 2z = 13 \\ 5x + 3y - 4z = -15 \end{cases}$ **61.** $\begin{cases} \dfrac{2}{x} + \dfrac{1}{y} = 8 \\ \dfrac{3}{x} - \dfrac{1}{y} = 7 \end{cases}$ **62.** $\begin{cases} \dfrac{1}{r} + \dfrac{2}{s} + \dfrac{3}{t} = 0 \\ \dfrac{2}{r} - \dfrac{3}{s} + \dfrac{1}{t} = 2 \\ \dfrac{5}{r} - \dfrac{4}{s} - \dfrac{2}{t} = -3 \end{cases}$

In problems 63–66, sketch the graph of each inequality.

63. $y < -2x + 4$ **64.** $y \geq 3x - 7$ **65.** $y - 3x \geq 1$ **66.** $4x - 2y > 3$

In problems 67–70, sketch the graph of each system of linear inequalities.

67. $\begin{cases} 2x + 4y < 3 \\ -2x + y > 4 \end{cases}$ **68.** $\begin{cases} 3x - 5y \geq 15 \\ -2x + 6y \geq 12 \end{cases}$ **69.** $\begin{cases} -3x + y \geq 2 \\ 2x - 3y \leq 1 \end{cases}$ **70.** $\begin{cases} \dfrac{x}{2} + \dfrac{y}{3} < 1 \\ \dfrac{x}{4} - \dfrac{y}{5} \leq 1 \end{cases}$

In problems 71–78, use a system of linear equations to solve each problem.

71. The sum of two numbers is 2. If one of the numbers is multiplied by 3 and the other is multiplied by 2, the sum of the products is 11. Find the numbers.

72. The sum of three numbers is 11. One number is twice the smallest number, and the remaining number is 3 more than the smallest number. Find the numbers.

73. A college mailed 100 letters, some requiring 20 cents postage and the rest requiring 30 cents postage. If the total bill was $23.70, find the number of letters sent at each rate.

74. The specific gravity of an object is defined to be its weight in air divided by its loss of weight when it is submerged in water. An object made partly of gold (which has specific gravity 16) and partly of silver (which has specific gravity 10.8) weighs 8 grams in air and 7.3 grams when it is submerged in water. How many grams of gold and how many grams of silver does the object contain?

75. A coin collection containing nickels and dimes consisted of 32 coins. If the total value of the coins was $2.65, how many coins of each kind were in the collection?

76. A piggy bank contained 35 coins, all nickels, dimes, and quarters. If there were twice as many nickels as quarters, and one-fourth as many dimes as nickels, how many coins of each kind were in the bank?

77. Sylvia invested a total of $30,000 in three different certificates, one paying 14% simple annual interest, one paying 12%, and the other paying 15%. If the total annual interest from the three certificates was $3,950, and if she invested twice as much in the 14% certificate as she did in the 15% certificate, how much did she invest in each kind of certificate?

78. Max and Joshua started a job that had to be finished in two days. The first day, Max worked 9 hours and Joshua worked 8 hours, and they finished half of the job. The next day, Max worked 6 hours and Joshua worked 12 hours to complete the job. How long would it take each, working alone, to do the job?

8 Logarithms

In this chapter we use the properties of exponents as the basis for establishing the properties of logarithms. Together, exponents and logarithms provide us with a tool to work applications in business, economics, engineering, and biology. They are also used to simplify many types of calculations.

8.1 Exponential Equations and Logarithms

Equations which contain a variable in an exponent are called **exponential equations.** Examples of exponential equations are

$$2^x = 8, \qquad 2^{t/5} = 32, \qquad \text{and}$$

$$5^{3y-7} = 125.$$

Some exponential equations may be solved by using the following property:

$$b^x = b^y \text{ if and only if } x = y \text{ for } b > 0 \text{ and } b \neq 1.$$

In Examples 1–4, solve each equation.

EXAMPLE 1 $\quad 2^x = 16$

SOLUTION $\quad$ First we express 16 as a power of 2, so that both sides of the equation will have the same base. Because $16 = 2^4$, we can rewrite

$$2^x = 16 \qquad \text{as} \qquad 2^x = 2^4.$$

Now we set the exponents equal to one another:

$$x = 4.$$

Therefore, 4 is the solution.

EXAMPLE 2 $9^t = 27^{4t-1}$

SOLUTION   Because $9 = 3^2$ and $27 = 3^3$, we can rewrite the given equation as

$$(3^2)^t = (3^3)^{4t-1} \qquad \text{or} \qquad 3^{2t} = 3^{3(4t-1)}.$$

We set the exponents equal to one another:

$$2t = 3(4t - 1)$$
$$2t = 12t - 3$$
$$-10t = -3$$
$$t = \frac{3}{10}.$$

Therefore, $\frac{3}{10}$ is the solution.

EXAMPLE 3 $(\frac{1}{5})^y = 25^{3y-1}$

SOLUTION   Because $\frac{1}{5} = 5^{-1}$ and $25 = 5^2$, we can rewrite the given equation as

$$(5^{-1})^y = (5^2)^{3y-1} \qquad \text{or} \qquad 5^{-y} = 5^{2(3y-1)}.$$

It follows that

$$-y = 2(3y - 1)$$
$$-y = 6y - 2$$
$$-7y = -2$$
$$y = \frac{2}{7}.$$

Therefore, $\frac{2}{7}$ is the solution.

EXAMPLE 4 $7^{w^2+w} = 49$

SOLUTION   Because $49 = 7^2$, we can rewrite the given equation as

$$7^{w^2+w} = 7^2,$$

so that,

$$w^2 + w = 2.$$

We solve this equation:

$$w^2 + w - 2 = 0 \qquad \text{or} \qquad (w + 2)(w - 1) = 0$$

so that

$$w + 2 = 0 \qquad \bigg| \qquad w - 1 = 0$$
$$w = -2 \qquad \bigg| \qquad w = 1.$$

Therefore, -2 and 1 are the solutions.

In the previous examples, we were able to express each side of the given equation as an exponential expression with the same base. Now suppose we are given an equation for which we are not able to do this. For example, consider the equation

$$10^x = 5.$$

We cannot simplify this equation so that both sides will be expressed in terms of the same base, as we have done previously. So, how do we solve this equation for x?

We can show that the equation $10^x = 5$ has a real solution, although the proof is beyond the scope of this text. We shall denote this solution by

$$x = \log_{10} 5,$$

which is read "x equals the logarithm of 5 to the base 10." That is, $\log_{10} 5$ is an exponent. It is the power to which we raise 10 in order to get 5. The actual value of $\log_{10} 5$ is an irrational number. (We explain how to obtain this irrational number later in the chapter.)

In general, we have the following definition:

DEFINITION 1   **Logarithm**

If $b > 0$, $b \neq 1$, and $c > 0$, then

$$x = \log_b c \text{ is equivalent to } b^x = c.$$

You can use the above definition to convert equations from exponential form to logarithmic form and vice versa, as the following table shows:

Exponential Form	Logarithmic Form
$3^2 = 9$	$2 = \log_3 9$
$5^4 = 625$	$4 = \log_5 625$
$64^{1/3} = 4$	$\frac{1}{3} = \log_{64} 4$
$1{,}000 = 10^3$	$\log_{10} 1{,}000 = 3$
$4 = (\frac{1}{2})^{-2}$	$\log_{1/2} 4 = -2$
$\frac{1}{49} = 7^{-2}$	$\log_7 \frac{1}{49} = -2$
$5^0 = 1$	$0 = \log_5 1$
$x^k = d$	$k = \log_x d \qquad (x > 0, x \neq 1, d > 0)$

EXAMPLE 5   Write each exponential equation in logarithmic form.

(a) $4^{3/2} = 8$ (b) $6^3 = 216$

SOLUTION   Using Definition 1, we have:

(a) $4^{3/2} = 8$ is equivalent to $\frac{3}{2} = \log_4 8$.

(b) $6^3 = 216$ is equivalent to $3 = \log_6 216$.

EXAMPLE 6   Write each equation in exponential form.

(a) $\log_8 4 = \frac{2}{3}$ (b) $\log_{10} 0.01 = -2$

SOLUTION   Using Definition 1, we have:

(a) $\log_8 4 = \frac{2}{3}$ is equivalent to $4 = 8^{2/3}$.

(b) $\log_{10} 0.01 = -2$ is equivalent to $0.01 = 10^{-2}$.

EXAMPLE 7   Find each value.

(a) $\log_3 81$ (b) $\log_2 32$ (c) $\log_8 \left(\frac{1}{64}\right)$

(d) $\log_{32} 8$ (e) $\log_7 7$ (f) $\log_5 1$

SOLUTION   (a) Let $x = \log_3 81$, so that $3^x = 81$. Then,

$$3^x = 3^4 \quad \text{or} \quad x = 4.$$

Therefore, $\log_3 81 = 4$.

(b) Let $x = \log_2 32$, so that $2^x = 32$. Then,

$$2^x = 2^5 \quad \text{or} \quad x = 5.$$

Therefore, $\log_2 32 = 5$.

(c) Let $x = \log_8 \left(\frac{1}{64}\right)$, so that $8^x = \frac{1}{64}$. Then

$$8^x = \frac{1}{8^2} = 8^{-2} \quad \text{or} \quad x = -2.$$

Therefore, $\log_8 \left(\frac{1}{64}\right) = -2$.

(d) Let $x = \log_{32} 8$, so that $32^x = 8$, or $(2^5)^x = 2^{5x} = 8$. Then,

$$2^{5x} = 2^3 \quad \text{or} \quad 5x = 3 \quad \text{or} \quad x = \frac{3}{5}.$$

Therefore, $\log_{32} 8 = \frac{3}{5}$.

(e) Let $x = \log_7 7$, so that $7^x = 7$. Then,

$$7^x = 7^1 \quad \text{or} \quad x = 1.$$

Therefore, $\log_7 7 = 1$.

(f) Let $x = \log_5 1$, so that $5^x = 1$. Then,

$$5^x = 1 = 5^0 \quad \text{or} \quad x = 0.$$

Therefore, $\log_5 1 = 0$.

Examples 7(e) and (f) can be generalized as follows:

If $b > 0$ and $b \neq 1$, then

$$\text{(i) } \log_b b = 1 \qquad \text{(ii) } \log_b 1 = 0$$

Definition 1, on page 312, can be used to solve some equations involving logarithms. Keep in mind that if $x = \log_b c$, the base b and the quantity c (whose logarithm we are taking) *must* be positive.

In Examples 8–10, solve each equation.

EXAMPLE 8 $\log_2(5x - 3) = 5$

SOLUTION   The equation $\log_2(5x - 3) = 5$ is equivalent to

$$2^5 = 5x - 3.$$

Because $2^5 = 32$, we have

$$5x - 3 = 32$$
$$5x = 35$$
$$x = 7$$

Therefore, 7 is the solution.

EXAMPLE 9 $\log_4 x^2 = 2$

SOLUTION   The equation $\log_4 x^2 = 2$ is equivalent to $4^2 = x^2$. Therefore,

$$x^2 = 16$$
$$x = \pm \sqrt{16} = \pm 4.$$

The solutions are $x = -4$ and $x = 4$. Notice that both numbers -4 and 4 satisfy the equation because the expression x^2 is positive when either value is substituted for x.

EXAMPLE **10** $\log_x 9 = 2$

SOLUTION   The equation $\log_x 9 = 2$ is equivalent to $x^2 = 9$. Definition 1 requires that the base of a logarithm be positive and not equal to 1. Therefore, we have the restriction that $x > 0$ and $x \neq 1$. Hence, the only solution is

$$x = \sqrt{9}$$
$$= 3.$$

PROBLEM SET 8.1

In problems 1–14, solve each exponential equation.

1. $5^x = 25$ **2.** $3^{2x} = 81$ **3.** $2^{2t} = 16$ **4.** $2^{-4p} = 64$

5. $3^{u-5} = 27$ **6.** $5^{2w-1} = 25$ **7.** $4^{3x} = 8^{x-1}$ **8.** $25^{3t} = 125^{2t-3}$

9. $3^{4c-5} = 81$ **10.** $5^{2x-1} = 125$ **11.** $3^{w^2+2w} = 27$ **12.** $2^{3u^2-2u} = 32$

13. $2^{x^2-6x} = (\frac{1}{2})^3$ **14.** $(\frac{1}{3})^{2x+7} = 9^2$

In problems 15–30, write each equation in logarithmic form.

15. $5^3 = 125$ **16.** $4^4 = 256$ **17.** $10^5 = 100,000$ **18.** $49^{0.5} = 7$

19. $4^{-2} = \frac{1}{16}$ **20.** $(\frac{1}{3})^{-2} = 9$ **21.** $6^{-2} = \frac{1}{36}$ **22.** $2^{-3} = 0.125$

23. $\sqrt{9} = 3$ **24.** $\sqrt[5]{32} = 2$ **25.** $(100)^{-3/2} = 0.001$ **26.** $(\frac{1}{8})^{-2/3} = 4$

27. $7^0 = 1$ **28.** $15^0 = 1$ **29.** $x^3 = a$ **30.** $\pi^t = z$

In problems 31–44, write each equation in exponential form.

31. $\log_9 81 = 2$ **32.** $\log_6 36 = 2$ **33.** $\log_{27} 9 = \frac{2}{3}$ **34.** $\log_{27} \frac{1}{9} = -\frac{2}{3}$

35. $\log_{10} 0.001 = -3$ **36.** $\log_{10} \frac{1}{10} = -1$ **37.** $\log_{1/3} 9 = -2$ **38.** $\log_{36} 216 = \frac{3}{2}$

39. $\log_{10} 4.35 = 0.64$ **40.** $\log_{10} 9.14 = 0.96$ **41.** $\log_{\sqrt{16}} 2 = \frac{1}{2}$ **42.** $\log_{4/9} \frac{27}{8} = -\frac{3}{2}$

43. $\log_x 1 = 0$ **44.** $\log_x 2 = 4$

In problems 45–64, find the value of each logarithm.

45. $\log_2 64$ **46.** $\log_4 \frac{1}{16}$ **47.** $\log_9 3$ **48.** $\log_4 8$

49. $\log_9 \frac{1}{3}$ **50.** $\log_{1/2} \frac{1}{8}$ **51.** $\log_{1/9} \frac{1}{81}$ **52.** $\log_2 \frac{1}{32}$

53. $\log_5 \frac{1}{125}$ **54.** $\log_{10} 0.00001$ **55.** $\log_3 9\sqrt{3}$ **56.** $\log_2 4\sqrt{2}$

57. $\log_{10} \frac{1}{10,000}$ **58.** $\log_7 343$ **59.** $\log_3 729$ **60.** $\log_5 5$

61. $\log_7 1$ **62.** $\log_b 1$ **63.** $\log_6 \frac{1}{216}$ **64.** $\log_b b^3$

In problems 65–76, solve each equation.

65. $\log_{10}(x + 1) = 1$ **66.** $\log_5(2y - 7) = 0$ **67.** $\log_4(3w + 1) = 2$ **68.** $\log_7(2u - 3) = 2$

69. $\log_5 N = 2$ **70.** $\log_b 36 = 2$ **71.** $\log_x 81 = 4$ **72.** $\log_c 16 = -\frac{4}{3}$

73. $\log_b 3 = \frac{1}{5}$ **74.** $\log_3 y = 4$ **75.** $\log_2(\log_5 5) = x$ **76.** $\log_4(\log_8 8) = y$

8.2 Basic Properties of Logarithms

In this section, we state the basic properties of logarithms and illustrate how these properties can be applied. We shall see that the properties of logarithms are based upon the properties of exponents. First we should note that it is possible to extend the definition of b^x so that all real numbers can be used as exponents. It is this that makes logarithms such a valuable computational tool. We conclude with proofs of some of the properties of logarithms.

Properties of Logarithms

Let M, N, and b be positive numbers, $b \neq 1$, and let r be any real number. Then

(i) $\log_b(MN) = \log_b M + \log_b N$ (ii) $\log_b\left(\dfrac{M}{N}\right) = \log_b M - \log_b N$

(iii) $\log_b N^r = r \log_b N$ (iv) $\log_b b^r = r$

Use the properties of logarithms to work Examples 1–4.

EXAMPLE 1   Write each expression as a sum or difference of multiples of logarithms.

(a) $\log_3 5y$ (b) $\log_8 \frac{17}{5}$ (c) $\log_5\left(\dfrac{x^2y}{2}\right)$ (d) $\log_2 \sqrt{\dfrac{t}{t+1}}$

SOLUTION   We assume that all quantities whose logarithms are taken are positive. Then:

(a) $\log_3 5y = \log_3 5 + \log_3 y$ [Property (i)]

(b) $\log_8 \frac{17}{5} = \log_8 17 - \log_8 5$ [Property (ii)]

(c) $\log_5\left(\dfrac{x^2y}{2}\right) = \log_5(x^2y) - \log_5 2$ [Property (ii)]

$$= \log_5 x^2 + \log_5 y - \log_5 2 \quad\quad \text{[Property (i)]}$$
$$= 2\log_5 x + \log_5 y - \log_5 2 \quad\quad \text{[Property (iii)]}$$

(d) $\log_2 \sqrt{\dfrac{t}{t+1}} = \log_2\left(\dfrac{t}{t+1}\right)^{1/2} = \dfrac{1}{2}\log_2\left(\dfrac{t}{t+1}\right)$ [Property (iii)]

$$= \frac{1}{2}[\log_2 t - \log_2(t+1)] \quad\quad \text{[Property (ii)]}$$

$$= \frac{1}{2}\log_2 t - \frac{1}{2}\log_2(t+1)$$

EXAMPLE 2   Write each expression as a single logarithm.

(a) $\log_3 15 + \log_3 13$ (b) $\log_5 216 - \log_5 54$
(c) $3 \log_2 x - \log_2 5x$ (d) $\log_4(t - 2) + \log_4(t + 2)$

SOLUTION   We assume that all quantities whose logarithms are taken are positive.

(a) $\log_3 15 + \log_3 13 = \log_3 15(13)$ [Property (i)]
$= \log_3 195$

(b) $\log_5 216 - \log_5 54 = \log_5 \frac{216}{54}$ [Property (ii)]
$= \log_5 4$

(c) $3 \log_2 x - \log_2 5x = \log_2 x^3 - \log_2 5x$ [Property (iii)]

$= \log_2 \frac{x^3}{5x} = \log_2 \frac{x^2}{5}$ [Property (ii)]

(d) $\log_4(t - 2) + \log_4(t + 2) = \log_4(t - 2)(t + 2)$ [Property (i)]
$= \log_4(t^2 - 4)$

EXAMPLE 3   Evaluate the following expressions.

(a) $\log_3 3^7$ (b) $\log_{11} \sqrt[3]{11}$

SOLUTION   We use Property (iv):

(a) $\log_3 3^7 = 7$ (b) $\log_{11} \sqrt[3]{11} = \log_{11} 11^{1/3} = \frac{1}{3}$

EXAMPLE 4   Use $\log_b 2 = 0.35$ and $\log_b 3 = 0.55$ to find the value of each expression.

(a) $\log_b 6$ (b) $\log_b \frac{2}{3}$ (c) $\log_b 8$ (d) $\log_b \sqrt{\frac{2}{3}}$ (e) $\log_b 24$ (f) $\dfrac{\log_b 2}{\log_b 3}$

SOLUTION   (a) $\log_b 6 = \log_b 2(3) = \log_b 2 + \log_b 3 = 0.35 + 0.55 = 0.90$

(b) $\log_b \frac{2}{3} = \log_b 2 - \log_b 3 = 0.35 - 0.55 = -0.20$

(c) $\log_b 8 = \log_b 2^3 = 3 \log_b 2 = 3(0.35) = 1.05$

(d) $\log_b \sqrt{\frac{2}{3}} = \log_b(\frac{2}{3})^{1/2} = \frac{1}{2} \log_b \frac{2}{3} = \frac{1}{2}(\log_b 2 - \log_b 3) = \frac{1}{2}(0.35 - 0.55)$
$= -0.10$

(e) $\log_b 24 = \log_b 8(3) = \log_b(2^3)(3) = \log_b 2^3 + \log_b 3$
$= 3 \log_b 2 + \log_b 3 = 3(0.35) + 0.55 = 1.05 + 0.55 = 1.60$

(f) $\dfrac{\log_b 2}{\log_b 3} = \dfrac{0.35}{0.55} = \dfrac{7}{11} = 0.64$

We can combine the basic properties of logarithms and Definition 1 of Section 8.1 to solve equations containing logarithms.

In Examples 5 and 6, solve each equation.

EXAMPLE 5 $\log_3(x + 1) + \log_3(x + 3) = 1$

SOLUTION   Because

$$\log_3(x + 1) + \log_3(x + 3) = \log_3[(x + 1)(x + 3)] \qquad \text{[Property (i)]}$$

we have

$$\log_3[(x + 1)(x + 3)] = 1$$

or

$$(x + 1)(x + 3) = 3^1 \qquad \text{[Definition 1,}$$
$$x^2 + 4x + 3 = 3 \qquad \text{Section 8.1]}$$
$$x^2 + 4x = 0$$
$$x(x + 4) = 0$$
$$x = 0 \quad | \quad x + 4 = 0$$
$$\qquad\qquad\qquad x = -4.$$

We cannot take the logarithm of a negative number. Therefore, -4 cannot be a solution. The only solution is 0.

EXAMPLE 6 $\log_4(x + 3) - \log_4 x = 1$

SOLUTION   Because

$$\log_4(x + 3) - \log_4 x = \log_4\left(\frac{x + 3}{x}\right) \qquad \text{[Property (ii)]}$$

we have

$$\log_4\left(\frac{x + 3}{x}\right) = 1$$

$$\frac{x + 3}{x} = 4^1 = 4$$

$$x + 3 = 4x$$

$$3 = 3x$$

$$1 = x.$$

Therefore, the solution is 1.

We now prove the first three properties of logarithms.

Proofs of the Properties of Logarithms

(i) To prove that

$$\log_b(MN) = \log_b M + \log_b N,$$

let

$$x = \log_b M \quad \text{and} \quad y = \log_b N,$$

so that

$$b^x = M \quad \text{and} \quad b^y = N.$$

It follows that

$$MN = b^x \cdot b^y = b^{x+y},$$

or

$$\log_b(MN) = x + y.$$

However, $x = \log_b M$ and $y = \log_b N$. Therefore,

$$\log_b(MN) = \log_b M + \log_b N.$$

(ii) To prove that

$$\log_b\left(\frac{M}{N}\right) = \log_b M - \log_b N,$$

we write

$$\frac{M}{N} \cdot N = M,$$

so that

$$\log_b\left(\frac{M}{N} \cdot N\right) = \log_b M.$$

We use Property (i):

$$\log_b\left(\frac{M}{N}\right) + \log_b N = \log_b M.$$

Thus,

$$\log_b\left(\frac{M}{N}\right) = \log_b M - \log_b N.$$

(iii) To prove that

$$\log_b N^r = r \log_b N,$$

we let

$$y = \log_b N.$$

Thus,

$$N = b^y.$$

It follows that

$$N^r = (b^y)^r = b^{yr},$$

or

$$\log_b N^r = yr.$$

Because $y = \log_b N$,

$$\log_b N^r = r \log_b N.$$

The proof of Property (iv) is straightforward and will be left as an exercise (Problem 73).

PROBLEM SET 8.2

In problems 1–28, use the properties of logarithms to write each expression as a sum or difference of multiples of logarithms. Assume that all variables represent positive real numbers.

1. $\log_4 5y$

2. $\log_5 7x$

3. $\log_3 uv$

4. $\log_7 cd$

5. $\log_5 \dfrac{x}{3}$

6. $\log_5 \dfrac{z}{11}$

7. $\log_2 \tfrac{7}{15}$

8. $\log_3 \dfrac{x}{y}$

9. $\log_7 3^5$

10. $\log_8 7^{1.4}$

11. $\log_3 c^5$

12. $\log_7 y^4$

13. $\log_4 \sqrt{w}$

14. $\log_5 \sqrt[3]{y}$

15. $\log_7 3^4 \cdot 5^2$

16. $\log_3 x^4 y^5$

17. $\log_{11} x^4 y^2$

18. $\log_7 \sqrt[3]{x} y^3$

19. $\log_2 \sqrt[5]{xy}$

20. $\log_4 \sqrt[7]{x^2 y^5}$

21. $\log_5 \dfrac{a^2}{b^4}$

22. $\log_5 \dfrac{x^7}{y^8}$

23. $\log_7 \dfrac{x^3 \sqrt[4]{y}}{z^3}$

24. $\log_5 \dfrac{\sqrt[5]{xy^4}}{w^4}$

25. $\log_4 \dfrac{u^4 v^5}{\sqrt[4]{z^3}}$

26. $\log_{10} \dfrac{c^7 \sqrt[9]{d^2}}{5\sqrt{f}}$

27. $\log_3 \sqrt[7]{\dfrac{y}{y+7}}$

28. $\log_{10} x(x+2)$

In problems 29–50, use the properties of logarithms to write each expression as a single logarithm. Assume that all variables represent positive real numbers.

29. $\log_5 4 - \log_5 3$

30. $\log_2 \tfrac{5}{7} + \log_2 \tfrac{14}{70}$

31. $\log_7 \tfrac{3}{8} - \log_7 \tfrac{9}{4}$

32. $\log_3 \tfrac{3}{4} - \log_3 \tfrac{5}{8}$

33. $\log_3 5 + \log_3 z - \log_3 y$

34. $\log_6 x + \log_6 y - \log_6 z$

35. $2 \log_5 \tfrac{4}{5} + 3 \log_3 \tfrac{1}{2}$

36. $3 \log_5 \tfrac{3}{4} + 2 \log_5 \tfrac{1}{5}$

37. $3 \log_2 x + 7 \log_2 y$

38. $3 \log_3 z - 2 \log_3 y$

39. $\log_a \dfrac{x}{y} + \log_a \dfrac{y^2}{3x}$

40. $\log_b \dfrac{x^2}{y} - \log_b \dfrac{x^4}{y^2}$

41. $\tfrac{1}{2} \log_4 a - 3 \log_4 b - 4 \log_4 z$

42. $5 \log_{10} x - \tfrac{5}{3} \log_{10} y - 7 \log_{10} z$

43. $\log_7(x-1) + \log_7(x+1)$

44. $\log_3 \dfrac{a}{a-1} + \log_3 \dfrac{a^2-1}{a}$

45. $\log_e(y^2 - 25) - \log_e(y - 5)$

46. $\log_e \dfrac{x+y}{z} - \log_e \dfrac{1}{x+y}$

47. $\log_7 7^4$

48. $\log_y y^5$

49. $\log_p \sqrt[5]{p}$

50. $\log_q \sqrt[7]{q}$

In problems 51–62, use $\log_{10} 2 = 0.3010$ and $\log_{10} 3 = 0.4771$ to find the value of each expression.

51. $\log_{10} 6$ **52.** $\log_{10} 12$ **53.** $\log_{10} 18$ **54.** $\log_{10} 24$

55. $\log_{10} \frac{3}{2}$ **56.** $\log_{10} \frac{1}{3}$ **57.** $\log_{10} 5$ **58.** $\log_{10} 32$

59. $\log_{10} 81$ **60.** $\log_{10} \sqrt[5]{2}$ **61.** $\log_{10} 0.5$ **62.** $\log_{10} 60$

In problems 63–72, solve each equation.

63. $\log_7 x + \log_7 14 = 1$ **64.** $\log_4 y + \log_4 3 = 2$ **65.** $\log_3 w - \log_3 2 = 2$

66. $\log_7 z - \log_7 2 = 2$ **67.** $\log_5 x + \log_5(x - 4) = 1$ **68.** $\log_2 x + \log_2(x - 1) = 1$

69. $\log_4(y + 2) - \log_4(y - 1) = 2$ **70.** $\log_3(u + 3) - \log_3(u - 1) = 2$

71. $\log_{10}(z^2 - 9) - \log_{10}(z + 3) = 2$ **72.** $\log_4 x + \log_4(6x + 11) = 1$

73. Prove that $\log_b b^r = r$, where $b > 0$, $b \neq 1$, and r is any real number.

8.3 **Common Logarithms**

In the preceding sections the bases of logarithms were *all* positive numbers such as 2, 3, 5, 6, etc. Since any positive number except the number 1 can serve as a base for logarithms, people have decided to simplify working with logarithms by choosing the same base. The usual positional system of writing numerals is based on 10. Therefore, logarithms to base 10 are the most useful for computational purposes. Logarithms with base 10 are called **common logarithms.** The symbol "log x" (with no subscript) is often used as an abbreviation for $\log_{10} x$. Thus, the expression

$$\log_{10} x \text{ is written as } \log x.$$

Scientific problems often involve very large or very small numbers. We express these numbers in a special notation to simplify matters. This notation is particularly appropriate to use with common logarithms.

Scientific Notation

A convenient way to write a very large or a very small number x is to express the number in the form

$$x = s \times 10^n$$

where s is a decimal between 1 and 10 ($1 \leq s < 10$) and n is an integer. This form of a number is referred to as **scientific notation.** For example, the speed c of light in a vacuum is written in scientific notation as

$$c = 3 \times 10^8 \text{ meters per second.}$$

The positive exponent 8 indicates that the decimal point should be moved 8 places to the *right* to change the scientific notation to ordinary notation:

$$c = 3 \times 10^8 = 300000000.0 \quad \text{or} \quad 300{,}000{,}000 \text{ meters per second.}$$

The same notation is used for very small numbers. For instance, in physics the average lifetime t of a lambda particle is estimated to be

$$t = 2.51 \times 10^{-10} \text{ second.}$$

The negative exponent -10 indicates that the decimal point should be moved 10 places to the *left* to change the scientific notation to ordinary notation. Thus,

$$t = 2.51 \times 10^{-10} = 0.000000000251 \text{ second.}$$

EXAMPLE 1   Rewrite each number in scientific notation.

(a) 3,951

(b) 21.33

(c) 0.0119

(d) 0.000573

SOLUTION   (a) $3{,}951 = 3.951 \times 10^3$

(b) $21.33 = 2.133 \times 10^1$

(c) $0.0119 = 1.19 \times 10^{-2}$

(d) $0.000573 = 5.73 \times 10^{-4}$

EXAMPLE 2   Rewrite each number in ordinary notation.

(a) 4.3×10^3

(b) 3.7×10^6

(c) 5.13×10^{-4}

(d) 2.73×10^{-1}

SOLUTION   (a) $4.3 \times 10^3 = 4{,}300$

(b) $3.7 \times 10^6 = 3{,}700{,}000$

(c) $5.13 \times 10^{-4} = 0.000513$

(d) $2.73 \times 10^{-1} = 0.273$

Using a Logarithmic Table

We use base 10 in computational work because every positive real number x can be written in scientific notation as

$$x = s \times 10^n$$

where $1 \leq s < 10$ and n is an integer. If we apply Property i of Section 8.2 to the above equation, we have

$$\begin{aligned} \log x &= \log[s \times 10^n] \\ &= \log s + \log 10^n. \end{aligned}$$

Since $\log 10^n = n$, we have

$$\log x = \log s + n.$$

This last equation is called the **standard form** of $\log x$. The number $\log s$, where $1 \leq s < 10$, is called the **mantissa** of $\log x$, and the integer n is called the **characteristic** of $\log x$.

Notice that for $1 \leq s < 10$, we have

$$\log 1 \leq \log s < \log 10,$$

or, equivalently,

$$0 \leq \log s < 1.$$

That is, the mantissa is either 0 or a positive number between 0 and 1. Therefore, to determine the value of $\log x$, we simply determine the value of $\log s$, where s is always between 1 and 10. We can obtain the approximate value of $\log s$ from a **table of common logarithms** (Table I in Appendix I).

The value of the characteristic of the logarithm of a number x is n. (In scientific notation we substitute $s \times 10^n$ for x.) Therefore, we know that $n \geq 0$ if $x \geq 1$, and that $n < 0$ if $0 < x < 1$. To illustrate, we list the values of n for several different values of x in the table below:

x	$s \times 10^n$	n
480	4.8×10^2	2
25	2.5×10^1	1
1.3	1.3×10^0	0
0.28	2.8×10^{-1}	-1
0.031	3.1×10^{-2}	-2

In Examples 3 and 4, use Table I in Appendix I to find the values of the given common logarithms. In each case, indicate the characteristic.

EXAMPLE 3   (a) log 53,900   (b) log 385   (c) log 28.4

SOLUTION   To find the characteristics n of the logarithms, we write the numbers x in scientific notation:

x	$s \times 10^n$	n
(a) 53,900	5.39×10^4	4
(b) 385	3.85×10^2	2
(c) 28.4	2.84×10^1	1

Thus, using Table I, we have

(a) $\log 53{,}900 = \log 5.39 + 4 = 0.7316 + 4 = 4.7316$

(b) $\log 385 = \log 3.85 + 2 = 0.5855 + 2 = 2.5855$

(c) $\log 28.4 = \log 2.84 + 1 = 0.4533 + 1 = 1.4533$

EXAMPLE 4   (a) log 4.06   (b) log 0.628   (c) log 0.0035

SOLUTION   The numbers x are written in scientific notation and their characteristics n are indicated below:

x	$s \times 10^n$	n
(a) 4.06	4.06×10^0	0
(b) 0.628	6.28×10^{-1}	-1
(c) 0.0035	3.5×10^{-3}	-3

Therefore, using Table I, we have

(a) $\log 4.06 = \log 4.06 + 0 = 0.6085 + 0 = 0.6085$

(b) $\log 0.628 = \log 6.28 + (-1) = 0.7980 + (-1) = -0.2020$

(c) $\log 0.0035 = \log 3.50 + (-3) = 0.5441 + (-3) = -2.4559$

The only logarithms that we can find directly from Table I, Appendix I, are logarithms of numbers that contain, at most, three significant digits.

If a number has four significant digits, it is possible to obtain an approximation of its logarithm by using a method known as **interpolation** or **linear interpolation.**

Interpolation depends on the following property:

If x, y, and z are positive numbers such that $z < x < y$, then

$$\log z < \log x < \log y.$$

Interpolation also depends on the following assumption: For small differences in numbers the change in the mantissas is proportional to the change in the numbers. (This assumption does not always hold, but the results are sufficiently accurate for our purposes, especially since the differences between consecutive mantissas in Table I are small.)

To illustrate the method of interpolation, let us find log 1.234.

We see from Table 1 that

$$\log 1.23 = 0.0899 \text{ and } \log 1.24 = 0.0934.$$

We selected these numbers because

$$1.23 < 1.234 < 1.24.$$

It follows that

$$\log 1.23 < \log 1.234 < \log 1.24.$$

The difference between the two consecutive logarithms in the table is

$$0.0934 - 0.0899 = 0.0035.$$

Also,

$$1.234 - 1.23 = 0.004 \text{ and } 1.24 - 1.23 = 0.01.$$

Because log 1.234 > log 1.23, there is a "correction" d such that

$$\log 1.234 = \log 1.23 + d.$$

The value of d can be determined by using the following ratio:

$$\frac{d}{0.0035} = \frac{0.004}{0.01}.$$

Hence,

$$d = \frac{0.0035(0.004)}{0.01} = 0.0014.$$

Therefore,

$$\log 1.234 = \log 1.23 + d$$
$$= 0.0899 + 0.014 = 0.0913.$$

EXAMPLE 5   Use interpolation to find log 52.33.

SOLUTION   We obtain the values for

$$\log 52.3 = \log 5.23 + 1 \quad \text{and} \quad \log 52.4 = \log 5.24 + 1,$$

from Table I:

$$\begin{aligned} \log 52.3 &= \log 5.23 + 1 \\ &= 1.7185 \end{aligned}$$

and

$$\begin{aligned} \log 52.4 &= \log 5.24 + 1 \\ &= 1.7193. \end{aligned}$$

We now find the difference between the two logarithms so that

$$1.7193 - 1.7185 = 0.0008.$$

We also find the difference between the two numbers whose logarithms these are,

$$52.4 - 52.3 = 0.1$$

and between the smaller number and the number whose logarithm we are seeking,

$$52.33 - 52.3 = 0.03.$$

Because log 52.3 < log 52.33 < log 52.4, we arrange our work as follows:

$$0.1 \begin{bmatrix} 0.03 \begin{bmatrix} \log 52.3 &= 1.7185 \\ \log 52.33 &= ? \end{bmatrix} d \\ \log 52.4 &= 1.7193 \end{bmatrix} 0.0008$$

Thus,

$$\frac{0.03}{0.1} = \frac{d}{0.0008} \quad \text{or}$$

$$d = \frac{0.03(0.0008)}{0.1}$$

$$= 0.0002.$$

Therefore,

$$\begin{aligned} \log 52.33 &= \log 52.3 + d \\ &= 1.7185 + 0.0002 \\ &= 1.7187. \end{aligned}$$

Antilogarithms

The process of finding a number whose logarithm is given is the reverse of finding the logarithm of a number. The number found is called the **antilogarithm** of the given logarithm. For instance, if we are given a number r, we can determine the value of x such that $\log x = r$. The number x is called the **antilogarithm of r** and is abbreviated **antilog r**. Thus:

> If $\log x = r$, then $x = $ antilog r.

To find the antilog of 4.4969 (or to determine the solution of the equation $\log x = 4.4969$), we write $\log x = 4.4969$ in standard form—that is, as the sum of a number between 0 and 1 and an integer. The given logarithm

$$4.4969 = 0.4969 + 4$$

has the mantissa 0.4969 and the characteristic 4. We use Table I in Appendix I to find a value s such that $\log s = 0.4969$. We find that $s = 3.14$. The characteristic of $\log x$ is 4. Therefore,

$$\begin{aligned}
\log x &= 0.4969 + 4 \\
&= \log 3.14 + 4 = \log 3.14 + \log 10^4 \\
&= \log (3.14 \times 10^4) = \log 31{,}400.
\end{aligned}$$

Therefore, $x = $ antilog $4.4969 = 31{,}400$.

In Examples 6–8 find the values of the given antilogarithms.

EXAMPLE 6   antilog 2.7210

SOLUTION   Let $x = $ antilog 2.7210, so that $\log x = 2.7210$. Then

$$\log x = 0.7210 + 2.$$

Using Table I, we find that $\log 5.26 = 0.7210$. Thus

$$\begin{aligned}
\log x &= \log 5.26 + 2 = \log 5.26 + \log 10^2 \\
&= \log (5.26 \times 10^2) = \log 526.
\end{aligned}$$

Therefore, $x = $ antilog $2.7210 = 526$.

EXAMPLE 7   antilog $[0.5105 + (-3)]$

SOLUTION   Let $x = $ antilog $[0.5105 + (-3)]$, so that

$$\log x = 0.5105 + (-3).$$

Using Table I, we find that $\log 3.24 = 0.5105$. Thus

$$\log x = \log 3.24 + (-3) = \log 3.24 + \log 10^{-3}$$
$$= \log (3.24 \times 10^{-3}) = \log 0.00324.$$

Therefore, $x = \text{antilog } [0.5105 + (-3)] = 0.00324.$

EXAMPLE 8   antilog (-2.0804)

SOLUTION   Let $x = \text{antilog } (-2.0804)$, so that $\log x = -2.0804$. The mantissa must always be positive. Therefore, we write

$$-2.0804 = (-2.0804 + 3) - 3 = 0.9196 + (-3).$$

Thus,

$$\log x = 0.9196 + (-3).$$

Using Table I, we find that $\log 8.31 = 0.9196$, and

$$\log x = \log 8.31 + (-3) = \log 8.31 + \log 10^{-3}$$
$$= \log (8.31 \times 10^{-3}) = \log 0.00831.$$

Therefore, $x = \text{antilog } (-2.0804) = 0.00831.$

The interpolation method can also be used to find antilogarithms.

EXAMPLE 9   Use interpolation to find antilog (-1.7186).

SOLUTION   Let $x = \text{antilog } (-1.7186)$, so that $\log x = -1.7186$. The mantissa must be positive. Therefore, we write

$$-1.7186 = (-1.7186 + 2) - 2$$
$$= 0.2814 + (-2).$$

Thus,

$$\log x = 0.2814 + (-2).$$

From Table I, we find that the values closest to 0.2814 are 0.2810 (log 1.91) and 0.2833 (log 1.92),

$$[0.2810 + (-2)] < [0.2814 + (-2)] < [0.2833 + (-2)].$$

Therefore,

$$\log 1.91 + (-2) < \log x < \log 1.92 + (-2)$$

or

$$\log 0.0191 < \log x < \log 0.0192.$$

We arrange the work as follows:

$$
0.0001 \left[\begin{array}{c} \begin{array}{c} \text{log } 0.0192 = 0.2833 + (-2) \\ d \left[\begin{array}{c} \text{log } x \quad\;\; = 0.2814 + (-2) \\ \text{log } 0.0191 = 0.2810 + (-2) \end{array} \right] 0.0004 \end{array} \end{array} \right] 0.0023
$$

The value of d can be determined by using the following ratio:

$$
\frac{d}{0.0001} = \frac{0.0004}{0.0023} \qquad \text{or} \qquad d = \frac{0.0001(0.0004)}{0.0023} = 0.00002.
$$

Therefore,

$$
\begin{aligned}
x &= \text{antilog } [0.2814 + (-2)] \\
&= \text{antilog } [0.2810 + (-2)] + d \\
&= 0.0191 + 0.00002 = 0.01912.
\end{aligned}
$$

PROBLEM SET 8.3

In problems 1–10, rewrite each number in scientific notation.

1. 3,782
2. 0.000132
3. 0.00381
4. 38,173
5. 375,000
6. 137,100,000
7. 0.0001321
8. 681,000,000
9. 0.000271312
10. 0.00127142281

In problems 11–20, rewrite each number in ordinary notation.

11. 2.1×10^2
12. 8.6×10^3
13. 3.14×10^{-5}
14. 7.5×10^{-3}
15. 1.13×10^4
16. 7.2×10^5
17. 5.41×10^{-6}
18. 1.871×10^{-4}
19. 3.127×10^{-4}
20. 1.94×10^{-5}

In problems 21–38, use Table I in Appendix I to find the value of each common logarithm.

21. log 317
22. log 3,910
23. log 53,400
24. log 348,000
25. log 17.1
26. log 5
27. log 6.81
28. log 7.59
29. log 1.18
30. log 9.81
31. log 0.315
32. log 0.712
33. log 0.0713
34. log 0.00512
35. log 0.000178
36. log 0.00081
37. log 0.000007
38. log 0.00000137

In problems 39–48, use Table I in Appendix I and interpolation to find the value of each common logarithm.

39. log 1,545
40. log 333.3
41. log 79.56
42. log 62.95
43. log 5.312
44. log 1.785
45. log 0.5725
46. log 0.7125
47. log 0.05342
48. log 0.006487

In problems 49–68, use Table I in Appendix I to find the value of each antilogarithm.

49. antilog 0.4133   **50.** antilog 0.4871   **51.** antilog 1.2945
52. antilog 1.7825   **53.** antilog 2.7427   **54.** antilog 2.9795
55. antilog 3.5514   **56.** antilog 3.8993   **57.** antilog $[0.7348 + (-1)]$
58. antilog $[0.8082 + (-2)]$ **59.** antilog $[0.8993 + (-3)]$ **60.** antilog $[0.5922 + (-4)]$
61. antilog (-1.6289) **62.** antilog (-2.4157) **63.** antilog (-3.4881)
64. antilog (-4.8153) **65.** antilog (-0.1574) **66.** antilog (-0.3251)
67. antilog (-2.1475) **68.** antilog (-3.1884)

In problems 69–76, use Table I in Appendix I and interpolation to find the value of each antilogarithm.

69. antilog 0.1452   **70.** antilog 1.5375
71. antilog 1.5425   **72.** antilog (-1.1275)
73. antilog $[0.2259 + (-2)]$ **74.** antilog $[0.4950 + (-2)]$
75. antilog (-4.4625) **76.** antilog (-4.565)

8.4 Using a Calculator to Evaluate Logarithms and Exponents

Before the development of calculators and computers, logarithms were used extensively to speed up certain numerical computations. Today, because of the wide availability of inexpensive and reliable calculators, common logarithms are rarely used for purposes of numerical calculation. Instead, scientific calculators that have both a y^x key and a log key are often employed.

For example, using a 10-digit calculator, we find log 364 by entering 364 and pressing the log key, to obtain

$$\log 364 = 2.561101384.$$

Similarly,

$$\log 478 = 2.679427897$$
$$\log 86.2 = 1.935507266$$
$$\log 0.568 = -0.2456516643$$
$$\log 0.0841 = -1.075204004$$
$$\log 0.34827 = -0.4580839341.$$

To find antilogarithms we use the y^x key. For example, using a 10-digit calculator and rounding off to four significant digits, we find antilog 1.7959 by entering 10, pressing the y^x key, entering 1.7959, and pressing the equals key, to obtain

$$\text{antilog } 1.7959 = 62.50.$$

We are commanding the calculator to evaluate $10^{1.7959}$. This is another way of expressing antilog 1.7959 (the number whose log is 1.7959). Other examples are:

$$\text{antilog } 2.6372 = 433.7$$

$$\text{antilog } 0.1381 = 1.374$$

$$\text{antilog } 0.0254 = 1.060.$$

ⓒ *In Examples 1 and 2, use a calculator with log and y^x keys to evaluate each of the following.*

EXAMPLE 1 (a) log 6.23

(b) log 8,921

(c) log 0.007316

SOLUTION On a 10-digit calculator, we obtain:

(a) log 6.23 = 0.7944880467

(b) log 8,921 = 3.950413539

(c) log 0.007316 = −2.135726303

EXAMPLE 2 (a) antilog 1.7782

(b) antilog 0.9741

(c) antilog (−1.8742)

SOLUTION We use a 10-digit calculator and round off to four significant digits:

(a) antilog 1.7782 = 60.01

(b) antilog 0.9741 = 9.421

(c) antilog (−1.8742) = 0.0134

In many applications of mathematics there are formulas which become much simpler if we use logarithms and exponential expressions with base

$$e \approx 2.718281828.$$

Logarithms with base e are called **natural logarithms.** The symbol $\ln x$, which is read "log of x to the base e" or "natural log of x," is often used as an abbreviation for $\log_e x$. Thus:

$$\ln x = \log_e x \text{ for } x > 0.$$

In other words, for $x > 0$,

$$\ln x = c \text{ if and only if } e^c = x.$$

EXAMPLE 3   Ⓒ Use a calculator ln key to evaluate each of the following:

(a) $\ln 8{,}132$ (b) $\ln 0.041326$

SOLUTION   We enter the number and press the ln key. On a 10-digit calculator we obtain:

(a) $\ln 8{,}132 = 9.003562175$

(b) $\ln 0.041326 = -3.186263437$

EXAMPLE 4   Ⓒ Use a calculator with an e^x key to evaluate each of the following:

(a) $e^{4.3}$ (b) $e^{\sqrt{5}}$ (c) $e^{-0.67}$

SOLUTION   We enter the exponent and press the e^x key. On a 10-digit calculator we obtain:

(a) $e^{4.3} = 73.6997937$

(b) $e^{\sqrt{5}} = 9.356469012$

(c) $e^{-0.67} = 0.5117085778$

EXAMPLE 5   Ⓒ Use a calculator with a y^x key to evaluate $(0.6423)^{-0.271}$.

SOLUTION   We enter 0.6423, press the y^x key, enter -0.271, and press the equals key. On a 10-digit calculator we obtain

$$(0.6423)^{-0.271} = 1.127464881.$$

EXAMPLE 6   Ⓒ Use a calculator with either an $\sqrt[x]{y}$ key or a y^x key to evaluate $\sqrt[5]{17}$.

SOLUTION   On a 10-digit calculator we obtain

$$\sqrt[5]{17} = (17)^{1/5} = (17)^{0.2} = 1.762340348.$$

Solving Exponential Equations by Using Logarithms

We can often solve an exponential equation by taking the logarithm of both sides of the equation and applying the properties of logarithms to simplify the results. For this purpose, we can use either the common or the natural logarithm.

ⓒ *In Examples 7–9, solve each equation. Round off your answers to four significant digits.*

EXAMPLE 7 $5^x = 7$

SOLUTION We take the common logarithm on both sides of the equation:

$$5^x = 7$$

$$\log 5^x = \log 7$$

$$x \log 5 = \log 7$$

$$x = \frac{\log 7}{\log 5}.$$

Therefore, using a 10-digit calculator we obtain

$$x = \frac{\log 7}{\log 5} = \frac{0.8450980400}{0.6989700041}$$

$$= 1.209.$$

EXAMPLE 8 $e^{3x+7} = 8$

SOLUTION Taking the natural logarithm on both sides of the equation, we have

$$e^{3x+7} = 8$$

$$\ln e^{3x+7} = \ln 8$$

$$(3x + 7)\ln e = \ln 8$$

$$3x + 7 = \ln 8 \qquad (\ln e = 1)$$

$$3x = \ln 8 - 7$$

$$x = \frac{\ln 8 - 7}{3}$$

$$= \frac{2.079441542 - 7}{3}$$

$$= -1.640.$$

EXAMPLE 9 $7^{2x} = 4^{x+1}$

SOLUTION   We take the common logarithms on both sides:

$$7^{2x} = 4^{x+1}$$

$$\log 7^{2x} = \log 4^{x+1}$$

$$2x \log 7 = (x + 1)\log 4$$

$$2x \log 7 = x \log 4 + \log 4$$

$$2x \log 7 - x \log 4 = \log 4$$

$$x(2 \log 7 - \log 4) = \log 4$$

$$x = \frac{\log 4}{2 \log 7 - \log 4}$$

$$= \frac{0.6020599913}{2(0.84509804) - 0.6020599913}$$

$$= \frac{0.6020599913}{1.088136089}$$

$$= 0.5533.$$

PROBLEM SET 8.4

© In problems 1–54, use a calculator to find the value of each of the following expressions. Round off your answers to four significant digits.

1. log 32.94
2. log 281.5
3. log 6.183
4. log 792.83
5. log 603.75
6. log 834.72
7. log 0.001342
8. log 0.005217
9. log 0.003561
10. log 0.0002794
11. log 0.00004175
12. log 0.000023517
13. antilog 1.9281
14. antilog 2.9741
15. antilog 0.09481
16. antilog 5.0546
17. antilog 1.47372
18. antilog 3.10375
19. antilog 4.40661
20. antilog 3.60437
21. antilog (−1.4837)
22. antilog (−1.0254)
23. antilog (−2.8459)
24. antilog (−3.20951)
25. ln 7,324
26. ln 543.1
27. ln 9.942
28. ln 0.6984
29. ln 0.5342
30. ln 0.90471
31. $e^{2.1}$
32. $e^{3.7}$
33. $e^{\sqrt{2}}$
34. $e^{\sqrt{3}}$
35. $e^{-0.15}$
36. $e^{-0.73}$
37. $e^{-3.1}$
38. $e^{-1.712}$
39. $e^{3.714}$
40. $e^{-\sqrt{7}}$
41. $(0.4014)^{3.2}$
42. $(4.81)^{1.5}$
43. $(5.977)^{-1.8}$
44. $(2.477)^{-3.7}$
45. $(3.912)^2$
46. $(21.85)^{-3}$
47. $(1.716)^{4.3}$
48. $(0.0763)^{0.34}$
49. $\sqrt[4]{7.18}$
50. $\sqrt[3]{5.32}$
51. $\sqrt[5]{91.81}$
52. $\sqrt[10]{2.63}$
53. $\sqrt[4]{0.293}$
54. $\sqrt[5]{(1.53)^3}$

© In problems 55–66, solve each equation. Round off your answers to four significant digits.

55. $2^x = 7$ **56.** $3^{2x} = 5$ **57.** $4^{-x} = 13$

58. $5^{-x} = 10$ **59.** $e^{2x} = 3$ **60.** $e^{-4x} = 7$

61. $e^{2x-1} = 6$ **62.** $7^{3x-1} = 9$ **63.** $3^{5-2t} = 8^{t-4}$

64. $4^{y+1} = 6^{8-3y}$ **65.** $3^{8p-1} = 5^{1-3p}$ **66.** $e^{3-7w} = 4^{1-5w}$

© **67.** In chemistry, the pH of a substance is defined by

$$pH = -\log[H^+],$$

where $[H^+]$ is the concentration of hydrogen ions in the substance measured in moles per liter. Find the pH of each substance (rounded off to two significant digits).

(a) Eggs: $[H^+] = 1.6 \times 10^{-8}$ moles per liter

(b) Tomatoes: $[H^+] = 6.3 \times 10^{-5}$ moles per liter

© **68.** On the **Richter scale,** the magnitude R of an earthquake of intensity I is given by the formula

$$R = \log\left(\frac{I}{I_0}\right),$$

where I_0 is a fixed minimum intensity. How many times greater than I_0 is the intensity of an earthquake that was measured as 6.4 on the Richter scale?

8.5 Applications

In this section, we examine some useful formulas that utilize exponential (or logarithmic) equations. For example, we show how exponential equations can be applied to calculate the growth of money accumulation due to interest payments. You will find the use of a calculator indispensable in this section.

Bankers use the **compound interest** formula

$$S = P\left(1 + \frac{r}{t}\right)^{nt}$$

to determine the accumulated amount of dollars S that will accrue from a principal of P dollars, invested for a term of n years at a nominal annual interest rate r compounded t times per year.

EXAMPLE **1**   c If you invest \$5,000 at a nominal annual interest rate of 14%, how much money do you accumulate after 4 years if the interest is compounded (a) annually and (b) quarterly?

SOLUTION   Here $P = 5,000$, $r = 0.14$, and $n = 4$.

(a) For interest compounded annually, $t = 1$. Therefore,

$$S = P\left(1 + \frac{r}{t}\right)^{nt}$$

$$= 5,000\left(1 + \frac{0.14}{1}\right)^{4(1)}$$

$$= 5,000(1.14)^4.$$

Using a calculator, we find that

$$S = 5,000(1.14)^4 = 8,444.80.$$

The amount accumulated is \$8,444.80.

(b) For interest compounded quarterly, $t = 4$. Therefore,

$$S = P\left(1 + \frac{r}{t}\right)^{nt}$$

$$= 5,000\left(1 + \frac{0.14}{4}\right)^{4(4)}$$

$$= 5,000(1.035)^{16} = 8,669.93.$$

The amount accumulated is \$8,669.93.

Effective Simple Annual Interest Rate

When a bank offers compound interest, it usually specifies not only the nominal annual interest rate r but also the **effective** simple annual interest rate R, that is, the rate of simple annual interest that would yield the same amount, accumulated over a 1-year term, as the compound interest yields. The formula

$$R = \left(1 + \frac{r}{t}\right)^t - 1$$

is used to calculate R in terms of r.

EXAMPLE **2**   c Find the effective simple annual interest rate R corresponding to a nominal annual interest rate of 14% compounded quarterly.

SOLUTION Here $r = 0.14$ and $t = 4$. Therefore,

$$R = \left(1 + \frac{r}{t}\right)^t - 1 = \left(1 + \frac{0.14}{4}\right)^4 - 1$$

$$= (1.035)^4 - 1 = 0.1475.$$

In other words, the effective simple annual interest rate is 14.75%.

Present Value

Money that you will receive in the future is worth *less* to you than the same amount of money received now. This is because you will miss out on the interest you could accumulate if you invested the money now. For this reason, the **present value** of money to be received in the future is the actual amount of money discounted by the prevailing interest rate during the period.

Suppose, for example, that you have an opportunity to invest P dollars at a nominal annual interest rate r, compounded t times a year. This principal plus the interest it earns will amount to S dollars after n years. This is stated in the compound interest formula:

$$S = P\left(1 + \frac{r}{t}\right)^{nt}.$$

Now, consider what this means. It means that P dollars today is worth S dollars to be received n years in the future. We can solve the compound interest equation for P to determine the **present value** of an offer of S dollars to be received n years in the future:

$$P = S\left(1 + \frac{r}{t}\right)^{-nt}.$$

EXAMPLE 3 © To dissolve a partnership that owns a sports arena, one partner buys the shares of the other partners for a total of $40,000, to be paid 2 years in the future. If, during this period, investments earn a nominal annual interest rate of 16% compounded quarterly, find the present value of the $40,000.

SOLUTION Here $S = 40,000$, $r = 0.16$, $t = 4$, and $n = 2$, so that

$$P = S\left(1 + \frac{r}{t}\right)^{-nt}$$

$$= 40,000\left(1 + \frac{0.16}{4}\right)^{-2(4)} = 40,000(1.04)^{-8} = 29,227.61.$$

Therefore, the present value of the $40,000 is $29,227.61.

Population Growth

We can also apply exponential equations to solve problems dealing with population growth. We use the formula

$$P = P_0 e^{kt}$$

where P is the population after time t, P_0 is the original population, and k is the rate of growth per unit of time.

EXAMPLE 4   ⓒ The population of a small country was 10 million in 1980, and has been growing according to the equation $P = P_0 e^{kt}$ at 3% per year. Predict the population in the year 2000.

SOLUTION   Here $P_0 = 10$ million, $k = 0.03$, and $t = 20$ years. Thus,

$$P = P_0 e^{kt}$$
$$= 10 e^{0.03(20)}$$
$$= 10 e^{0.6} = 18.22.$$

Therefore, the population will be approximately 18.22 million in the year 2000.

Science

One example of the use of exponentials is found in an application of Boyle's law, an important law in chemistry.

EXAMPLE 5   ⓒ Boyle's law for adiabatic expansion of air is given by the equation $PV^{1.4} = C$, where P is the pressure of the air, V is its volume, and C is a constant. At a certain instant, the volume of the air is 75.2 cubic inches and $C = 12,600$ pounds per inch. Find the pressure P.

SOLUTION   We have $PV^{1.4} = C$ or $P = \dfrac{C}{V^{1.4}} = CV^{-1.4}$

Here $V = 75.2$ and $C = 12,600$ so that

$$P = 12,600(75.2)^{-1.4}$$
$$= 29.76.$$

Therefore, the pressure is 29.76 pounds per square inch.

PROBLEM SET 8.5

[C] **1.** If you invest $7,000 in a savings certificate at a nominal annual interest rate of 13% compounded annually, how much money is accumulated after 3 years?

[C] **2.** If $8,000 is invested at a nominal annual interest rate of 14% compounded quarterly, how much money is accumulated after 4 years?

[C] **3.** If $10,000 is invested at a nominal annual interest rate of 11.8% compounded monthly, how much money is accumulated after 6 years?

[C] **4.** Find the amount of interest earned on $4,000 that was placed in a bank at a nominal annual interest rate of $7\frac{1}{4}$% compounded semiannually for 5 years.

[C] **5.** Find the effective simple annual interest rate R corresponding to a nominal annual interest rate of 11% compounded annually.

[C] **6.** Find the effective simple annual interest rate R corresponding to a nominal annual interest rate of 12% compounded semiannually.

[C] **7.** Find the effective simple annual interest rate R corresponding to a nominal annual interest rate of 8% compounded quarterly.

[C] **8.** Find the effective simple annual interest rate R corresponding to a nominal annual interest rate of 10% compounded monthly.

[C] **9.** Find the present value of $10,000, to be paid 3 years in the future, if money could be invested at 14.5% compounded annually during that time.

[C] **10.** Find the present value of $5,000 to be paid to you 7 years in the future, if investments during this period will earn a nominal annual interest rate of 10% compounded quarterly.

[C] **11.** On Gus's 18th birthday, his father promises to give him $25,000 when he turns 21, to help set him up in business. Savings certificates are available at a nominal annual interest rate of 12.8% compounded quarterly. How much does his father need to invest on Gus's 18th birthday in order to fulfill his promise?

[C] **12.** Suppose that someone owes you $130,000 to be paid to you 3 years in the future. What is the present value of this money to you if investments are earning a nominal interest rate of 10.5% compounded annually?

[C] **13.** The population of a small country was 2 million in 1983 and has been growing according to the equation $P = P_0 e^{kt}$ at 4% per year. Predict the population in the year 1993.

[C] **14.** A biologist finds that there are 2,000 bacteria in a culture and that the culture has been growing according to the equation $P = P_0 e^{kt}$ at 7% per hour. How many bacteria will be present after 12 hours?

[C] **15.** A bank offers a savings account with continuously compounded interest at a nominal annual rate of 7%. Such an account is worth an amount $P = P_0 e^{kt}$ after t years. If a principal $P = $1,000$ is deposited when $t = 0$, what is the final value of this investment after 10 years?

[C] **16.** A manufacturing plant estimates that the value V, in dollars, of a machine is decreasing according to the equation $V = 45,000 e^{-0.13t}$, where t is the number of years since the machine was placed in service. Find the value of the machine after 8 years.

C **17.** According to Newton's law of cooling, under certain conditions the temperature T (in degrees Celsius) of an object is given by the equation $T = 75e^{-2t}$, where t is the time in hours. Find the temperature of an object after 2.5 hours.

C **18.** The required area A of the cross section of a chimney is given by $A = 0.06ph^{-1.2}$, where p is the number of pounds of coal burned each hour and h is the height of the chimney in feet. Find the required cross-sectional area (in square feet) of a chimney 72 feet high if 750 pounds of coal are burned each hour.

C **19.** A fully charged electrical condenser is allowed to discharge. After t seconds the remaining charge Q (in coulombs) is given by the equation $Q = 750(2.7)^{-0.4t}$. What is the charge Q after 23 seconds?

C **20.** Under certain conditions, the atmospheric pressure P, in inches of mercury, at altitude h, in feet, is given by the equation $P = 29(2.6)^{-0.000034h}$. What is the pressure at an altitude of 35,000 feet?

REVIEW PROBLEM SET

In problems 1–8, solve each exponential equation.

1. $3^{2x-1} = 9$

2. $4^{1-t} = 2^{t+2}$

3. $8^{y+1} = 4^y$

4. $5^{u+2} = 625$

5. $27^{v-1} = 9$

6. $6^{3x+7} = 216^{3-x}$

7. $3^{2x+1} = 27^{x-1}$

8. $(1.2)^{2c+1} = 1.44$

In problems 9–16, write each equation in logarithmic form.

9. $2^5 = 32$

10. $9^{3/2} = 27$

11. $8^{2/3} = 4$

12. $32^{-4/5} = \frac{1}{16}$

13. $27^{-2/3} = \frac{1}{9}$

14. $13^0 = 1$

15. $z^n = w$

16. $c^t = d$

In problems 17–24, write each equation in exponential form.

17. $\log_2 8 = 3$

18. $\log_2 64 = 6$

19. $\log_{10} 100 = 2$

20. $\log_9 27 = \frac{3}{2}$

21. $\log_{17} 1 = 0$

22. $\log_{125} 625 = \frac{4}{3}$

23. $\log_{10} \frac{1}{10} = -1$

24. $\log_a z = w$

In problems 25–36, find the value of each logarithm.

25. $\log_3 9$

26. $\log_4 8$

27. $\log_6 1$

28. $\log_5 0.04$

29. $\log_{100} 0.001$

30. $\log_9 \frac{1}{3}$

31. $\log_4 \frac{1}{128}$

32. $\log_2 16^{-2}$

33. $\log_5 625^{-1}$

34. $\log_2 1024^{-1}$

35. $\log_4 8\sqrt{2}$

36. $\log_{6/5} \frac{25}{36}$

In problems 37–42, solve each equation.

37. $\log_4 16 = t$

38. $\log_3 9\sqrt{3} = y$

39. $\log_2(7x - 1) = 4$

40. $\log_5(3u - 11) = 2$

41. $\log_9(17z - 33) = 0$

42. $\log_5\left(\frac{x}{2} - \frac{3}{2}\right) = 1$

In problems 43–50, use the properties of logarithms to write each expression as a sum or difference of multiples of logarithms. Assume that all variables represent positive real numbers.

43. $\log_6 7u$

44. $\log_2 4x^3$

45. $\log_2 3^6 \cdot 4^7$

46. $\log_8 \dfrac{5^7}{9^3}$

47. $\log_4 xy^5$

48. $\log_3 \sqrt[7]{5^3 \cdot 8^6}$

49. $\log_a \dfrac{w^2}{z^4}$

50. $\log_b x^3 y^2 z^4$

In problems 51–58, use the properties of logarithms to write each expression as a single logarithm or number. Assume that all variables represent positive real numbers.

51. $\log_2 \frac{3}{7} + \log_2 \frac{14}{27}$

52. $\log_3 \frac{5}{12} + \log_3 \frac{4}{15}$

53. $\log_5 \frac{6}{7} - \log_5 \frac{27}{4} + \log_5 \frac{21}{16}$

54. $\log_9 \frac{11}{5} + \log_9 \frac{14}{3} - \log_9 \frac{22}{15}$

55. $5\log_a x - 3\log_a y$

56. $2\log_c x^3 + \log_c \dfrac{2}{x} - \log_c \dfrac{2}{x^4}$

57. $\log_9 9^5 + \log_9 3^{-7}$

58. $\log_t \sqrt[3]{t} + \log_t \sqrt[3]{t^2}$

In problems 59–64, use the properties of logarithms to solve each equation.

59. $\log_3 z + \log_3 4 = 2$

60. $\log_2(t + 1) + \log_2 3 = 4$

61. $\log_5 x - \log_5 3 = 1$

62. $\log_3 4y - \log_3 2 = 5$

63. $\log_5(2x - 1) + \log_5(2x + 1) = 2$

64. $\log_{1/2}(4t^2 - 1) - \log_{1/2}(2t + 1) = 1$

In problems 65–68, express each number in scientific notation.

65. 46,800,000

66. 432,000,000

67. 0.0000012

68. 0.0000000326

In problems 69–72, express each number in ordinary notation.

69. 5.6×10^3

70. 3.21×10^4

71. 1.92×10^{-4}

72. 8.37×10^{-6}

In problems 73–82, use Table I in Appendix I to find the value of each common logarithm. Use interpolation when necessary.

73. log 846

74. log 55.6

75. log 75.2

76. log 7.38

77. log 3.184

78. log 0.5315

79. log 0.473

80. log 0.0392

81. log 0.005867

82. log 0.0009254

In problems 83–92, use Table I in Appendix I to find the value of each antilogarithm. Interpolate when necessary.

83. antilog 0.7466

84. antilog 0.5514

85. antilog 1.9533

86. antilog 2.9243

87. antilog 3.2375

88. antilog 4.1152

89. antilog [0.4518 + (−2)]

90. antilog [0.9289 + (−3)]

91. antilog (−3.4076)

92. antilog (−2.3478)

C In problems 93–108, use a calculator to find the value of each of the following expressions. Round off your answers to four significant digits.

93. log 14.73

94. log 289.7

95. log 0.000315

96. log 0.000008792

97. antilog 3.4917

98. antilog 4.2163

99. antilog (−2.3168)

100. antilog (−1.9721)

101. ln 45.31   **102.** ln 178.9   **103.** $e^{\sqrt{5}}$ **104.** $e^{-0.51}$

105. $(2.91)^{-1.7}$ **106.** $(0.87)^{-2.3}$ **107.** $\sqrt[4]{23.89}$ **108.** $\sqrt[5]{173.8}$

C In problems 109–114, solve each equation.

109. $2^x = 5$ **110.** $5^{2x} = 3$ **111.** $4^{3-t} = 5^{t+1}$ **112.** $2^{2v+1} = 4^{v-2}$

113. $e^{3u-2} = 7$ **114.** $e^{4x+2} = 6^{3-2x}$

C **115.** A sum of $5,000 is invested in a stock whose average growth rate is 15% compounded annually. Assuming that the rate of growth continues, find the investment's value after 6 years.

C **116.** If $10,000 is invested at a nominal annual interest rate of 12% compounded quarterly, how much money is accumulated after 8 years?

C **117.** Find the effective simple annual interest rate R corresponding to a nominal annual interest rate of 14% compounded semiannually.

C **118.** Find the effective simple annual interest rate R corresponding to a nominal annual interest rate of 12.6% compounded quarterly.

C **119.** Find the present value of $7,000 to be paid to you 5 years in the future if investments during this period will be earning a nominal annual interest rate of 11.6% compounded every 2 months.

C **120.** Find the present value of the $20,000 you will receive in 10 years if investments during this period will be earning a nominal annual interest rate of 14% compounded quarterly.

C **121.** The population of a city was 520,000 in 1983, and it has been growing according to the equation $P = P_0 e^{kt}$ at 3.5% per year. Predict the city's population in the year 1990.

C **122.** At the start of an experiment, the number of bacteria present in a culture is 3,000. If the culture is growing according to the equation $P = P_0 e^{kt}$ at 6% per hour, how many bacteria will be present after 8 hours?

9 Functions and Related Curves

Scientific advances often result from the discovery that things depend upon one another in definite ways. An example is Einstein's famous equation, $E = mc^2$, which relates mass m to energy E, using the speed of light, c, squared. Such relationships among quantities are usually expressed by *functions.* In this chapter, we discuss briefly some functions and their graphs. We also learn how to graph special curves, known as conic sections, which may be used to solve nonlinear systems of equations.

9.1 Functions

You have encountered the concept of a function many times in everyday life. For example, the amount of sales tax charged on a purchase is a function of the amount of the purchase; the number of textbooks to be ordered for a course is a function of the number of students enrolled in the course; and the number of congressional representatives for a particular state is a function of the population of the state. A function suggests some kind of **correspondence.** In each of the preceding examples, there is an established correspondence between numbers: between the amount of the sales tax and the amount of the purchase; between the number of books and the number of students; and between the number of congressional representatives and the number of people in each state.

In mathematics, the general idea of a function is simple. Suppose that one variable quantity—say, y—depends, in a definite way, on another variable quantity—say, x. Then for each particular value of x, there is *one* corresponding value of y. Such a correspondence defines a **function,** and we say that (the variable) y is a function of (the variable) x.

For example, if x is used to represent the length of the side of a square and y is used to represent the area of this square, then y depends on x in a definite way, namely, $y = x^2$. Therefore, we say that the area y of the square is a function of the length of its side x.

In a sense, the value of y *depends* on the value assigned to x. For this reason, we sometimes refer to x as the **independent variable** and to y as the **dependent variable.** Thus, if $y = x^2$, then $y = 4$ when $x = 2$, $y = 25$ when $x = 5$, and $y = 121$ when $x = 11$. More formally, we have the following definition:

DEFINITION **1**   **A Function as a Correspondence**

A **function** is a correspondence that assigns to each member in a certain set, called the *domain* of the function, one and only one member in a second set, called the *range* of the function.

The **independent variable** of the function can take on any value in the domain of the function. The range of the function is the set of all possible corresponding values that the **dependent variable** can assume.

Suppose that a function is denoted by the letter f and is determined by the equation $3r + 5t = 3$. If r represents the independent variable for the function f determined by $3r + 5t = 3$, we say that t is a function of r. We indicate this fact by the function notation $t = f(r)$. This equation, which is read "t equals f of r", means that t is a function of r, in other words that r represents the independent variable and t represents the dependent variable. If we solve for t in terms of r, we obtain $t = f(r) = -\frac{3}{5}r + \frac{3}{5}$.

On the other hand, if we write $r = f(t)$, we are saying that t represents the independent variable and r the dependent variable. If we then solve for r in terms of t, we have $r = f(t) = -\frac{5}{3}t + 1$.

We often use letters other than f to represent functions. For example, g and h as well as F, G, and H are favorites for this purpose. *If f is a function and x represents a member of the domain, then $f(x)$ represents the corresponding member of the range.* Note that $f(x)$ is *not* the function f. However, to save time, we frequently use the phrase "the function $y = f(x)$." There is no great harm in this practice. Indeed, we use this phrase whenever it is convenient. However, it is important to remember that "the function $y = f(x)$" actually means "the function f determined by the equation $y = f(x)$."

Thus

$$g(t) = t^2, \qquad h(x) = x + 7, \qquad V(r) = \frac{4}{3}\pi r^3, \qquad \text{and} \qquad F(s) = \sqrt{s}$$

represent functions with the independent variables $t, x, r,$ and s, respectively.

If $g(t) = t^2$, then we can determine $g(3)$ by substituting 3 for t:

$$g(3) = 3^2 = 9.$$

Similarly, we can find $g(x + h)$ by substituting $x + h$ for t:

$$g(x + h) = (x + h)^2$$
$$= x^2 + 2xh + h^2.$$

EXAMPLE 1   Let $f(x) = 5x + 6$. Find the following values.

(a) $f(1)$ (b) $f(-3)$

(c) $f(\sqrt{2})$ (d) $\sqrt{f(2)}$

(e) $[f(-4)]^2$ (f) $f(a)$

(g) $f(a + 4)$ (h) $f(3b)$

(i) $f(-c)$ (j) $f(3t + 7)$

SOLUTION   (a) $f(1) = 5(1) + 6 = 11$

(b) $f(-3) = 5(-3) + 6 = -9$

(c) $f(\sqrt{2}) = 5\sqrt{2} + 6$

(d) $\sqrt{f(2)} = \sqrt{5(2) + 6} = \sqrt{16} = 4$

(e) $[f(-4)]^2 = [5(-4) + 6]^2 = (-14)^2 = 196$

(f) $f(a) = 5a + 6$

(g) $f(a + 4) = 5(a + 4) + 6 = 5a + 20 + 6 = 5a + 26$

(h) $f(3b) = 5(3b) + 6 = 15b + 6$

(i) $f(-c) = 5(-c) + 6 = -5c + 6$

(j) $f(3t + 7) = 5(3t + 7) + 6 = 15t + 35 + 6 = 15t + 41$

EXAMPLE 2   Let $f(x) = \sqrt{25 - x^2}$. Find the following values.

(a) $f(0)$ (b) $f(3)$

(c) $f(4)$ (d) $f(5)$

SOLUTION   (a) $f(0) = \sqrt{25 - 0^2} = \sqrt{25} = 5$

(b) $f(3) = \sqrt{25 - 3^2} = \sqrt{25 - 9} = \sqrt{16} = 4$

(c) $f(4) = \sqrt{25 - 4^2} = \sqrt{25 - 16} = \sqrt{9} = 3$

(d) $f(5) = \sqrt{25 - 5^2} = \sqrt{25 - 25} = \sqrt{0} = 0$

EXAMPLE 3   Let $f(x) = |x|$. Find the following values.

(a) $f(-2)$

(b) $f(0)$

(c) $f(2)$

SOLUTION   (a) $f(-2) = |-2| = 2$

(b) $f(0) = |0| = 0$

(c) $f(2) = |2| = 2$

Whenever a function f is defined by an equation $y = f(x)$, you may assume (unless you are told otherwise) that its domain consists of all values of x for which the equation makes sense and determines a corresponding real number y. The range of the function is then automatically determined, since it consists of the set of all values of y that correspond, by the equation that defines the function, to values of x in the domain.

In Examples 4–6, find the domain of the function determined by each equation.

EXAMPLE 4 $f(x) = 2x + 1$

SOLUTION   The expression $2x + 1$ is defined for all real values of x. Therefore, the domain of f is the set of real numbers.

EXAMPLE 5 $g(x) = \dfrac{1}{x + 2}$

SOLUTION   The domain of g is the set of all real numbers except -2 because $x + 2 = 0$ for $x = -2$, and division by 0 is undefined.

EXAMPLE 6 $h(x) = \sqrt{1 - x}$

SOLUTION   The expression $\sqrt{1 - x}$ represents a real number if and only if $1 - x \geq 0$, that is, if and only if $x \leq 1$. Therefore, the domain of h is the set of all real numbers x such that $x \leq 1$.

The **graph** of a function f is the graph of the corresponding equation $y = f(x)$. The graph of f is the set of all points (x, y) in the Cartesian plane such that x is in the domain of f and $y = f(x)$.

In Examples 7 and 8, sketch the graph of each function.

EXAMPLE **7** $f(x) = x$ (the identity function)

SOLUTION   The domain of f consists of all real numbers. The graph of the funtion f is the graph of the equation $y = x$. The graph of $y = x$ is the line with slope 1 and y intercept 0 (Figure 1).

Figure 1

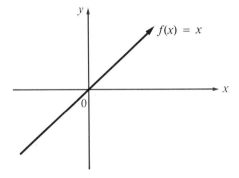

EXAMPLE **8** $f(x) = |x|$ (the absolute value function)

SOLUTION   The domain of f consists of all real numbers. The graph of the function f is the graph of the equation $y = |x|$. If we apply the definition of absolute value,

$$y = \begin{cases} x & \text{for } x \geq 0 \\ -x & \text{for } x < 0. \end{cases}$$

The graphs are portions of straight lines with slopes 1 and -1, respectively (Figure 2).

Figure 2

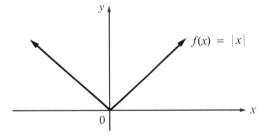

It is important to realize that *not every curve* in the Cartesian plane is the *graph* of a function. Indeed, the definition of a function requires that there be one and only one value of y corresponding to each value of x in the domain. Thus, we cannot have two points (x_1, y_1) and (x_1, y_2) on the graph of a function with the same abscissa x_1 and different ordinates y_1 and y_2.

EXAMPLE 9   Which of the curves in Figure 3 is the graph of a function?

SOLUTION   Notice that the dashed vertical line in Figure 3a intersects the curve in exactly one point. This is true of every vertical line that would intersect this curve. However, the dashed vertical line in Figure 3b intersects the curve in two points. Thus, the curve in Figure 3a is the graph of a function, but the curve in Figure 3b is not, because on the graph of a function we cannot have two points (x, y_1) and (x, y_2) with the same abscissa x and different ordinates y_1 and y_2.

Figure 3

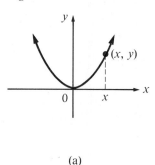

(a)

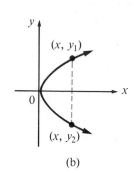

(b)

PROBLEM SET 9.1

In problems 1–10, let $f(x) = 3x + 1$ and find the following values.

1. $f(1)$ **2.** $f(\frac{2}{3})$ **3.** $f(-2)$ **4.** $f(u + v)$

5. $f(0)$ **6.** $f(2z)$ **7.** $f(a + z)$ **8.** $f(a + b) - f(a)$

9. $[f(4)]^2$ **10.** $\sqrt{f(3)}$

In problems 11–18, let $g(x) = \sqrt{16 - x^2}$ and find the following values.

11. $g(0)$ **12.** $g(2)$ **13.** $g(4)$ **14.** $g(-3)$

15. $g(-4)$ **16.** $g(x + 4)$ **17.** $g(\sqrt{7})$ **18.** $[g(2)]^2$

In problems 19–26, let $h(x) = |x - 2|$ and find the following values.

19. $h(7)$ **20.** $h(w + 2)$ **21.** $h(-3)$ **22.** $h(b + 3)$

23. $h(0)$ **24.** $h(2\frac{1}{2})$ **25.** $h(-a)$ **26.** $\sqrt{h(6)}$

In problems 27–36, find the domain of the function determined by each equation.

27. $f(x) = -3x + 2$ **28.** $f(x) = \frac{2}{3}x + \frac{1}{3}$ **29.** $f(x) = \dfrac{1}{x}$ **30.** $f(x) = \sqrt[3]{-2x + 1}$

31. $f(x) = \dfrac{1}{\sqrt{x - 1}}$ **32.** $f(x) = 2x^2 - 5$ **33.** $f(x) = \sqrt{2 - x}$ **34.** $f(x) = \dfrac{4}{\sqrt{x^2 - 1}}$

35. $f(x) = x^2$ **36.** $f(x) = x^3 - 1$

In problems 37–40, sketch the graph of each function.

37. $f(x) = x + 3$ **38.** $f(x) = x - 2$ **39.** $f(x) = |x - 1|$ **40.** $f(x) = |x + 3|$

In problems 41–46, determine which curve is the graph of a function.

41.

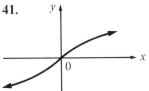

42.

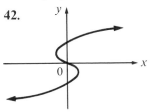

43.

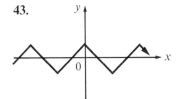

44.

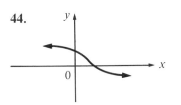

45.

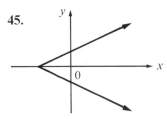

46.

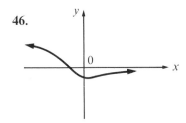

9.2 **Variation**

We often encounter a function described by an equation of the form $y = kx$, where k is some constant number. When this occurs, we have what is called a **direct variation.**

DEFINITION 1   **Direct Variation**

> Let x and y be two variable quantities. We say that y **varies directly with** x or that y **is directly proportional to** x, if there is a constant k such that
>
> $$y = kx.$$
>
> The number k is called the **constant of variation** or the **constant of proportionality.**

For example, if an airplane is flying at a rate of 650 miles per hour, then the distance d that it travels in t hours is $d = 650t$. The distance d is directly proportional to the time t, and the constant of proportionality is 650.

At times, we deal with two variables x and y, related by an equation of the form

$$y = kx^n$$

for some constant number k and some positive rational number n. In this case we say that y **is directly proportional to the nth power of x.**

EXAMPLE 1   Express y as a function of x if y is directly proportional to x, and if $y = 4$ when $x = 1$.

SOLUTION   Since y is directly proportional to x, there is some constant k such that

$$y = kx.$$

Because $y = 4$ when $x = 1$, we have $4 = k(1)$ or $k = 4$, so that the equation becomes

$$y = 4x.$$

EXAMPLE 2   The surface area of a sphere is directly proportional to the square of its radius. If a sphere with a radius of 4 inches has a surface area of 64π square inches, express the surface area of a sphere as a function of its radius.

SOLUTION   Let S represent the surface area of a sphere, and let r represent the radius, in inches. Because S is directly proportional to r^2, there is a constant k such that

$$S = kr^2.$$

Because $S = 64\pi$ when $r = 4$, then

$$64\pi = 16k,$$

and

$$k = \frac{64}{16}\pi = 4\pi.$$

Hence,

$$S = 4\pi r^2.$$

EXAMPLE 3   It has been estimated that the amount of pollution A entering the atmosphere is directly proportional to the number of people N living in a certain area. If a city with a population of 140,000 people produces 100,000 tons of atmospheric pollutants in a year, how many tons of pollutants are likely to enter the atmosphere annually in a city with a population of 1,350,000 people?

SOLUTION   Because the pollution A is directly proportional to the number of people N, there is some constant k such that

$$A = kN.$$

Because $A = 100,000$ when $N = 140,000$, we have

$$100,000 = 140,000k.$$

We solve for k: $k = \frac{5}{7}$, so that the function becomes

$$A = \frac{5}{7}N.$$

Hence, if $N = 1{,}350{,}000$, then

$$A = \frac{5}{7}(1{,}350{,}000) = 964{,}285.71.$$

This means that about 964,286 tons of pollutants will probably enter the atmosphere.

Consider the function determined by the equation

$$y = \frac{k}{x},$$

where k is a positive constant. Notice that as the value of x increases, the value of y decreases. This is an example of an **inverse variation.**

DEFINITION 2   **Inverse Variation**

> Let x and y be two variable quantities. We say that **y varies inversely with x** or that **y is inversely proportional to x** if there is a constant k such that
>
> $$y = \frac{k}{x} \qquad \text{for } x \neq 0.$$

For example, if y is inversely proportional to x, and $y = 0.4$ when $x = 0.8$, then

$$y = \frac{k}{x}$$

and

$$0.4 = \frac{k}{0.8} \qquad \text{or} \qquad k = 0.32.$$

Therefore, the function that relates y to x is given by the equation

$$y = \frac{0.32}{x}.$$

If

$$y = \frac{k}{x^n}$$

for some constant number k and some positive rational number n, we say that **y is inversely proportional to the nth power of x.**

EXAMPLE 4   Express y as a function of x if y is inversely proportional to x^2, and if $y = 12$ when $x = 2$.

SOLUTION   Because y is inversely proportional to x^2, there is a number k such that

$$y = \frac{k}{x^2}.$$

Given that $y = 12$ when $x = 2$, then $12 = k/2^2$ and $k = 48$. Thus,

$$y = \frac{48}{x^2}.$$

EXAMPLE 5   Boyle's law states that the pressure P of an ideal gas at a constant temperature is inversely proportional to its volume V. Find the constant of variation if the pressure P of a gas is 30 pounds per square inch when the volume V is 100 cubic inches.

SOLUTION   Since the pressure P is inversely proportional to V, there is a number k such that

$$P = \frac{k}{V} \quad \text{or} \quad k = PV.$$

At $P = 30$ pounds per square inch, $V = 100$ cubic inches, so that we have $k = 30(100) = 3,000$ pounds inches.

Consider the area of a rectangle with length l units and width w units. The area A is given by the equation $A = lw$. In this situation, we say that A varies jointly with l and w. We have the following formal definition:

DEFINITION 3   **Joint Variation**

Let x, y, and z be variable quantities. We say that z **varies jointly with x and y** if there is a constant k such that

$$z = kxy.$$

We call the relationship between z and the product xy **joint variation.**

Several kinds of variations can occur together. For example, if

$$w = \frac{kx^2y^3}{z^4},$$

then w varies jointly with the square of x and with the cube of y, and inversely with the fourth power of z.

EXAMPLE 6   Suppose that z varies directly with the cube of x and inversely with y. If $z = 8$ when $x = 2$ and $y = 4$, find z when $x = 5$ and $y = 10$.

SOLUTION   Because z varies directly with the cube of x and inversely with y, there is some constant k such that

$$z = \frac{kx^3}{y}.$$

We substitute for z, x, and y:

$$8 = \frac{k(2)^3}{4} \quad \text{or} \quad k = 4$$

so that $z = 4x^3/y$. Thus, for $x = 5$ and $y = 10$,

$$z = \frac{4(5)^3}{10} = 50.$$

EXAMPLE 7   The volume V of a right circular cone varies jointly with its altitude h and the square of its base radius r. If $V = 12\pi$ cubic inches when $r = 3$ inches and $h = 4$ inches, find V in terms of r and h.

SOLUTION   Because V varies jointly with r^2 and h, there is a real number k such that

$$V = kr^2h.$$

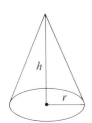

We substitute for V, r, and h:

$$12\pi = k(9)(4) \quad \text{or} \quad k = \frac{12\pi}{36} = \frac{\pi}{3}.$$

Hence, the required formula is

$$V = \frac{1}{3}\pi r^2 h.$$

PROBLEM SET 9.2

1. Let y be directly proportional to x, and let $y = f(x)$. If $y = 8$ when $x = 4$, find a formula for $f(x)$. Also find $f(x + 2)$, $f(2) + f(3)$, and $[f(x + h) - f(x)]/h$, where $h \neq 0$.

2. If y is directly proportional to x^2 and $y = f(x)$, does $f(ax) = af(x)$?

In problems 3–8, y is directly proportional to x^3. Express y as a function of x in each case.

3. $y = 4$ when $x = 2$ 4. $y = 12$ when $x = -2$ 5. $y = 3$ when $x = 1$

6. $y = -2$ when $x = 3$ 7. $y = 14$ when $x = 11$ 8. $y = 10$ when $x = -3$

9. If y is inversely proportional to x^2, and if $y = 9$ when $x = 2$, find y when $x = 3$.

10. If y is inversely proportional to $\sqrt[3]{x}$ and $y = 9$ when $x = 8$, find y when $x = 216$.

11. If T is directly proportional to x and inversely proportional to y, and if $T = 0.01$ when $x = 20$ and $y = 20$, express T as a function of x and y.

12. If y is inversely proportional to x^2, and if $y = 8$ when $x = 10$, find y when $x = 2$.

13. If y is inversely proportional to x^3, and if $y = 3$ when $x = 4$, express y as a function of x.

14. If V varies directly with T and inversely with P, and if $V = 40$ when $T = 300$ and $P = 30$, find V when $T = 324$ and $P = 24$.

15. The surface areas of two spheres have the ratio of 9 to 4. (The *ratio* 9 to 4 may be written as the fraction $\frac{9}{4}$ or as $9:4$.) What is the ratio of their radii? their volumes? (Hint: $S = 4\pi r^2$ and $V = \frac{4}{3}\pi r^3$.)

16. Coulomb's law states that the magnitude of the force F (in newtons) that acts on two charges q_1 and q_2 varies directly with the product of q_1 and q_2 (in coulombs) and inversely with the square of the distance r (in meters) between them. If two charges, each having a magnitude of 1 coulomb, are separated in air by a distance of 0.1 meters and if the force on the two charges is 9×10^{11} newtons, find the force when the charges are separated by 0.2 meters.

17. The total surface area S of a cube is directly proportional to the square of an edge x. If the cube with an edge of 3 inches has a surface area of 54 square inches, express the surface area S as a function of x. Then find the surface area of a cube with an edge of 12 inches.

18. Newton's law of gravitational attraction states that the force F with which two particles of mass m_1 and m_2 attract each other varies directly with the product of the masses and inversely with the square of the distance r between them. If one of the masses is tripled, and the distance between the masses is also tripled, what happens to the force?

9.3 Linear and Quadratic Functions

In this section, we explore a variety of *linear* and *quadratic* functions by examining their graphs. These functions have many important applications.

Linear Functions

If m and b are constants, then the graph of the function f defined by

$$f(x) = mx + b$$

is the same as the graph of the equation

$$y = mx + b.$$

As we saw in Section 7.4, $y = mx + b$ is a straight line with slope m and y intercept b. Thus, we have the following definition:

DEFINITION 1   **Linear Function**

> A function f of the form
> $$f(x) = mx + b$$
> is called a **linear function.**

For example,

$$f(x) = 2x - 1, \qquad g(x) = -3x + 7, \qquad \text{and} \qquad h(x) = 1 - 7x$$

are linear functions.

If we substitute $m = 0$ in the equation $f(x) = mx + b$, we have a specific type of linear function:

DEFINITION 2   **Constant Function**

> A function of the form
> $$f(x) = b$$
> is called a **constant function.**

The graph of a constant function $f(x) = b$ is a horizontal straight line (Figure 1).

Figure 1

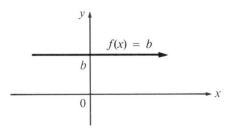

In Examples 1 and 2, sketch the graph of each linear function. Find the slope of the graph, and the domain and range of each function.

EXAMPLE 1 $f(x) = 2x + 5$

SOLUTION   The graph of f is the same as the graph of the equation

$$y = 2x + 5.$$

This graph is a line with slope $m = 2$ and y intercept 5. The x intercept is $-\frac{5}{2}$ (Figure 2). We see from the graph that the domain of f is the set $\mathbb{R}$ of all real numbers, and that the range of f is also $\mathbb{R}$.

Figure 2

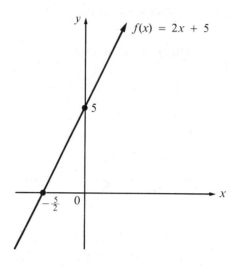

EXAMPLE 2 $g(x) = -\frac{9}{5}x + \frac{7}{5}$

SOLUTION   The graph of g is the same as the graph of the equation

$$y = -\frac{9}{5}x + \frac{7}{5}.$$

This graph is a line with slope $m = -\frac{9}{5}$ and y intercept $\frac{7}{5}$. The x intercept is $\frac{7}{9}$ (Figure 3). We see from the graph that the domain and the range of g are both $\mathbb{R}$.

Figure 3

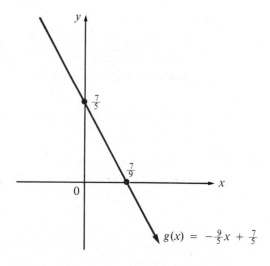

Quadratic Functions

Quadratic functions are used often in applied mathematics. This type of function is defined as follows:

DEFINITION 3   **Quadratic Function**

> A function f of the form
>
> $$f(x) = ax^2 + bx + c,$$
>
> where $a, b,$ and c are real numbers and $a \neq 0$, is called a **quadratic function.**

The simplest quadratic function is

$$f(x) = ax^2.$$

The graph of this function is called a **parabola** with **vertex** at the origin. It opens upward if $a > 0$ (Figure 4a) and downward if $a < 0$ (Figure 4b).

Figure 4

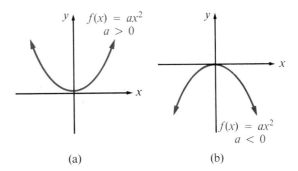

(a)   (b)

Graphs of quadratic functions resemble the graphs in Figure 4. We can obtain these graphs by plotting some points and connecting these points with a smooth curve.

In Examples 3 and 4, sketch the graph of each function, and find the domain and range of the function.

EXAMPLE 3 $f(x) = x^2$

SOLUTION   The function f is a quadratic function of the form $f(x) = ax^2$ with $a > 0$. The graph of f is a parabola which has its vertex at the origin and opens upward. We calculate values of $f(x) = x^2$ for some integer values of x, as shown in the table in Figure 5. Then we plot these points and connect them with a smooth curve to obtain the graph of f (Figure 5). The graph of f reveals that the domain is $\mathbb{R}$, and that the range consists of all nonnegative real numbers.

Figure 5

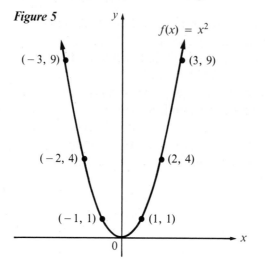

x	$y = f(x) = x^2$
-3	9
-2	4
-1	1
0	0
1	1
2	4
3	9

EXAMPLE 4 $g(x) = -x^2$

SOLUTION   The function g is of the form $f(x) = ax^2$ with $a < 0$. The graph of g is a parabola which has its vertex at the origin and opens downward. We calculate values of $g(x) = -x^2$ for some integer values of x, as shown in the table in Figure 6. Then we plot the corresponding points and connect them by a smooth curve to obtain the graph of g (Figure 6). The graph of g reveals that the domain is $\mathbb{R}$ and that the range consists of all nonpositive real numbers.

Figure 6

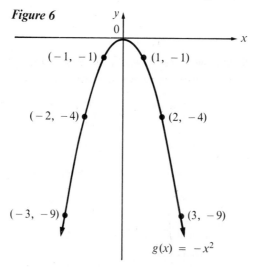

x	$y = g(x) = -x^2$
-3	-9
-2	-4
-1	-1
0	0
1	-1
2	-4
3	-9

The graph of any quadratic function of the form

$$f(x) = ax^2 + bx + c$$

will have the same general shape as the curves in Figure 4. If $a > 0$, the graph of f is a parabola that opens upward. If $a < 0$, the graph of f is a parabola that opens downward. The location of the graph will vary, depending upon specific values of a, b, and c.

The vertex of the parabola is one of the key points that we locate when we graph a quadratic function: $f(x) = ax^2 + bx + c$.

If we designate the point (h, k) as the vertex, we can find the values for the coordinates h and k by using the procedure given in Section 6.2 for completing the squares. We will find (see problem 34) that the vertex has the coordinates

$$(h, k) = \left(-\frac{b}{2a}, f\left(-\frac{b}{2a}\right)\right) = \left(-\frac{b}{2a}, \frac{4ac - b^2}{4a}\right).$$

The vertex (h, k) is the lowest point on the graph if $a > 0$ (Figure 5) and the highest point on the graph if $a < 0$ (Figure 6). Therefore, if $a > 0$, the number $f(-b/2a)$ is the **minimum** value of f, and if $a < 0$, the number $f(-b/2a)$ is the **maximum** value of f.

In Examples 5 and 6, find the vertex and the x and y intercepts of the graph of each function and sketch the graph. What are the domain and the range of the function?

EXAMPLE 5 $f(x) = x^2 - 3x + 2$

SOLUTION   Here we have $a = 1$, $b = -3$, and $c = 2$. The graph is a parabola that opens upward. The vertex (h, k) has coordinates

$$h = -\frac{b}{2a} = -\frac{(-3)}{2(1)} = \frac{3}{2}$$

and

$$k = f\left(-\frac{b}{2a}\right) = \frac{4ac - b^2}{4a} = \frac{4(1)(2) - (-3)^2}{4(1)} = -\frac{1}{4}$$

so that

$$(h, k) = \left(\frac{3}{2}, -\frac{1}{4}\right).$$

Figure 7

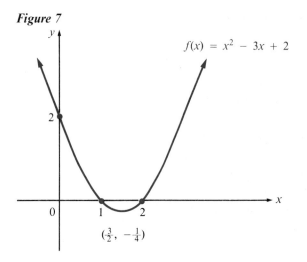

The y intercept is obtained when $x = 0$. Therefore, the y intercept is 2. The x intercepts are obtained by solving the equation

$$x^2 - 3x + 2 = 0$$

or

$$(x - 1)(x - 2) = 0.$$

Thus, the x intercepts are 1 and 2. We can plot these four points and connect them with a smooth curve to obtain the graph of f (Figure 7). The graph of f reveals that the domain of f is $\mathbb{R}$, and that the range consists of all real numbers $\geq -\frac{1}{4}$.

EXAMPLE 6 $g(x) = -x^2 + 2x$

SOLUTION   Here $a = -1$, $b = 2$, and $c = 0$. The graph is a parabola that opens downward. Its vertex (h, k) has the coordinates

Figure 8

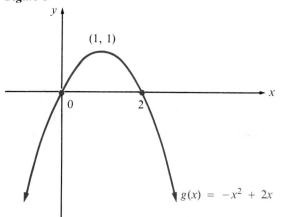

$$h = -\frac{b}{2a} = -\frac{2}{2(-1)} = 1$$

and

$$k = f\left(-\frac{b}{2a}\right) = \frac{4ac - b^2}{4a} = \frac{0 - 2^2}{4(-1)} = 1$$

so that

$$(h, k) = (1, 1).$$

The y intercept is 0, and the x intercepts are 0 and 2. (Why?) The graph of g shows that the domain of g is $\mathbb{R}$ and that the range consists of all real numbers ≤ 1 (Figure 8).

PROBLEM SET 9.3

In problems 1–10, sketch the graph of each linear function. Find the slope of the graph and the domain and the range of the function.

1. $f(x) = -3x + 5$ **2.** $f(x) = 5x + 1$ **3.** $f(x) = -\frac{3}{4}x + 1$ **4.** $f(x) = \frac{1}{4}x + 3$
5. $f(x) = -1$ **6.** $f(x) = -3x$ **7.** $f(x) = 2(x - 2) + 1$ **8.** $f(x) = 7$
9. $f(x) = 4x$ **10.** $f(x) = 3 - 2(1 - x)$

In problems 11–16, find a linear function f such that each condition is satisfied.

11. $f(1) = 3$ and $f(2) = 5$ **12.** $f(-2) = 7$ and $f(3) = -5$

13. $f(0) = 4$ and $f(3) = 0$ **14.** $f(1) = -8$ and $f(\frac{1}{2}) = -6$

15. $f(-7) = 3$ and $f(5) = 3$ **16.** $f(2) = -4$ and $f(-10) = -4$

17. Find a linear function f such that $2f(x) = f(2x)$ for every real number x.

18. Find a linear function f such that $f(3x + 2) = f(3x) + 2$.

19. Find a linear function f such that $f(1) = 3$ and the graph of f is parallel to the graph of the line determined by the points $(-2, 1)$ and $(3, 2)$.

20. Find a linear function f such that $f(1) = 3$ and the graph of f is perpendicular to the graph of the line determined by the points $(-2, 1)$ and $(3, 2)$.

In problems 21–32, find the vertex and the x and y intercepts of the graph of each quadratic function, and sketch the graph. What are the domain and the range of each function?

21. $f(x) = 2x^2$ **22.** $f(x) = -\frac{1}{2}x^2$

23. $f(x) = 2x^2 - 3$ **24.** $f(x) = x^2 - 3$

25. $f(x) = -x^2 - 2x - 1$ **26.** $f(x) = (x - 5)^2$

27. $f(x) = x^2 + 5x + 6$ **28.** $f(x) = -x^2 - 1$

29. $f(x) = 2x^2 - 3x$ **30.** $f(x) = -(x + 1)^2$

31. $f(x) = x^2 + 4x + 3$ **32.** $f(x) = -x^2 + x - 5$

33. Sketch the graph of each of the following functions on the same coordinate system: $f(x) = x^2 - 1$, $f(x) = x^2 + 1$, $f(x) = x^2 - 2$, and $f(x) = x^2 + 2$.

34. Show that the coordinates of the vertex of $y = ax^2 + bx + c, a \neq 0$, are

$$(h, k) = \left(-\frac{b}{2a}, \frac{4ac - b^2}{4a}\right).$$

(Hint: Use the method of completing the square, page 209.)

9.4 Exponential and Logarithmic Functions

In this section, we introduce two additional functions that are of great practical importance in many applications of mathematics: exponential and logarithmic functions. Bankers use exponential functions, for instance, to compute compound interest, and scientists use them to determine the rate of radioactive decay of material in order to determine its age.

Exponential Functions

In Section 9.3, we studied quadratic functions which take the form

$$p(x) = ax^2,$$

where the base is the variable x and the exponent 2 is constant. We now consider functions of the form

$$f(x) = b^x,$$

where the exponent is the variable x and the base b is constant. Such a function is called an **exponential function.**

DEFINITION 1   **Exponential Function**

> If b is a positive number, the function f defined by
>
> $$f(x) = b^x,$$
>
> where $b \neq 1$, is called an **exponential function** with base b.

Examples of exponential functions are

$$f(x) = 2^x,$$

$$g(x) = \left(\frac{1}{3}\right)^x,$$

and

$$h(x) = (5^{-1})^x.$$

If b is a positive constant, and if you plot some points (x, b^x) for rational values of x, you will notice that these points lie along a smooth curve. The curve is the graph of the function $f(x) = b^x$ with base b. For example, if $b = 2$, and if you plot a few points $(x, 2^x)$ for rational values of x, you obtain a collection of points similar to those in Figure 1a. In Figure 1b, we have connected these points with a smooth curve to obtain the graph of $f(x) = 2^x$. This graph reveals that the domain of the function $f(x) = 2^x$ is the set of real numbers $\mathbb{R}$ and that the range is the set of all positive real numbers.

x	$f(x)$
-3	$\frac{1}{8}$
-2	$\frac{1}{4}$
-1	$\frac{1}{2}$
0	1
1	2
2	4
3	8

Figure 1

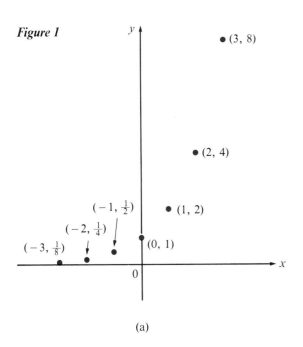

(a)

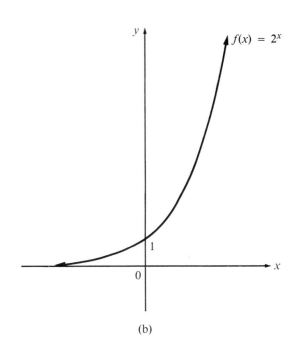

(b)

In Examples 1–3, sketch the graph of each function. Indicate the domain and the range of the function.

EXAMPLE 1 $f(x) = 3^x$

SOLUTION   We begin by calculating values of $f(x) = 3^x$ for some integer values of x, as shown in the table in Figure 2. Then we plot the corresponding points and connect them with a smooth curve to obtain the graph of f (Figure 2). From the graph we see that the domain of f is the set of real numbers $\mathbb{R}$, and that the range consists of all positive real numbers.

Figure 2

x	$f(x)$
-3	$\frac{1}{27}$
-2	$\frac{1}{9}$
-1	$\frac{1}{3}$
0	1
1	3
2	9
3	27

EXAMPLE **2** $g(x) = (\tfrac{1}{4})^x$

SOLUTION   We calculate values of $g(x) = (\tfrac{1}{4})^x$ for some integer values of x and obtain the table in Figure 3. We plot the corresponding points and connect them by a smooth curve. We obtain the graph of g (Figure 3). The graph reveals that the domain of g is $\mathbb{R}$, and that the range consists of all positive real numbers.

Figure 3

$f(x) = (\tfrac{1}{4})^x$

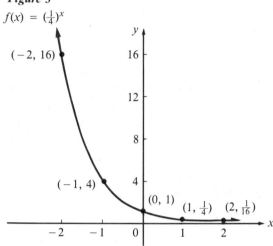

x	$g(x)$
-2	16
-1	4
0	1
1	$\tfrac{1}{4}$
2	$\tfrac{1}{16}$

EXAMPLE **3**   ⓒ $h(x) = \pi^x$

SOLUTION   Using a calculator with a y^x key, we evaluate $h(x) = \pi^x$ for some integer values of x, as shown in the table in Figure 4. We plot the corresponding points and connect them with a smooth curve to obtain the graph of h (Figure 4). From the graph we see that the domain of h is $\mathbb{R}$, and that the range consists of all positive real numbers.

Generally, if $b > 0$, the domain of the exponential function $f(x) = b^x$ is $\mathbb{R}$, and the range is the set of positive real numbers.

Figure 4

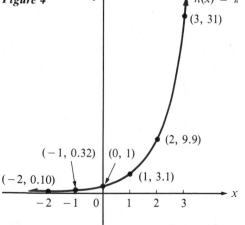

x	$h(x)$
-2	0.10
-1	0.32
0	1
1	3.1
2	9.9
3	31

Logarithmic Functions

In Section 8.1, we showed the connection between the equations $x = b^y$ and $y = \log_b x$, that is,

$$x = b^y \quad \text{if and only if} \quad y = \log_b x.$$

A function f of the form

$$f(x) = \log_b x,$$

where $b > 0$ and $b \neq 1$, is called a **logarithmic function** with base b. We must also have $x > 0$. Therefore, the domain of f consists of all positive real numbers. The range of f is $\mathbb{R}$. In other words, the domain of the exponential function is the range of the logarithmic function, and the range of the exponential function is the domain of the logarithmic function.

To sketch the graph of $f(x) = \log_b x$, we use the equivalent equation

$$x = b^{f(x)}$$

to locate some points on the graph.

For example, to sketch the graph of

$$f(x) = \log_3 x,$$

we use the equivalent equation

$$x = 3^{f(x)} = 3^y, \text{ where } y = f(x).$$

Thus,

if $f(x) = 0$, then $x = 3^0 = 1$
if $f(x) = 1$, then $x = 3^1 = 3$
if $f(x) = 2$, then $x = 3^2 = 9$.

Note that we have reversed the usual technique for finding points on the graph: We have selected values of y first and then determined the corresponding values of x, as shown in the table in Figure 5. The graph shows that the domain of $f(x) = \log_3 x$ consists of all positive real numbers and that the range of f is the set of real numbers $\mathbb{R}$.

In Examples 4 and 5, sketch the graph of each function. Indicate the domain and the range of the function.

EXAMPLE 4 $f(x) = \log_2 x$

SOLUTION The equation

$$f(x) = \log_2 x$$

Figure 5

$x = 3^y$	y
$\frac{1}{9}$	-2
$\frac{1}{3}$	-1
1	0
3	1
9	2
27	3

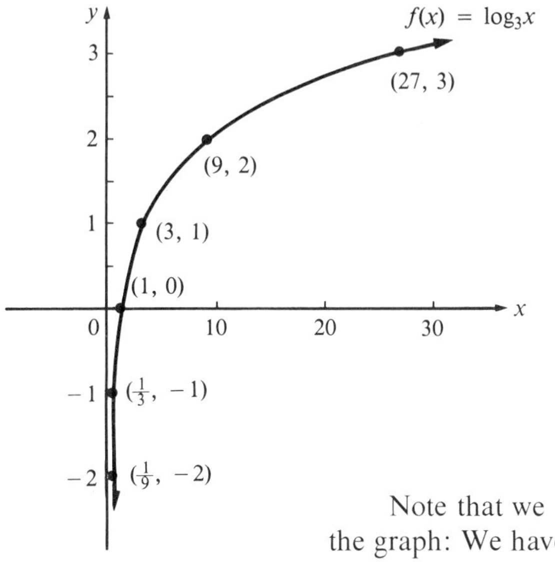

is equivalent to the equation

$$x = 2^{f(x)} = 2^y, \qquad \text{where} \qquad y = f(x).$$

We calculate values of $x = 2^y$ for some integer values of y, as shown in the table in Figure 6. Then we plot the corresponding points and connect them by a smooth curve to obtain the graph of $f(x) = \log_2 x$ (Figure 6). The graph of f reveals that the domain consists of all positive real numbers and that the range is $\mathbb{R}$.

Figure 6

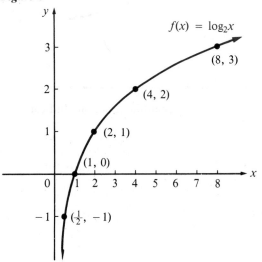

$x = 2^y$	y
$\frac{1}{2}$	-1
1	0
2	1
4	2
8	3

EXAMPLE 5 $f(x) = \log_{1/4} x$

SOLUTION   The equation

$$f(x) = \log_{1/4} x$$

is equivalent to the equation

$$x = \left(\frac{1}{4}\right)^{f(x)}$$

$$= \left(\frac{1}{4}\right)^y, \text{ where } y = f(x).$$

We calculate values of $x = (\frac{1}{4})^y$ for some values of y as shown in the table in Figure 7. Then we plot the corresponding points and connect them by a smooth curve to obtain the graph of $f(x) = \log_{1/4} x$ (Figure 7). From the graph of f we see that the domain consists of all positive real numbers, and that the range is $\mathbb{R}$.

Figure 7

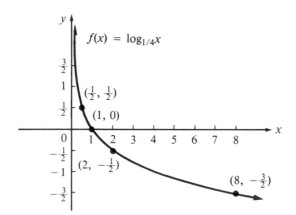

$$f(x) = \log_{1/4} x$$

$x = (\frac{1}{4})^y$	y
$\frac{1}{2}$	$\frac{1}{2}$
1	0
2	$-\frac{1}{2}$
8	$-\frac{3}{2}$

PROBLEM SET 9.4

In problems 1–10, sketch the graph of each exponential function. Indicate the domain and the range.

1. $f(x) = 4^x$ **2.** $f(x) = 4^{x-1}$ **3.** $f(x) = 3^{x+1}$ **4.** $f(x) = -2^x$

5. $f(x) = -(\frac{1}{3})^x$ **6.** $f(x) = 2^{-x}$ **7.** $f(x) = (\frac{1}{5})^{-x}$ **8.** $f(x) = (0.1)^x$

9. $f(x) = 5(3^x)$ **10.** $f(x) = -(4)^x$

In problems 11–16, determine the base of the exponential function $f(x) = b^x$ if the graph of $f(x)$ contains the given points.

11. $(2, 9)$ **12.** $(3, 27)$ **13.** $(2, 16)$ **14.** $(3, 125)$

15. $(0, 1)$ **16.** $(\frac{1}{2}, \sqrt{10})$

17. Let f be an exponential function with base b. Show that $f(u - v) = f(u) \div f(v)$ for any real numbers u and v.

18. Use the graph in Figure 2 (page 363) to approximate the value of $3^{3/2}$.

In problems 19–24, sketch the graph of each function. Indicate the domain and the range.

19. $f(x) = -\log_2 x$ **20.** $f(x) = \log_5 x$ **21.** $f(x) = \log_{1/2} x$ **22.** $f(x) = \log_4 x$

23. $f(x) = \log_6 x$ **24.** $f(x) = \log_{1/3} x$

In problems 25–30, determine the base of the logarithmic function $f(x) = \log_b x$ if the graph of $f(x)$ contains the given points.

25. $(8, 3)$ **26.** $(125, 3)$ **27.** $(\frac{1}{16}, -2)$ **28.** $(8, \frac{3}{2})$

29. $(3, \frac{1}{2})$ **30.** $(c^{3/2}, 3)$

31. Let f be a logarithmic function with base b. Show that $f(u) + f(v) = f(uv)$ for any positive real numbers u and v.

32. Let f be a logarithmic function with base b. Show that $f(u) - f(v) = f(u/v)$ for any positive real numbers u and v.

9.5 Graphs of Special Curves—Conic Sections

The **graph** of an equation in two unknowns x and y is defined to be the set of all points $P = (x, y)$ in the Cartesian plane whose coordinates x and y satisfy the equation. Many (but not all) equations in x and y have graphs that are smooth curves in the plane. If we have an equation whose graph is a given curve in the Cartesian plane, the equation is called an **equation of the curve.** In this section, we discuss certain types of equations that play a special role in geometry and in many other applications. We also examine certain types of curves. We introduce these curves by establishing the standard forms for the equations of their graphs. The curves that we consider are **circles, ellipses, hyperbolas,** and **parabolas.** Each of these curves is obtained by sectioning or cutting a circular cone with a plane. Therefore, the curves are called **conic sections.**

Circles and Ellipses

Consider the equation

$$x^2 + y^2 = r^2, \qquad \text{for } r > 0.$$

We can rewrite this equation as

$$\sqrt{x^2 + y^2} = \sqrt{r^2} \quad \text{or} \quad \sqrt{(x - 0)^2 + (y - 0)^2} = r.$$

According to the distance formula, the last equation is true if and only if the point $P = (x, y)$ on the curve is r units from the origin $0 = (0, 0)$. Therefore, the graph of

$$\boxed{x^2 + y^2 = r^2}$$

is a circle of radius r units with its center at the origin 0 (Figure 1).

The equation of a circle $x^2 + y^2 = r^2$ can be rewritten in the form

$$\frac{x^2}{r^2} + \frac{y^2}{r^2} = 1.$$

We can modify this last equation to read:

$$\boxed{\frac{x^2}{a^2} + \frac{y^2}{b^2} = 1,}$$

Figure 1

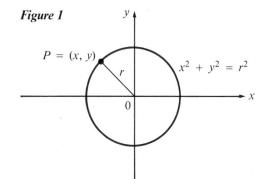

$$x^2 + y^2 = r^2$$

where $a > 0$, $b > 0$, and $a \neq b$. This produces an equation of an ellipse (Figure 2). The ellipse intersects the x axis at the points $(-a, 0)$ and $(a, 0)$. It intersects the y axis at the points $(0, -b)$ and $(0, b)$. These four points are called the **vertices** of the ellipse. The two line segments joining opposite pairs of vertices are called the **axes** of the ellipse. The longer axis is called the **major axis,** the shorter axis is called the **minor axis,** and the two axes intersect at the center of the ellipse. If $a > b > 0$, the major axis is horizontal (Figure 2a). If $0 < a < b$, the major axis is vertical (Figure 2b).

Figure 2

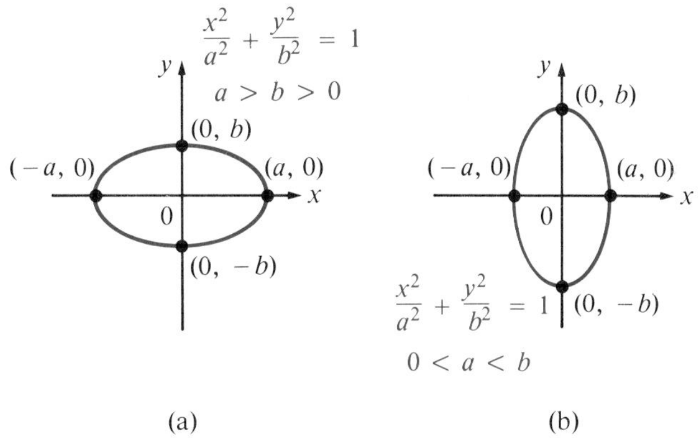

$$\frac{x^2}{a^2} + \frac{y^2}{b^2} = 1$$

$a > b > 0$

$(0, b)$

$(-a, 0)$ $(a, 0)$

$(0, -b)$

(a)

$(0, b)$

$(-a, 0)$ $(a, 0)$

$$\frac{x^2}{a^2} + \frac{y^2}{b^2} = 1$$ $(0, -b)$

$0 < a < b$

(b)

In Examples 1–3, sketch the graph of each equation.

EXAMPLE 1 $x^2 + y^2 = 4$

SOLUTION   This equation has the form

$$x^2 + y^2 = r^2,$$

so the graph is a circle whose center is at the origin and whose radius is 2 (Figure 3).

Figure 3

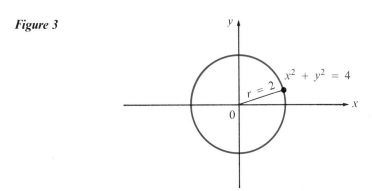

$x^2 + y^2 = 4$

$r = 2$

EXAMPLE 2 $9x^2 + 25y^2 = 225$

SOLUTION   We divide both sides of the equation by 225 to obtain

$$\frac{x^2}{25} + \frac{y^2}{9} = 1.$$

This equation has the form

$$\frac{x^2}{a^2} + \frac{y^2}{b^2} = 1$$

with $a = 5$ and $b = 3$. Thus the graph is an ellipse with its center at the origin and its vertices at $(-5, 0)$, $(5, 0)$, $(0, -3)$, and $(0, 3)$ (Figure 4).

Figure 4

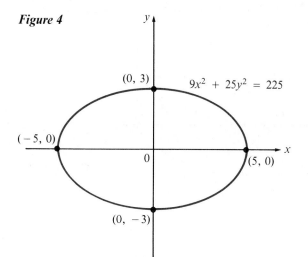

Figure 5

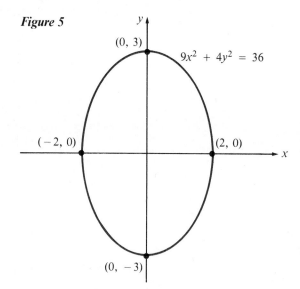

EXAMPLE 3 $9x^2 + 4y^2 = 36$

SOLUTION   We divide both sides of the equation by 36:

$$\frac{x^2}{4} + \frac{y^2}{9} = 1.$$

This equation has the form

$$\frac{x^2}{a^2} + \frac{y^2}{b^2} = 1$$

with $a = 2$ and $b = 3$. Therefore, the graph is an ellipse with its center at the origin and its vertices at $(-2, 0)$, $(2, 0)$, $(0, -3)$, and $(0, 3)$ (Figure 5).

Parabolas

In Section 9.3, we saw that the graph of the equation $y = ax^2 + bx + c$, $a \neq 0$, is a parabola that opens upward if $a > 0$ and downward if $a < 0$. If we interchange the variables x and y, we obtain the equation

$$x = ay^2 + by + c.$$

If $b = c = 0$, then

$$x = ay^2, \quad a \neq 0.$$

The graph of this equation is a **parabola** that has its vertex at the origin, and that opens to the right if $a > 0$ (Figure 6a) and to the left if $a < 0$ (Figure 6b).

Figure 6

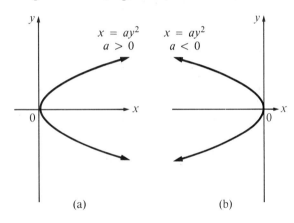

In Examples 4 and 5, sketch the graph of each equation.

EXAMPLE 4 $x = 4y^2$

SOLUTION   The equation has the form $x = ay^2$ with $a > 0$, so the graph is a parabola that has its vertex at the origin and that opens to the right (Figure 7).

Figure 7

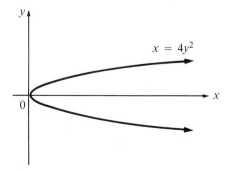

EXAMPLE 5 $x = -3y^2$

SOLUTION   The equation has the form $x = ay^2$ with $a < 0$, so the graph is a parabola that has its vertex at the origin and that opens to the left (Figure 8).

Figure 8

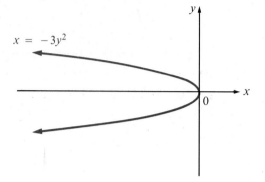

Hyperbolas

The graph of the equation

$$\frac{x^2}{a^2} - \frac{y^2}{b^2} = 1, \qquad a > 0, b > 0$$

is a **hyperbola** with its center at the origin (Figure 9). The hyperbola has two branches—one opening to the right and one opening to the left. The two points $(-a, 0)$ and $(a, 0)$, where these branches intersect the x axis, are called the **vertices** of the hyperbola. The line segment between the two vertices is called the **transverse** axis. The midpoint of the transverse axis is called the **center** of the hyperbola.

If we position the hyperbola so that the y axis contains the transverse axis, the equation of the graph is:

$$\frac{y^2}{b^2} - \frac{x^2}{a^2} = 1, \qquad a > 0, b > 0.$$

Figure 9

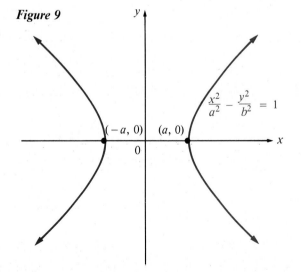

The center of this graph is at the origin, and the vertices are now the points $(0, -b)$ and $(0, b)$ (Figure 10).

Figure 10

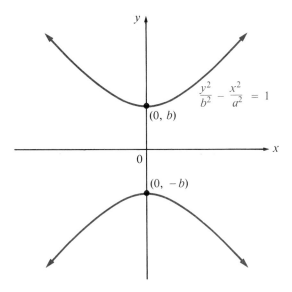

In Examples 6 and 7, sketch the graph of each equation.

EXAMPLE **6** $4x^2 - 9y^2 = 36$

SOLUTION   We divide both sides of the equation by 36 to obtain

$$\frac{x^2}{9} - \frac{y^2}{4} = 1.$$

This equation has the form

$$\frac{x^2}{a^2} - \frac{y^2}{b^2} = 1$$

with $a = 3$ and $b = 2$. Thus the graph is a hyperbola with a horizontal transverse axis and vertices $(-3, 0)$ and $(3, 0)$ (Figure 11).

Figure 11

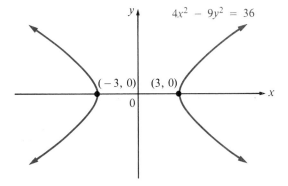

EXAMPLE 7 $4y^2 - 25x^2 = 100$

SOLUTION   We divide both sides of the equation by 100 to obtain

$$\frac{y^2}{25} - \frac{x^2}{4} = 1.$$

The equation has the form

$$\frac{y^2}{b^2} - \frac{x^2}{a^2} = 1$$

with $a = 2$ and $b = 5$. Therefore, the graph is a hyperbola with vertical transverse axis and vertices $(0, -5)$ and $(0, 5)$ (Figure 12).

Figure 12

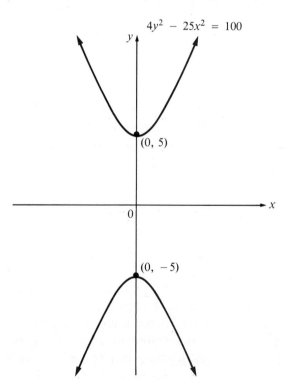

$4y^2 - 25x^2 = 100$

PROBLEM SET 9.5

In problems 1–6, sketch the graph of each circle and find its radius.

1. $x^2 + y^2 = 9$ **2.** $x^2 + y^2 = 16$ **3.** $4x^2 + 4y^2 = 25$ **4.** $9x^2 + 9y^2 = 36$

5. $\dfrac{x^2}{3} + \dfrac{y^2}{3} = 2$ **6.** $x^2 + y^2 = k, k > 0$

In problems 7–12, sketch the graph of each ellipse and find its vertices.

7. $\dfrac{x^2}{9} + \dfrac{y^2}{4} = 1$
8. $\dfrac{x^2}{49} + \dfrac{y^2}{81} = 1$
9. $25x^2 + 4y^2 = 100$
10. $16x^2 + 25y^2 = 400$

11. $7x^2 + 8y^2 = 56$
12. $11x^2 + 5y^2 = 55$

In problems 13–18, sketch the graph of each parabola.

13. $x = 2y^2$
14. $x = -3y^2$
15. $x = -\tfrac{1}{2}y^2$
16. $x = \tfrac{4}{5}y^2$

17. $4x = 3y^2$
18. $2x = 7y^2$

In problems 19–24, sketch the graph of each hyperbola and find its vertices.

19. $\dfrac{x^2}{16} - \dfrac{y^2}{7} = 1$
20. $\dfrac{y^2}{49} - \dfrac{x^2}{81} = 1$
21. $\dfrac{y^2}{4} - \dfrac{x^2}{16} = 1$
22. $\dfrac{x^2}{64} - \dfrac{y^2}{25} = 1$

23. $5x^2 - 9y^2 = 45$
24. $7y^2 - 4x^2 = 28$

In problems 25–40, sketch the graph of the given conic.

25. $y^2 - x^2 = 1$
26. $y^2 + 2x = 0$
27. $4x^2 + 9y^2 = 1$
28. $x^2 + y^2 = 81$

29. $x^2 - 4y = 0$
30. $16y^2 - 4x^2 = 48$
31. $x^2 + y^2 = 49$
32. $y^2 + 4x^2 = 16$

33. $36x^2 - 9y^2 = 1$
34. $3x^2 + 3y^2 = 24$
35. $3x^2 + 2y = 0$
36. $9x^2 - y^2 = 9$

37. $4x^2 + 4y^2 = 9$
38. $-2x^2 + 5y = 0$
39. $16x^2 + 4y^2 = 64$
40. $9x^2 + 25y^2 = 1$

9.6 Systems Containing Quadratic Equations

In this section, we consider systems containing quadratic equations in two unknowns. An approximate solution of such a system can be found by sketching graphs of the two equations on the same coordinate system, and then determining the points where the two graphs intersect.

The substitution and elimination methods, introduced in Section 7.5, can often be used to solve systems containing quadratic equations. Even then, graphs can be used to determine the number of solutions to the system and to provide a rough check on the calculations.

In Examples 1–3, sketch graphs to determine the number of solutions to each system, and then solve the system.

EXAMPLE 1
$\begin{cases} 2x + 3y = 8 \\ 2x^2 - 3y^2 = -10 \end{cases}$

SOLUTION
The graph of $2x + 3y = 8$ is a line. The graph of $2x^2 - 3y^2 = -10$ is a hyperbola that intersects the line at two points (Figure 1). Thus, there are

Figure 1

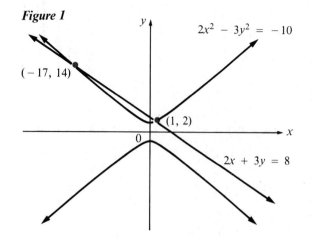

two solutions to the system. To find these solutions algebraically, we use the method of substitution. We solve the first equation for x:

$$x = \frac{8 - 3y}{2}.$$

Substituting $(8 - 3y)/2$ for x in the second equation, we obtain

$$2\left(\frac{8 - 3y}{2}\right)^2 - 3y^2 = -10.$$

We simplify this equation:

$$3y^2 - 48y + 84 = 0 \qquad \text{or} \qquad y^2 - 16y + 28 = 0.$$

Factoring, we have

$$(y - 2)(y - 14) = 0.$$

We set each factor equal to zero and solve for y:

$$y = 2 \qquad \text{or} \qquad y = 14.$$

For each of these values of y there is a corresponding value for x given by

$$x = \frac{8 - 3y}{2}.$$

That is, for $y = 2$, $x = 1$ and for $y = 14$, $x = -17$. Therefore, the solutions are $(1, 2)$ and $(-17, 14)$.

EXAMPLE 2 $\begin{cases} 4x^2 + 7y^2 = 32 \\ -3x^2 + 11y^2 = 41 \end{cases}$

SOLUTION   The graph of $4x^2 + 7y^2 = 32$ is an ellipse and the graph of $-3x^2 + 11y^2 = 41$ is a hyperbola that intersects the ellipse at four points (Figure 2). Thus, there are four solutions to the system. We use the method of elimination to find these solutions algebraically.

$$\begin{cases} 4x^2 + 7y^2 = 32 \\ -3x^2 + 11y^2 = 41 \end{cases} \quad \xrightarrow[\text{We multiply each side by 4.}]{\text{We multiply each side by 3.}} \quad \begin{cases} 12x^2 + 21y^2 = 96 \\ \underline{-12x^2 + 44y^2 = 164} \quad \text{add} \\ 65y^2 = 260 \end{cases}$$

so that $y^2 = 4$. Hence,

$$y = -2 \qquad \text{or} \qquad y = 2.$$

We substitute $y = -2$ into the first equation $4x^2 + 7y^2 = 32$:

$$4x^2 + 7(-2)^2 = 32 \quad \text{or} \quad x^2 = 1$$

so that

$$x = -1 \quad \text{or} \quad x = 1.$$

We substitute $y = 2$ into $4x^2 + 7y^2 = 32$:

$$4x^2 + 7(2)^2 = 32 \quad \text{or} \quad x^2 = 1$$

so that

$$x = -1 \quad \text{or} \quad x = 1.$$

Therefore, the solutions are $(1, 2)$, $(-1, 2)$, $(1, -2)$, and $(-1, -2)$.

Figure 2

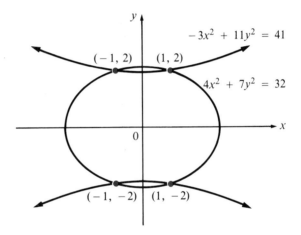

EXAMPLE 3

$$\begin{cases} x^2 + 2y^2 = 22 \\ 2x^2 + y^2 = 17 \end{cases}$$

SOLUTION

The graph of $x^2 + 2y^2 = 22$ is an ellipse and the graph of $2x^2 + y^2 = 17$ is an ellipse that intersects the first ellipse at four points (Figure 3). Thus, there are four solutions, which can be found algebraically by the method of elimination.

$$\begin{cases} x^2 + 2y^2 = 22 \\ 2x^2 + y^2 = 17 \end{cases} \xrightarrow[\text{We multiply each side by 2.}]{} \begin{cases} 2x^2 + 4y^2 = 44 \\ \underline{2x^2 + \ y^2 = 17} \quad \text{subtract} \\ 3y^2 = 27 \end{cases}$$

so that $y^2 = 9$. Hence,

$$y = -3 \quad \text{or} \quad y = 3.$$

Figure 3

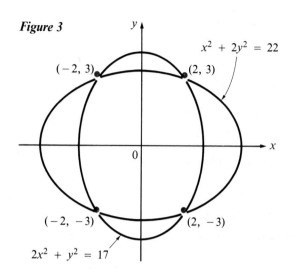

We substitute $y = 3$ into the equation $x^2 + 2y^2 = 22$:

$$x^2 + 2(3)^2 = 22 \quad \text{or} \quad x^2 = 4$$

so that

$$x = -2 \quad \text{or} \quad x = 2.$$

We substitute $y = -3$ into $x^2 + 2y^2 = 22$:

$$x^2 + 2(-3)^2 = 22 \quad \text{or} \quad x^2 = 4$$

so that

$$x = -2 \quad \text{or} \quad x = 2.$$

Therefore, the solutions are $(2, 3)$, $(2, -3)$, $(-2, 3)$, and $(-2, -3)$.

Systems of nonlinear equations are used in certain applications. The next example illustrates one such application.

EXAMPLE 4   A manufacturer of art supplies makes templates by cutting out right triangles with perimeters of 60 centimeters from plastic sheets. If the hypotenuse of each triangle is 25 centimeters, find the lengths of the two sides of the triangle.

SOLUTION   Let x and y represent the lengths of the other two sides of the right triangle (Figure 4). Then

$$x + y + 25 = 60 \qquad \text{(perimeter of triangle)}$$

and

$$x^2 + y^2 = 25^2 \qquad \text{(Pythagorean theorem)}.$$

This gives us the system of equations

$$\begin{cases} x^2 + y^2 = 625 \\ x + y = 35. \end{cases}$$

Figure 4

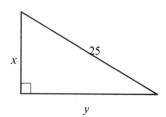

We solve the second equation for y, and we obtain $y = 35 - x$. We substitute this result into the first equation:

$$x^2 + (35 - x)^2 = 625$$

$$2x^2 - 70x + 1{,}225 = 625$$

$$2x^2 - 70x + 600 = 0$$

$$x^2 - 35x + 300 = 0$$

$$(x - 15)(x - 20) = 0$$

so that

$$x = 15 \quad \text{or} \quad x = 20.$$

When $x = 15$, then $y = 35 - 15 = 20$; when $x = 20$, then $y = 35 - 20 = 15$. Hence, the lengths of the other two sides of each triangle are 15 and 20 centimeters.

PROBLEM SET 9.6

In problems 1–6, use the graphs of the equations to determine the number of solutions of each system of equations. Then solve the system.

1. $\begin{cases} x - y = 1 \\ x^2 + y^2 = 5 \end{cases}$

2. $\begin{cases} x - 2y = 3 \\ x^2 - y^2 = 24 \end{cases}$

3. $\begin{cases} 3x - y = 2 \\ x^2 + y^2 = 20 \end{cases}$

4. $\begin{cases} x + y = 3 \\ 3x^2 - y^2 = \frac{9}{2} \end{cases}$

5. $\begin{cases} 3x + 2y = 1 \\ 3x^2 - y^2 = -4 \end{cases}$

6. $\begin{cases} x + y = 6 \\ x^2 + y^2 = 20 \end{cases}$

In problems 7–30, solve each system of equations by the elimination or substitution method.

7. $\begin{cases} 5x - 3y = 10 \\ x^2 - y^2 = 6 \end{cases}$

8. $\begin{cases} 2x + y = 10 \\ xy = 12 \end{cases}$

9. $\begin{cases} 2x + 3y = 7 \\ x^2 + y^2 + 4y + 4 = 0 \end{cases}$

10. $\begin{cases} x - y + 4 = 0 \\ x^2 + 3y^2 = 12 \end{cases}$

11. $\begin{cases} 5x - y = 21 \\ y = x^2 - 5x + 4 \end{cases}$

12. $\begin{cases} x^2 - 25y^2 = 20 \\ 2x^2 + 25y^2 = 88 \end{cases}$

13. $\begin{cases} x - y^2 = 0 \\ x^2 + 2y^2 = 24 \end{cases}$

14. $\begin{cases} 3x^2 - 8y^2 = 40 \\ 5x^2 + y^2 = 81 \end{cases}$

15. $\begin{cases} 2x^2 - 3y^2 = 6 \\ 3x^2 + 2y^2 = 35 \end{cases}$

16. $\begin{cases} x^2 - y^2 = 7 \\ x^2 + y^2 = 25 \end{cases}$

17. $\begin{cases} x^2 + 9y^2 = 33 \\ x^2 + y^2 = 25 \end{cases}$

18. $\begin{cases} x^2 + 5y^2 = 70 \\ 3x^2 - 5y^2 = 30 \end{cases}$

19. $\begin{cases} 4x^2 - y^2 = 4 \\ 4x^2 + \frac{5}{3}y^2 = 36 \end{cases}$

20. $\begin{cases} x^2 - 2y^2 = 17 \\ 2x^2 + y^2 = 54 \end{cases}$

21. $\begin{cases} 2x^2 - 3y^2 = 20 \\ x^2 + 2y = 20 \end{cases}$

22. $\begin{cases} 4x^2 + 3y^2 = 43 \\ 3x^2 - y^2 = 3 \end{cases}$

23. $\begin{cases} x^2 - 2y^2 = 1 \\ x^2 + 4y^2 = 25 \end{cases}$

24. $\begin{cases} 2x^2 - 5y^2 + 8 = 0 \\ x^2 - 7y^2 + 4 = 0 \end{cases}$

25. $\begin{cases} x^2 + 4y = 8 \\ x^2 + y^2 = 5 \end{cases}$

26. $\begin{cases} 3x - 2y = 9 \\ 9x = y^2 \end{cases}$

27. $\begin{cases} x^2 + y^2 = 16 \\ x^2 - y^2 = -34 \end{cases}$

28. $\begin{cases} x^2 - 4y^2 = -15 \\ -x^2 + 3y^2 = 11 \end{cases}$

29. $\begin{cases} x^2 + y^2 = 25 \\ (x - 5)^2 + y^2 = 9 \end{cases}$

30. $\begin{cases} x^2 - y = 0 \\ x^2 + (y - 6)^2 = 36 \end{cases}$

31. Find the dimensions of a rectangle whose area is 96 square centimeters and whose perimeter is 40 centimeters.

32. Suppose that the demand and supply curves of a certain product are given by the equations $p + 2q^2 = 8$ and $p - q = 5$, respectively, where p is the price and q is the quantity. Find the (point of) equilibrium of price and quantity (the point of intersection of the two curves where both p and q are nonnegative).

33. A woman receives $170 interest on a sum of money she lent at simple annual interest. If the interest rate had been 1 percent higher, she would have received $238. What was the amount lent and what was the interest rate?

34. The sum of the squares of two positive numbers is 73 and the difference of their squares is 5. What are the numbers?

35. A rectangular parking lot has an area of 27,000 square feet. If a strip 10 feet wide is eliminated on each of the ends and on each of the sides, the available parking space is reduced to 20,800 square feet. What are the dimensions of the original parking lot?

REVIEW PROBLEM SET

In problems 1–12, let $f(x) = 2x + 3$, $g(x) = \sqrt{4 - x^2}$, and $h(x) = |x + 5|$. Find the following values.

1. $f(3)$

2. $g(0)$

3. $g(-2)$

4. $\sqrt{h(-14)}$

5. $h(-8)$

6. $f(u + v) - f(u)$

7. $[f(1)]^2$

8. $\dfrac{1}{g(0)}$

9. $f(u + 1)$

10. $4f(4)$

11. $h(2) - h(-2)$

12. $h(-5) - h(0)$

In problems 13–18, find the domain of the function determined by each equation.

13. $f(x) = 7x - 2$

14. $f(x) = 2x^2$

15. $f(x) = \dfrac{2}{(x - 1)(x + 2)}$

16. $f(x) = \sqrt[3]{3x + 1}$

17. $f(x) = \sqrt{3 - 2x}$

18. $f(x) = \dfrac{x}{\sqrt{2 - x}}$

In problems 19–22, determine whether the curve is the graph of a function.

19.

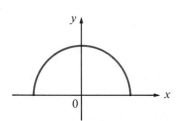

20.

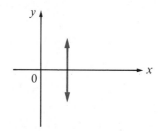

21.

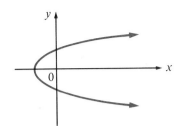

22.

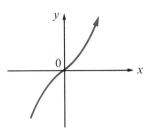

In problems 23–28, express y as a function of x, that is, $y = f(x)$, and graph the function.

23. If y is directly proportional to x and if $y = 8$ when $x = 12$.

24. If y is directly proportional to x^2 and if $y = 18$ when $x = 3$.

25. If y is directly proportional to $\sqrt{x}$ and if $y = 16$ when $x = 16$.

26. If y is directly proportional to $\sqrt{x}$ and if $y = 9$ when $x = 16$.

27. If y is inversely proportional to x and if $y = 4$ when $x = 5$.

28. If y is inversely proportional to x and if $y = 12$ when $x = \frac{3}{4}$.

29. Hooke's law states that the extension of an elastic spring beyond its natural length is directly proportional to the force applied. If a weight of 8 pounds causes a spring to stretch from a length of 9 inches to a length of 9.5 inches, what weight will cause it to stretch to a length of 1 foot?

30. The power required to operate a fan is directly proportional to the speed of the fan. If 1 horsepower will drive the fan at a speed of 480 revolutions per minute, how fast will 8 horsepower drive it? What power will be required to give the fan a speed of 600 revolutions per minute?

In problems 31–34, sketch the graph of each linear function. Find the slope of the graph and the domain and the range of the function.

31. $f(x) = 2x - 2$ **32.** $f(x) = \frac{3}{4}x + 7$ **33.** $f(x) = \frac{3}{2}x$ **34.** $f(x) = -3(x + 1) + 4$

In problems 35–40, find the vertex and the x and y intercepts of the graph of each quadratic function. Sketch the graph of the function and determine the domain and the range of the function.

35. $f(x) = 6x^2 - 5x - 4$ **36.** $f(x) = 2x^2 - x - 6$ **37.** $f(x) = x^2 + 6x + 9$

38. $f(x) = x^2 - 8x + 16$ **39.** $f(x) = -3 - 10x - 8x^2$ **40.** $f(x) = 10 + 3x - x^2$

In problems 41–46, sketch the graph of each exponential function. Indicate the domain and the range.

41. $f(x) = 5^x$ **42.** $f(x) = 7^x$ **43.** $f(x) = 3(2^x)$ **44.** $f(x) = -2(3^x)$

45. $f(x) = (\frac{1}{2})^x$ **46.** $f(x) = (\frac{1}{3})^x$

In problems 47–50, sketch the graph of each logarithmic function. Indicate the domain and the range.

47. $g(x) = \log_{1/2} x$ **48.** $g(x) = \log_2(x + 1)$ **49.** $f(x) = \log_3(x + 2)$ **50.** $f(x) = \log_3 |x|$

In problems 51–54, let $f(x) = 4^x$. Find the values.

51. $f(0)$ **52.** $f(2)$ **53.** $f(-\frac{1}{2})$ **54.** $f(\frac{5}{2})$

In problems 55–58, let $f(x) = \log_{16} x$. Find the values.

55. $f(32)$ **56.** $f(64)$ **57.** $f(\sqrt[5]{2})$ **58.** $f(\sqrt[3]{4})$

In problems 59–66, sketch the graph of each equation. Find the x and y intercepts and identify the graph.

59. $2x^2 + 2y^2 = 50$ **60.** $2x^2 + 3y^2 = 18$ **61.** $9x^2 + 16y^2 = 36$ **62.** $y^2 - x^2 = 4$

63. $y^2 = 4x$ **64.** $x^2 = 2y$ **65.** $16x^2 - 11y^2 = 64$ **66.** $\dfrac{x^2}{10} + \dfrac{y^2}{10} = 1$

In problems 67–72, solve each system of equations algebraically. Check your solutions by sketching the graph of each equation and approximating the points of intersection.

67. $\begin{cases} 3x - 4y = 25 \\ x^2 + y^2 = 25 \end{cases}$ **68.** $\begin{cases} 2x - y = 2 \\ x^2 + 2y^2 = 12 \end{cases}$ **69.** $\begin{cases} x + y^2 = 6 \\ x^2 + y^2 = 36 \end{cases}$ **70.** $\begin{cases} 3x^2 - 2y^2 = 27 \\ 7x^2 + 5y^2 = 63 \end{cases}$

71. $\begin{cases} x^2 + y^2 = 29 \\ x^2 - y^2 = 21 \end{cases}$ **72.** $\begin{cases} 3x^2 - 2y^2 = 190 \\ 2x^2 + 5y^2 = 133 \end{cases}$

73. The diagonal of a rectangle is 34 feet and the perimeter is 92 feet. Find the dimensions of the rectangle.

74. Suppose that the demand and supply curves of a certain product are given by the equations $p - q^2 + 16q = 35$ and $p - q^2 - 4q = -5$, respectively, where p is the price and q is the quantity. Find the (point of) equilibrium of price and quantity (the point of intersection of the two curves where both p and q are nonnegative).

75. A rectangle has an area of 20 square centimeters and a perimeter of 18 centimeters. Find the length and width of the rectangle.

76. The sum of the squares of two numbers is 117, and the difference of the two numbers is 3. What are the numbers?

10

Topics in Algebra

This chapter provides a brief introduction to a variety of topics that supplement ideas presented in previous chapters. These topics include sequences, series, the binomial theorem, determinants, and Cramer's rule.

10.1 Sequences

A **sequence** is a function whose domain is the set of positive integers. We often represent the values of the function with the symbol a_n. We call these values the **terms** of the sequence. Thus, a_1 is the **first term,** a_2 is the **second term,** and a_n is the **general term,** or **nth term,** of the sequence.

The symbols

$$a_1, a_2, a_3, \ldots, a_n, \ldots$$

are typically used to denote the sequence. This notation suggests a never-ending list, in which the terms appear in order, with the nth term in the nth position for each positive integer n. The set of three dots is read "and so on." We also use the notation $\{a_n\}$ to denote the sequence whose nth term is a_n.

We can specify a particular sequence by giving a rule for determining the nth term a_n. This is often done by using a formula.

Examples of sequences are $\{a_n\}$ and $\{b_n\}$, where

$$a_n = n^2 - n \quad \text{and} \quad b_n = \frac{2}{n}.$$

EXAMPLE 1   Find the first five terms of each sequence.

(a) $a_n = (-1)^n$ (b) $a_n = 3 - \dfrac{1}{n}$

SOLUTION   To find the first five terms of each sequence we substitute the positive integers 1, 2, 3, 4, and 5 in turn for n in the formula for the general term.

383

(a) We have

$$a_1 = (-1)^1 = -1, \qquad a_2 = (-1)^2 = 1, \qquad a_3 = (-1)^3 = -1,$$

$$a_4 = (-1)^4 = 1, \qquad \text{and} \qquad a_5 = (-1)^5 = -1.$$

Thus, the first five terms of the sequence $\{a_n\}$ are $-1, 1, -1, 1,$ and -1.

(b) We have

$$a_1 = 3 - \frac{1}{1} = 2, \qquad a_2 = 3 - \frac{1}{2} = \frac{5}{2}, \qquad a_3 = 3 - \frac{1}{3} = \frac{8}{3},$$

$$a_4 = 3 - \frac{1}{4} = \frac{11}{4}, \qquad \text{and} \qquad a_5 = 3 - \frac{1}{5} = \frac{14}{5}.$$

Therefore, the first five terms of the sequence $\{a_n\}$ are $2, \frac{5}{2}, \frac{8}{3}, \frac{11}{4},$ and $\frac{14}{5}$.

Arithmetic and Geometric Sequences

We now consider special types of sequences. The first type we consider, the arithmetic sequence, is defined as follows:

DEFINITION 1   **Arithmetic Sequence**

> A sequence $a_1, a_2, a_3, \ldots, a_n, \ldots$ is called an **arithmetic sequence** (or an **arithmetic progression**) if each term (after the first term) differs from the preceding term by a fixed amount.

For example, the sequence $\{a_n\}$ whose general term a_n is given by

$$a_n = 1 + 2n$$

is an arithmetic sequence. The terms of $\{a_n\}$ are:

$$a_1 = 3, a_2 = 5, a_3 = 7, a_4 = 9, \ldots, a_n = 1 + 2n, \ldots$$

Note that after the first term a_1, each term in the sequence is always 2 more than the preceding term:

$$a_1 = 3 \qquad\qquad\qquad = 3$$
$$a_2 = 3 + 2 \qquad\qquad\qquad = 3 + 1 \cdot 2$$
$$a_3 = (3 + 2) + 2 \qquad\qquad\qquad = 3 + 2 \cdot 2$$
$$a_4 = [(3 + 2) + 2] + 2 \qquad\qquad\qquad = 3 + 3 \cdot 2$$
$$\vdots$$
$$a_n = [(3 + 2) + 2] + 2 + \cdots + 2 = 3 + (n - 1) \cdot 2.$$

In general, a sequence

$$a_1, a_2, a_3, \ldots, a_n, \ldots$$

is an **arithmetic sequence** (or an **arithmetic progression**) if it can be expressed in the form

$$a_1, \quad a_1 + d, \quad a_1 + 2d, \quad a_1 + 3d, \ldots, a_1 + (n - 1)d, \ldots$$

for every positive integer n. The number d is called the **common difference** associated with the arithmetic sequence. The nth term a_n of such a sequence is given by the following equation:

$$a_n = a_1 + (n - 1)d.$$

EXAMPLE 2   Find the tenth term of the arithmetic progression whose first four terms are $2, -1, -4$, and -7.

SOLUTION   The common difference is $d = -3$. We substitute $a_1 = 2$, $d = -3$, and $n = 10$ in the formula $a_n = a_1 + (n - 1)d$:

$$\begin{aligned} a_{10} &= 2 + (10 - 1)(-3) \\ &= 2 - 27 \\ &= -25. \end{aligned}$$

EXAMPLE 3   If the third term of an arithmetic progression is 7 and the seventh term is 15, find the fifth term.

SOLUTION   We substitute $n = 3$ and $n = 7$ in the formula $a_n = a_1 + (n - 1)d$. We use the fact that $a_3 = 7$ and $a_7 = 15$ to obtain the following system of linear equations in the unknowns a_1 and d:

$$7 = a_1 + (3 - 1)d \quad \text{or} \quad a_1 + 2d = 7$$
$$15 = a_1 + (7 - 1)d \quad \text{or} \quad a_1 + 6d = 15.$$

We solve for d and a_1:

$$4d = 8 \quad \text{or} \quad d = 2 \quad \text{and}$$
$$a_1 = 7 - 2(2) = 3.$$

Therefore,

$$\begin{aligned} a_5 &= 3 + (5 - 1)(2) \\ &= 3 + 8 \\ &= 11. \end{aligned}$$

Another important type of sequence is defined as follows:

Geometric Sequence

A sequence $a_1, a_2, a_3, \ldots, a_n, \ldots$ is called a **geometric sequence** (or a **geometric progression**) if each term (after the first) is obtained by multiplying the preceding term by a fixed amount.

For example, the sequence $\{a_n\}$ whose general term a_n is given by

$$a_n = 3(2)^{n-1}$$

is a geometric sequence. The terms of $\{a_n\}$ are:

$$3, 3(2), 3(2^2), 3(2^3), \ldots$$

Note that each term in this geometric sequence (after the first term) is obtained by multiplying the preceding term by 2.

In general, a sequence of the form

$$a_1, a_1r, a_1r^2, \ldots, a_1r^{n-1}, \ldots$$

for every positive integer n is a **geometric sequence** or a **geometric progression.** The number r is called the **common ratio** associated with the geometric progression, and the nth term a_n is given by

$$a_n = a_1 r^{n-1}.$$

EXAMPLE 4   Find the tenth term of the geometric progression having the first term $a_1 = \frac{1}{2}$ and common ratio $r = \frac{1}{2}$.

SOLUTION   Substituting $a_1 = \frac{1}{2}$, $r = \frac{1}{2}$, and $n = 10$ in the formula

$$a_n = a_1 r^{n-1},$$

we have

$$a_{10} = \frac{1}{2}\left(\frac{1}{2}\right)^{10-1}$$

$$= \frac{1}{2}\left(\frac{1}{2}\right)^{9} = \frac{1}{1,024}.$$

EXAMPLE 5   If the first two terms of a geometric sequence are 2 and 4, respectively, which term of the sequence is equal to 512?

SOLUTION   Since $a_1 = 2$ and $a_2 = 4$, we conclude that $r = 2$. If $a_n = 512$, then

$$512 = 2(2^{n-1})$$

or

$$2^9 = 2^{1+n-1} = 2^n,$$

so that

$$n = 9.$$

Therefore, the ninth term of the sequence is 512.

PROBLEM SET 10.1

In problems 1–6, find the first five terms in each sequence.

1. $a_n = \dfrac{n(n + 2)}{2}$ **2.** $b_n = \dfrac{n + 4}{n}$ **3.** $c_n = \dfrac{n(n - 3)}{2}$ **4.** $a_n = \dfrac{3}{n(n + 1)}$

5. $a_n = (-1)^n + 3$ **6.** $c_n = \dfrac{n^2 - 2}{2}$

In problems 7–14, determine which sequences are arithmetic progressions, and find the common difference d for each arithmetic progression.

7. $2, 5, 8, 11, \ldots$

8. $3, 5, 7, 9, \ldots$

9. $7, 12, 17, 22, \ldots$

10. $11a + 7b, 7a + 2b, 3a - 3b, \ldots$

11. $67, 54, 41, 28, \ldots$

12. $9a^2, 16a^2, 23a^2, 30a^2, \ldots$

13. $5.7, 6.9, 8.1, 9.3, \ldots$

14. $1.4, 4.5, 7.6, 10.7, \ldots$

15. Find the tenth and fifteenth terms of the arithmetic progression $-13, -6, 1, 8, \ldots$.

16. Find the twelfth and thirty-fifth terms of the arithmetic progression $19, 17, 15, 13, \ldots$.

17. Find the sixth and ninth terms of the arithmetic progression $a + 24b, 4a + 20b, 7a + 16b, \ldots$.

18. Find the third and sixteenth terms of the arithmetic progression $7a^2 - 4b, 2a^2 + 7b, -3a^2 + 18b, \ldots$.

In problems 19–26, determine which sequences are geometric progressions and give the value of the common ratio r for each geometric progression.

19. $2, 6, 18, \ldots$ **20.** $1, \frac{1}{5}, \frac{1}{25}, \ldots$ **21.** $1, -2, 4, \ldots$ **22.** $\frac{4}{9}, \frac{1}{6}, \frac{1}{16}, \ldots$

23. $81, 54, 36, \ldots$ **24.** $147, -21, 3, \ldots$ **25.** $9, -6, 4, \ldots$ **26.** $64, -32, 16, \ldots$

In problems 27–32, find the indicated term of each geometric progression.

27. The tenth term of $-4, 2, -1, \frac{1}{2}, \ldots$

28. The eighth term of $\frac{1}{8}, \frac{1}{4}, \frac{1}{2}, \ldots$

29. The fifth term of $32, 16, 8, \ldots$

30. The eleventh term of $1, 1.03, (1.03)^2, \ldots$

31. The nth term of $1, 1 + a, (1 + a)^2, \ldots$

32. The twelfth term of $10^{-5}, 10^{-7}, 10^{-9}, \ldots$

33. Find the sixth and tenth terms of the geometric progression $6, 12, 24, 48, \ldots$.

34. Find the sixth and eighth terms of the geometric progression $2, 6, 18, \ldots$.

35. Find the fifth term of the geometric progression $3, 6, 12, \ldots$.

36. Find the eleventh term of the geometric progression $10, 10^2, 10^3, \ldots$.

37. If the first two terms of a geometric sequence are 2 and 1, respectively, which term of the sequence is equal to $\frac{1}{16}$?

38. If the first two terms of a geometric sequence are $3\sqrt{3}$ and 9, respectively, which term of the sequence is equal to $243\sqrt{3}$?

10.2 Series

Some applications of mathematics involve finding the sum of the terms of a sequence. This sum is called a **series.**

Summation Notation

Although the sum of the first n terms of a sequence

$$a_1, a_2, a_3, \ldots, a_n, \ldots$$

can be written as

$$a_1 + a_2 + a_3 + \cdots + a_n,$$

a more compact notation is useful. The Greek capital letter Σ (sigma) is used for this purpose. We write the sum in **sigma notation** as

$$\sum_{k=1}^{n} a_k = a_1 + a_2 + a_3 + \cdots + a_n.$$

Here Σ indicates a sum, and the symbols above and below the Σ indicate that k is an integer from 1 to n inclusive. k is called the **index of summation.** There is no particular reason to use k for the index of summation, any letter will do; however, i, j, k, and n are the most commonly used indices. For instance,

$$\sum_{k=1}^{n} 5^k = \sum_{i=1}^{n} 5^i$$

$$= \sum_{j=1}^{n} 5^j = 5^1 + 5^2 + 5^3 + \cdots + 5^n.$$

In Examples 1 and 2, evaluate each sum.

EXAMPLE 1 $\displaystyle\sum_{k=1}^{3} (4k^2 - 3k)$

SOLUTION Here we have $a_k = 4k^2 - 3k$. To find the indicated sum, we substitute the integers 1, 2, and 3 for k in succession, and then add the resulting numbers. Thus,

$$\sum_{k=1}^{3} (4k^2 - 3k) = [4(1^2) - 3(1)] + [4(2^2) - 3(2)] + [4(3^2) - 3(3)]$$
$$= 1 + 10 + 27 = 38.$$

EXAMPLE 2 $\displaystyle\sum_{k=2}^{5} \frac{k-1}{k+1}$

SOLUTION Here we have $a_k = (k-1)/(k+1)$, and

$$\sum_{k=2}^{5} \frac{k-1}{k+1} = \left(\frac{2-1}{2+1}\right) + \left(\frac{3-1}{3+1}\right) + \left(\frac{4-1}{4+1}\right) + \left(\frac{5-1}{5+1}\right)$$
$$= \frac{1}{3} + \frac{2}{4} + \frac{3}{5} + \frac{4}{6} = \frac{21}{10}.$$

The Sum of the First *n* Terms of Arithmetic and Geometric Sequences

We now derive formulas for the sum of the first n terms of an arithmetic sequence or a geometric sequence. We begin with an arithmetic progression. Let

$$a_1, a_2, a_3, \ldots, a_n, \ldots$$

be an arithmetic sequence with a first term a_1 and a common difference d. Let S_n represent the sum of the first n terms, that is,

$$S_n = \sum_{k=1}^{n} a_k.$$

The sum can be written out as

$$S_n = a_1 + (a_1 + d) + (a_1 + 2d) + (a_1 + 3d) + \cdots + [a_1 + (n-3)d]$$
$$+ [a_1 + (n-2)d] + [a_1 + (n-1)d].$$

If we reverse the order of these terms, we have

$$S_n = [a_1 + (n-1)d] + [a_1 + (n-2)d] + [a_1 + (n-3)d] + \cdots + (a_1 + 3d)$$
$$+ (a_1 + 2d) + (a_1 + d) + a_1.$$

Now observe what happens when we add the two representations of S_n term by term:

$$
\begin{array}{rllllll}
S_n = & a_1 & + & (a_1 + d) & + \cdots + & [a_1 + (n-2)d] & + [a_1 + (n-1)d] \\
+ \; S_n = & [a_1 + (n-1)d] & + & [a_1 + (n-2)d] & + \cdots + & (a_1 + d) & + \quad a_1 \\
\hline
2S_n = & [2a_1 + (n-1)d] & + [2a_1 + d + (n-2)d] & + \cdots & + [2a_1 + d + (n-2)d] & + [2a_1 + (n-1)d]
\end{array}
$$

n times

$$
\begin{array}{rl}
= & \overbrace{[2a_1 + (n-1)d] \; + \; [2a_1 + (n-1)d] \; + \cdots + \; [2a_1 + (n-1)d] \; + [2a_1 + (n-1)d]} \\
2S_n = & n[2a_1 + (n-1)d].
\end{array}
$$

We divide both sides of the equation by 2:

$$S_n = \frac{n}{2}[2a_1 + (n-1)d].$$

Therefore, the formula for the sum of the first n terms of an arithmetic sequence is:

$$
\begin{aligned}
S_n &= \sum_{k=1}^{n} a_k \\
&= \frac{n}{2}[2a_1 + (n-1)d].
\end{aligned}
$$

Using the fact that $a_n = a_1 + (n-1)d$, we can rewrite this formula as follows:

$$
\begin{aligned}
S_n &= \frac{n}{2}[2a_1 + (n-1)d] \\
&= \frac{n}{2}[a_1 + a_1 + (n-1)d] \\
&= \frac{n}{2}(a_1 + a_n).
\end{aligned}
$$

Therefore,

$$
\begin{aligned}
S_n &= \sum_{k=1}^{n} a_k \\
&= \frac{n}{2}(a_1 + a_n).
\end{aligned}
$$

EXAMPLE 3   Find the sum of the first twenty terms of an arithmetic sequence whose first term is 2 and whose common difference is 4.

SOLUTION   We substitute $a_1 = 2$, $d = 4$, and $n = 20$ in $S_n = (n/2)[2a_1 + (n - 1)d]$:

$$S_{20} = \frac{20}{2}[2(2) + (20 - 1)4]$$

$$= 10(4 + 76)$$

$$= 800.$$

EXAMPLE 4   The sum of the first ten terms of an arithmetic sequence is 351 and the tenth term is 51. Find the first term and the common difference.

SOLUTION   Using the formula $S_n = (n/2)(a_1 + a_n)$, we have

$$S_{10} = \frac{10}{2}(a_1 + a_{10}).$$

We know the values of S_{10} and a_{10}, so the equation is

$$351 = \frac{10}{2}(a_1 + 51) = 5a_1 + 255.$$

We solve for a_1:

$$5a_1 = 96 \quad \text{or} \quad a_1 = 19.2.$$

To solve for d, we use the formula $a_n = a_1 + (n - 1)d$:

$$a_{10} = a_1 + 9d.$$

Thus,

$$51 = 19.2 + 9d, \quad 9d = 31.8, \quad \text{and} \quad d = \frac{31.8}{9} = \frac{53}{15}.$$

Therefore, the first term is 19.2 and the common difference is $\frac{53}{15}$.

EXAMPLE 5   How many terms are there in the arithmetic sequence for which $a_1 = 3$, $d = 5$, and $S_n = 255$?

SOLUTION   Using the formula

$$S_n = \frac{n}{2}(a_1 + a_n),$$

we have

$$255 = \frac{n}{2}(3 + a_n) \quad \text{or} \quad 510 = n(3 + a_n).$$

We can obtain another equation in the variables n and a_n by using the formula

$$a_n = a_1 + (n - 1)d.$$

We substitute the given values for a_1 and d in this formula:

$$a_n = 3 + (n - 1)5 = 5n - 2.$$

We can now solve the system

$$\begin{cases} 510 = n(3 + a_n) \\ a_n = 5n - 2 \end{cases}$$

by using the substitution method:

$$510 = n[3 + (5n - 2)] = n(5n + 1) = 5n^2 + n$$

or

$$5n^2 + n - 510 = 0.$$

This equation can be factored as

$$(5n + 51)(n - 10) = 0.$$

This gives us $n = 10$ or $n = -\frac{51}{5}$. Hence, the sequence has ten terms (n must be a positive integer).

We now consider a geometric sequence with first term a_1 and a common ratio r,

$$a_1, a_1r, a_1r^2, \ldots, a_1r^{n-1}, \ldots.$$

To find a formula for the sum

$$S_n = \sum_{k=1}^{n} a_1 r^{k-1},$$

we start with the expression for S_n in expanded form:

$$S_n = a_1 + a_1r + a_1r^2 + \cdots + a_1r^{n-1}.$$

We then multiply both sides of the equation by r:

$$rS_n = a_1r + a_1r^2 + a_1r^3 + \cdots + a_1r^n.$$

Next we subtract rS_n from S_n:

$$S_n - rS_n = a_1 - a_1r^n,$$

so that the equation becomes

$$(1 - r)S_n = a_1(1 - r^n) \qquad \text{or} \qquad S_n = \frac{a_1(1 - r^n)}{1 - r}.$$

Therefore, the formula for the sum of the first n terms of a geometric sequence is:

$$S_n = \sum_{k=1}^{n} a_1 r^{k-1} = \frac{a_1(1 - r^n)}{1 - r}, \qquad r \neq 1.$$

EXAMPLE 6   Find the sum of the first ten terms of the geometric sequence whose first term is $\frac{1}{2}$ and whose common ratio is 2.

SOLUTION   We substitute $n = 10$, $a_1 = \frac{1}{2}$, and $r = 2$ in the formula

$$S_n = \frac{a_1(1 - r^n)}{1 - r}$$

and obtain

$$S_{10} = \frac{\frac{1}{2}(1 - 2^{10})}{1 - 2} = \frac{\frac{1}{2}(-1{,}023)}{-1} = 511.5.$$

EXAMPLE 7   The sum of the first five terms of a geometric sequence is $2\frac{7}{27}$ and the common ratio is $-\frac{1}{3}$. Find the first four terms of the sequence.

SOLUTION   Using the formula for S_n, we have

$$2\frac{7}{27} = \frac{a_1[1 - (-\frac{1}{3})^5]}{1 - (-\frac{1}{3})}$$

so that

$$2\frac{7}{27} = \left(\frac{\frac{244}{243}}{\frac{4}{3}}\right)a_1, \qquad \frac{61}{27} = \frac{61}{81}a_1, \qquad \text{and} \qquad a_1 = \frac{\frac{61}{27}}{\frac{61}{81}} = 3.$$

Hence, the first four terms of the sequence are 3, $3(-\frac{1}{3})$, $3(-\frac{1}{3})^2$, and $3(-\frac{1}{3})^3$, or 3, -1, $\frac{1}{3}$, and $-\frac{1}{9}$.

PROBLEM SET 10.2

In problems 1–14, evaluate each sum.

1. $\displaystyle\sum_{k=1}^{5} k$

2. $\displaystyle\sum_{k=0}^{4} \frac{2^k}{k + 1}$

3. $\displaystyle\sum_{i=1}^{10} 2i(i - 1)$

4. $\displaystyle\sum_{k=0}^{4} 3^{2k}$

5. $\displaystyle\sum_{k=2}^{5} 2^{k-2}$

6. $\displaystyle\sum_{i=2}^{6} \frac{1}{i}$

7. $\displaystyle\sum_{k=1}^{3} (2k + 1)$

8. $\displaystyle\sum_{k=1}^{5} (3k^2 - 5k + 1)$

9. $\displaystyle\sum_{i=1}^{4} \frac{i}{i+1}$

10. $\displaystyle\sum_{k=1}^{4} k^k$

11. $\displaystyle\sum_{k=1}^{100} 5$

12. $\displaystyle\sum_{i=3}^{7} (i+2)$

13. $\displaystyle\sum_{k=1}^{5} \frac{1}{k(k+1)}$

14. $\displaystyle\sum_{k=1}^{4} \frac{3}{k}$

15. Find the sum of the first ten terms of an arithmetic sequence whose first term is 1 and whose common difference is 3.

16. Find the sum of the first fifteen terms of an arithmetic sequence whose first term is $\frac{1}{2}$ and whose common difference is $\frac{1}{3}$.

17. Find the sum of the first eight terms of an arithmetic sequence whose first term is -5 and whose common difference is $\frac{3}{7}$.

18. Find the sum of the first twelve terms of an arithmetic sequence whose first term is 11 and whose common difference is -2.

19. Find S_7 for the arithmetic sequence $6, 3b + 1, 6b - 4, \ldots$.

20. Find S_{10} for the arithmetic sequence $x + 2y, 3y, -x + 4y, \ldots$.

In problems 21–26, certain elements of an arithmetic sequence are given. Find the indicated unknown elements.

21. $a_1 = 6$; $d = 3$; a_{10} and S_{10}

22. $a_1 = 38$; $d = -2$; $n = 25$; S_n

23. $a_1 = 17$; $S_{18} = 2{,}310$; d and a_{18}

24. $d = 3$; $S_{25} = 400$; a_1 and a_{25}

25. $a_1 = 27$; $a_n = 48$; $S_n = 1{,}200$; n and d

26. $a_1 = -3$; $d = 2$; $S_n = 140$; n

27. Find the sum of the first six terms of the geometric sequence whose first term is $\frac{3}{2}$ and whose common ratio is 2.

28. Find the sum of the first ten terms of the geometric sequence whose first term is 6 and whose common ratio is $\frac{1}{2}$.

29. Find the sum of the first twelve terms of the geometric sequence whose first term is -4 and whose common ratio is -2.

30. Find the sum of the first eight terms of the geometric sequence whose first term is 5 and whose common ratio is $-\frac{1}{2}$.

31. Find S_6 for the geometric sequence $10, 10a, 10a^2, 10a^3, \ldots$.

32. Find S_8 for the geometric sequence $k, \dfrac{k}{b}, \dfrac{k}{b^2}, \dfrac{k}{b^3}, \ldots$.

In problems 33–40 find the indicated element in each geometric progression with the given elements.

33. $a_1 = 2$; $n = 3$; $S_n = 26$; r

34. $r = 2$; $n = 5$; $a_n = -48$; a_1 and S_n

35. $a_1 = 3$; $a_n = 192$; $n = 7$; r

36. $a_6 = 3$; $a_9 = -81$; r and a_1

37. $a_5 = \frac{1}{8}$; $r = -\frac{1}{2}$; a_9 and S_8

38. $a_1 = 1$; $r = (1.03)^{-1}$; $a_9 = (1.03)^{-8}$; S_8

39. $a_1 = \frac{1}{16}$; $r = 2$; $a_n = 32$; n and S_n

40. $a_1 = 250$; $r = \frac{3}{5}$; $a_n = 32\frac{2}{5}$; n and S_n

10.3 The Binomial Theorem

In Chapter 2, Section 2.6, we considered the special products $(a + b)^2$ and $(a + b)^3$. We often work with expressions of the form $(a + b)^n$, where n is a positive integer. Since the expression $a + b$ is a binomial, the formula for expanding $(a + b)^n$ is called the **binomial theorem.**

We can expand $(a + b)^n$, for small values of n, by using direct calculation. For instance,

$$(a + b)^1 = a + b$$

$$(a + b)^2 = a^2 + 2ab + b^2$$

$$(a + b)^3 = a^3 + 3a^2b + 3ab^2 + b^3$$

$$(a + b)^4 = a^4 + 4a^3b + 6a^2b^2 + 4ab^3 + b^4$$

$$(a + b)^5 = a^5 + 5a^4b + 10a^3b^2 + 10a^2b^3 + 5ab^4 + b^5$$

This pattern holds for the expansion of $(a + b)^n$, where n is any positive integer. The following rules are used for this expansion:

1. There are $n + 1$ terms. The first term is a^n and the last term is b^n.

2. The powers of a decrease by 1 and the powers of b increase by 1 for each term. The sum of the exponents of a and b is n for each term.

One way to display the coefficients in the expansion of $(a + b)^n$ for $n = 1, 2, 3, \ldots$ is the following array of numbers, known as **Pascal's triangle:**

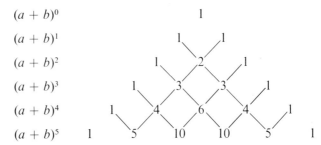

The first and last numbers in each line are always 1. The other numbers can be found by adding the pair of numbers from the preceding line, as indicated by the V's. For example,

indicates that 10 was obtained by adding 4 and 6.

It is easier to detect the pattern for determining the coefficients in the expansion of $(a + b)^n$ if we use the following notation for the product of all positive integers from 1 to n inclusive. The symbol $n!$ (read "n factorial" or "factorial n") is defined by:

$$n! = 1 \cdot 2 \cdot 3 \cdots (n - 1)n$$

or

$$n! = n(n - 1)(n - 2) \cdots 2 \cdot 1.$$

Thus, $4! = 4 \cdot 3 \cdot 2 \cdot 1 = 24$ and $6! = 6 \cdot 5 \cdot 4 \cdot 3 \cdot 2 \cdot 1 = 720$.

We have defined $n!$ for positive integers n as

$$n! = n(n - 1)(n - 2) \cdots 4 \cdot 3 \cdot 2 \cdot 1.$$

Therefore,

$$(n - 1)! = (n - 1)(n - 2)(n - 3) \cdots 4 \cdot 3 \cdot 2 \cdot 1.$$

If we multiply both sides of this equation by n, we find that

$$n! = n \, (n - 1)! \qquad \text{if} \qquad n \neq 1.$$

We want this relationship to hold for $n = 1$:

$$1! = 1 \, (1 - 1)! \qquad \text{or} \qquad 1! = 1 \cdot 0!,$$

therefore, we define

$$0! = 1.$$

We can use this definition together with the above formula, $n! = n(n - 1)!$, to make the table of values of $n!$ shown in Table 1.

Expressions involving factorial notation may be simplified as follows:

Table 1

$0! = 1$
$1! = 1$
$2! = 2$
$3! = 6$
$4! = 24$
$5! = 120$
$6! = 720$
$7! = 5040$

EXAMPLE 1 Simplify:

(a) $\dfrac{7!}{5!}$ (b) $\dfrac{8!}{3! \cdot 5!}$ (c) $\dfrac{(n + 1)!}{(n - 1)!}$

SOLUTION (a) $\dfrac{7!}{5!} = \dfrac{7 \cdot 6 \cdot \cancel{5} \cdot \cancel{4} \cdot \cancel{3} \cdot \cancel{2} \cdot \cancel{1}}{\cancel{5} \cdot \cancel{4} \cdot \cancel{3} \cdot \cancel{2} \cdot \cancel{1}} = 7 \cdot 6 = 42$

(b) $\dfrac{8!}{3! \cdot 5!} = \dfrac{8 \cdot 7 \cdot \cancel{6} \cdot \cancel{5} \cdot \cancel{4} \cdot \cancel{3} \cdot \cancel{2} \cdot \cancel{1}}{(\cancel{3} \cdot \cancel{2} \cdot \cancel{1})(\cancel{5} \cdot \cancel{4} \cdot \cancel{3} \cdot \cancel{2} \cdot 1)} = 8 \cdot 7 = 56$

(c) $\dfrac{(n + 1)!}{(n - 1)!} = \dfrac{(n + 1)(n)[\cancel{(n - 1)!}]}{\cancel{(n - 1)!}} = (n + 1)n = n^2 + n$

It is helpful to introduce factorial notation to provide a general description of the binomial expansion $(a + b)^n$ without relying on Pascal's triangle.

The variables in the expansion of $(a + b)^n$ have the following pattern:

$$a^n, a^{n-1}b, a^{n-2}b^2, a^{n-3}b^3, \ldots, ab^{n-1}, b^n.$$

Notice that the sum of the exponents of a and b is n for each term. In addition:

1. The first term is a^n, and the coefficient is 1.

2. The second term contains $a^{n-1}b$, and the coefficient is

$$\frac{n}{1!}.$$

3. The third term contains $a^{n-2}b^2$, and the coefficient is

$$\frac{n(n-1)}{2!}.$$

4. The fourth term contains $a^{n-3}b^3$, and the coefficient is

$$\frac{n(n-1)(n-2)}{3!}.$$

The following formula provides the general expansion of $(a + b)^n$:

THEOREM 1 **The Binomial Theorem**

$$(a + b)^n = a^n + \frac{n}{1!}a^{n-1}b + \frac{n(n-1)}{2!}a^{n-2}b^2$$

$$+ \frac{n(n-1)(n-2)}{3!}a^{n-3}b^3 + \cdots$$

$$+ \frac{n(n-1)(n-2) \cdots (n-k+2)}{(k-1)!}a^{n-k+1}b^{k-1} + \cdots + b^n,$$

where k is an integer such that $1 \leq k \leq n + 1$.

For example, if we substitute x for a, $2y^2$ for b, and 5 for n in the above theorem, we have

$$(x + 2y^2)^5 = [x + (2y^2)]^5$$

$$= x^5 + \frac{5}{1!}x^4(2y^2) + \frac{5 \cdot 4}{2!}x^3(2y^2)^2 + \frac{5 \cdot 4 \cdot 3}{3!}x^2(2y^2)^3$$

$$+ \frac{5 \cdot 4 \cdot 3 \cdot 2}{4!}x(2y^2)^4 + \frac{5 \cdot 4 \cdot 3 \cdot 2 \cdot 1}{5!}(2y^2)^5$$

$$= x^5 + 10x^4y^2 + 40x^3y^4 + 80x^2y^6 + 80xy^8 + 32y^{10}.$$

Note that the kth term of the binomial expansion $(a + b)^n$, denoted by u_k, is given by the equation

$$u_k = \frac{n(n-1)(n-2)\cdots(n-k+2)}{(k-1)!}a^{n-k+1}b^{k-1}.$$

This information will help us write a particular term of the binomial expansion or to find the term where b has a particular exponent. For example, the sixth term u_6 of $(x^2 + 2y)^{12}$ is

$$\frac{\cancel{12}\cdot 11 \cdot \cancel{10}\cdot 9 \cdot 8}{\cancel{5}\cdot\cancel{4}\cdot\cancel{3}\cdot\cancel{2}\cdot 1}(x^2)^7(2y)^5 = 25{,}344x^{14}y^5.$$

If we replace $k - 1$ by k, in the expression for u_k, we obtain an expression for the $(k + 1)$th term u_{k+1} which contains the factor b^k:

$$u_{k+1} = \frac{n(n-1)(n-2)\cdots(n-k+1)}{k!}a^{n-k}b^k$$

$$= \frac{n(n-1)(n-2)\cdots(n-k+1)(n-k)!}{k!(n-k)!}a^{n-k}b^k \qquad \text{[We multiplied the numerator and denominator of the coefficient by } (n-k)!]$$

$$= \frac{n(n-1)(n-2)\cdots(n-k+1)(n-k)(n-k-1)(n-k-2)\cdots 3\cdot 2\cdot 1}{k!(n-k)!}a^{n-k}b^k$$

$$= \frac{n!}{k!(n-k)!}a^{n-k}b^k.$$

Thus:

$$u_{k+1} = \frac{n!}{k!(n-k)!}a^{n-k}b^k.$$

For example, to find the term involving y^4 in the expansion of $(x^2 + 2y)^{12}$, we have $b^k = (2y)^k$. We choose a value of k that will give us the variable factor y^4: $k = 4$. We substitute 4 for k in the above formula:

$$u_5 = \frac{12!}{4!8!}(x^2)^8(2y)^4 = \frac{12\cdot 11\cdot 10\cdot 9\cdot 8!}{4!8!}x^{16}(16y^4)$$

$$= \frac{12\cdot 11\cdot 10\cdot 9}{4\cdot 3\cdot 2\cdot 1}(16)x^{16}y^4$$

$$= 7{,}920x^{16}y^4.$$

EXAMPLE 2 Expand $(x + y)^7$

SOLUTION We substitute $a = x$, $b = y$, and $n = 7$ in the binomial theorem:

$$(x + y)^7$$

$$= x^7 + \frac{7}{1!}x^6y + \frac{7 \cdot 6}{2!}x^5y^2 + \frac{7 \cdot 6 \cdot 5}{3!}x^4y^3 + \frac{7 \cdot 6 \cdot 5 \cdot 4}{4!}x^3y^4$$

$$+ \frac{7 \cdot 6 \cdot 5 \cdot 4 \cdot 3}{5!}x^2y^5 + \frac{7 \cdot 6 \cdot 5 \cdot 4 \cdot 3 \cdot 2}{6!}xy^6 + y^7$$

$$= x^7 + 7x^6y + 21x^5y^2 + 35x^4y^3 + 35x^3y^4 + 21x^2y^5 + 7xy^6 + y^7.$$

EXAMPLE 3 Find the eighth term in the expansion of $(x - y)^{12}$.

SOLUTION The kth term u_k of $(a + b)^n$ is given by

$$u_k = \frac{n(n - 1)(n - 2) \cdots (n - k + 2)}{(k - 1)!}a^{n-k+1}b^{k-1}.$$

Substituting $a = x$, $b = -y$, $k = 8$, and $n = 12$ in the above formula, we have

$$n - k + 2 = 12 - 8 + 2 = 6$$

$$n - k + 1 = 12 - 8 + 1 = 5$$

$$k - 1 = 8 - 1 = 7.$$

Therefore,

$$u_8 = \frac{12 \cdot 11 \cdot 10 \cdot 9 \cdot 8 \cdot 7 \cdot 6}{7!}x^5(-y)^7$$

$$= -792x^5y^7.$$

EXAMPLE 4 Find and simplify the term involving x^7 in the expansion of $(2 - x)^{12}$.

SOLUTION We use the formula for the $(k + 1)$th term,

$$u_{k+1} = \frac{n!}{k!(n - k)!}a^{n-k}b^k,$$

with $k = 7$, $a = 2$, $b = -x$, and $n = 12$. We have

$$\frac{12!}{7!(12 - 7)!}(2)^5(-x)^7 = \frac{\cancel{12} \cdot 11 \cdot \cancel{10} \cdot 9 \cdot 8 \cdot \cancel{7!}}{\cancel{7!} \cdot \cancel{5} \cdot \cancel{4} \cdot \cancel{3} \cdot \cancel{2} \cdot 1}(-32x^7)$$

$$= -25,344x^7.$$

PROBLEM SET 10.3

In problems 1–10, write each expression in expanded form and simplify the results.

1. $\dfrac{4!}{6!}$

2. $\dfrac{10!}{5! \cdot 7!}$

3. $\dfrac{2!}{4! - 3!}$

4. $\dfrac{1}{4!} + \dfrac{1}{3!}$

5. $\dfrac{3! \cdot 8!}{4! \cdot 7!}$

6. $\dfrac{4! \cdot 6!}{8! - 5!}$

7. $\dfrac{0}{0!}$

8. $\dfrac{(n - 2)!}{(n + 1)!}$

9. $\dfrac{(n + 1)!}{(n - 3)!}$

10. $\dfrac{(n + k)!}{(n + k - 2)!}$

In problems 11–18, expand each expression by using the binomial theorem and simplify each term.

11. $(x + 2)^5$

12. $(a - 2b)^4$

13. $(x^2 + 4y^2)^3$

14. $(1 - a^{-1})^5$

15. $(a^3 - a^{-1})^6$

16. $\left(1 - \dfrac{x}{y^2}\right)^5$

17. $\left(2 + \dfrac{x}{y}\right)^5$

18. $(x + y + z)^3$

In problems 19–22, use the binomial theorem to expand each expression and check the results using Pascal's triangle.

19. $(2z + x)^8$

20. $(x - 3)^8$

21. $(y^2 - 2x)^5$

22. $\left(\dfrac{1}{a} + \dfrac{x}{2}\right)^5$

In problems 23–26, find the first four terms of each expansion.

23. $(x^2 - 2a)^{10}$

24. $\left(2a - \dfrac{1}{b}\right)^6$

25. $\left(\sqrt{\dfrac{x}{2}} + 2y\right)^7$

26. $\left(\dfrac{1}{a} + \dfrac{x}{2}\right)^{11}$

In problems 27–34, find the first five terms in each expansion and simplify.

27. $(x + y)^{16}$

28. $(a^2 + b^2)^{12}$

29. $(a - 2b^2)^{11}$

30. $(a + 2y^2)^8$

31. $(x - 2y)^7$

32. $\left(1 - \dfrac{x}{y^2}\right)^8$

33. $(a^3 - a^2)^9$

34. $\left(x + \dfrac{1}{2y}\right)^{15}$

In problems 35–40, find the indicated term for each expression.

35. $\left(\dfrac{x^2}{2} + a\right)^{15}$, fourth term

36. $(y^2 - 2z)^{10}$, sixth term

37. $\left(2x^2 - \dfrac{a^2}{3}\right)^9$, seventh term

38. $(x + \sqrt{a})^{12}$, middle term

39. $\left(a + \dfrac{x^3}{3}\right)^9$, term containing x^{12}

40. $\left(2\sqrt{y} - \dfrac{x}{2}\right)^{10}$, term containing y^4

10.4 **Determinants**

Consider the following system of two linear equations in two variables:

$$\begin{cases} a_1 x + b_1 y = c_1 \\ a_2 x + b_2 y = c_2. \end{cases}$$

We can solve this system by using the substitution method or the elimination method (Section 7.5) to obtain the solution (x, y), where

$$x = \frac{b_2 c_1 - b_1 c_2}{a_1 b_2 - a_2 b_1} \quad \text{and} \quad y = \frac{a_1 c_2 - a_2 c_1}{a_1 b_2 - a_2 b_1},$$

provided that $a_1 b_2 - a_2 b_1 \neq 0$.

Any system of two linear equations in two variables can be arranged in the above form. Note that the solution appears as fractions with a common denominator. We can represent this denominator by the symbol

$$\begin{vmatrix} a_1 & b_1 \\ a_2 & b_2 \end{vmatrix}$$

which is called a **determinant.**

We have the following formal definition:

$$\begin{vmatrix} a_1 & b_1 \\ a_2 & b_2 \end{vmatrix} = a_1 b_2 - a_2 b_1.$$

The numbers a_1, b_1, a_2, and b_2 are called the **elements** of the determinant. The elements of the two **rows** are a_1, b_1 and a_2, b_2. The elements of the two **columns** are a_1, a_2 and b_1, b_2. Therefore, this is called a **two-by-two** (2×2) determinant, or a determinant of **order 2.** The expression $a_1 b_2 - a_2 b_1$ is referred to as the **expansion** of the determinant.

In Examples 1 and 2, find the value of the following 2×2 determinants.

EXAMPLE 1

$$\begin{vmatrix} 1 & -2 \\ 3 & 4 \end{vmatrix}$$

SOLUTION

Here we have $a_1 = 1$, $b_1 = -2$, $a_2 = 3$, and $b_2 = 4$. So by substitution in the expression $a_1 b_2 - a_2 b_1$, we have

$$\begin{vmatrix} 1 & -2 \\ 3 & 4 \end{vmatrix} = 1(4) - 3(-2) = 4 + 6 = 10.$$

EXAMPLE 2
$$\begin{vmatrix} 1 & 0 \\ 2 & -1 \end{vmatrix}$$

SOLUTION
$$\begin{vmatrix} 1 & 0 \\ 2 & -1 \end{vmatrix} = 1(-1) - 2(0) = -1 - 0 = -1$$

The value of a 3×3 determinant is defined in terms of 2×2 determinants as follows:

$$\begin{vmatrix} a_1 & b_1 & c_1 \\ a_2 & b_2 & c_2 \\ a_3 & b_3 & c_3 \end{vmatrix} = a_1 \begin{vmatrix} b_2 & c_2 \\ b_3 & c_3 \end{vmatrix} - a_2 \begin{vmatrix} b_1 & c_1 \\ b_3 & c_3 \end{vmatrix} + a_3 \begin{vmatrix} b_1 & c_1 \\ b_2 & c_2 \end{vmatrix}.$$

We can use the definition of a 2×2 determinant to obtain:

$$\begin{vmatrix} a_1 & b_1 & c_1 \\ a_2 & b_2 & c_2 \\ a_3 & b_3 & c_3 \end{vmatrix} = a_1 b_2 c_3 - a_1 b_3 c_2 - a_2 b_1 c_3 + a_2 b_3 c_1 + a_3 b_1 c_2 - a_3 b_2 c_1.$$

The right-hand side of this equation is called the **expansion** of a 3×3 determinant. We can express the expansions of 2×2 determinants and 3×3 determinants in terms of sums and differences of the products of their elements.

In Examples 3 and 4, find the value of the 3×3 determinants.

EXAMPLE 3
$$\begin{vmatrix} 1 & 0 & 2 \\ 4 & 6 & -1 \\ -1 & 0 & -1 \end{vmatrix}$$

SOLUTION   By definition,

$$\begin{vmatrix} 1 & 0 & 2 \\ 4 & 6 & -1 \\ -1 & 0 & -1 \end{vmatrix} = 1 \begin{vmatrix} 6 & -1 \\ 0 & -1 \end{vmatrix} - 4 \begin{vmatrix} 0 & 2 \\ 0 & -1 \end{vmatrix} + (-1) \begin{vmatrix} 0 & 2 \\ 6 & -1 \end{vmatrix}$$

$$= 1[6(-1) - 0(-1)] - 4[0(-1) - 0(2)] - 1[0(-1) - 6(2)]$$

$$= 1(-6) - 4(0) - 1(-12) = -6 + 12 = 6.$$

EXAMPLE **4**

$$\begin{vmatrix} 3 & 1 & -1 \\ 0 & 2 & 4 \\ -1 & 4 & 2 \end{vmatrix}$$

SOLUTION

$$\begin{vmatrix} 3 & 1 & -1 \\ 0 & 2 & 4 \\ -1 & 4 & 2 \end{vmatrix} = 3 \begin{vmatrix} 2 & 4 \\ 4 & 2 \end{vmatrix} - 0 \begin{vmatrix} 1 & -1 \\ 4 & 2 \end{vmatrix} + (-1) \begin{vmatrix} 1 & -1 \\ 2 & 4 \end{vmatrix}$$

$$= 3(-12) - 0 - 1(6) = -42$$

PROBLEM SET 10.4

In problems 1–10, evaluate each 2×2 determinant.

1. $\begin{vmatrix} 7 & 1 \\ -5 & 3 \end{vmatrix}$

2. $\begin{vmatrix} 0 & 1 \\ 1 & 0 \end{vmatrix}$

3. $\begin{vmatrix} 6 & 0 \\ -3 & 4 \end{vmatrix}$

4. $\begin{vmatrix} 1 & 2 \\ 3 & 5 \end{vmatrix}$

5. $\begin{vmatrix} -3 & -1 \\ -5 & \frac{1}{2} \end{vmatrix}$

6. $\begin{vmatrix} 2 & 1 \\ -10 & 4 \end{vmatrix}$

7. $\begin{vmatrix} 3 & -1 \\ 6 & -2 \end{vmatrix}$

8. $\begin{vmatrix} 3 & -2 \\ 3 & 2 \end{vmatrix}$

9. $\begin{vmatrix} 1 & 0 \\ 0 & 1 \end{vmatrix}$

10. $\begin{vmatrix} 7 & 14 \\ 3 & 6 \end{vmatrix}$

In problems 11–20, evaluate each 3×3 determinant.

11. $\begin{vmatrix} -3 & 1 & 7 \\ 0 & 2 & 6 \\ -4 & 5 & 1 \end{vmatrix}$

12. $\begin{vmatrix} 2 & 1 & 1 \\ 9 & 3 & 6 \\ 0 & 0 & 1 \end{vmatrix}$

13. $\begin{vmatrix} 1 & 0 & 0 \\ 0 & 1 & 0 \\ 0 & 0 & 1 \end{vmatrix}$

14. $\begin{vmatrix} -10 & -1 & 5 \\ -7 & 8 & 2 \\ 3 & -6 & 0 \end{vmatrix}$

15. $\begin{vmatrix} -1 & 3 & 5 \\ -7 & 4 & 2 \\ -6 & 2 & 0 \end{vmatrix}$

16. $\begin{vmatrix} 3 & -1 & 2 \\ 0 & 1 & -5 \\ 6 & 7 & 4 \end{vmatrix}$

17. $\begin{vmatrix} 2 & 3 & 5 \\ 9 & 4 & 2 \\ 11 & -6 & 2 \end{vmatrix}$

18. $\begin{vmatrix} 2 & 2 & 2 \\ 3 & 3 & 3 \\ 4 & 4 & 4 \end{vmatrix}$

19. $\begin{vmatrix} -2 & -1 & 3 \\ 7 & -7 & 4 \\ 8 & -6 & 2 \end{vmatrix}$

20. $\begin{vmatrix} \frac{1}{2} & 4 & 7 \\ 1 & -1 & 2 \\ 3 & 2 & 5 \end{vmatrix}$

In problems 21–25, solve each equation for x.

21. $\begin{vmatrix} x & x \\ 5 & 3 \end{vmatrix} = 2$

22. $\begin{vmatrix} x+1 & x \\ x & x-2 \end{vmatrix} = -6$

23. $\begin{vmatrix} x & 4 & 5 \\ 0 & 1 & x \\ 5 & 2 & 0 \end{vmatrix} = 7$

24. $\begin{vmatrix} x & 0 & 1 \\ 4x & 1 & 2 \\ 3x & 1 & 3 \end{vmatrix} = 4$

25. $\begin{vmatrix} x & 5 \\ 4 & 2-x \end{vmatrix} = -x^2 + 3$

26. Show that the equation

$$\begin{vmatrix} 0 & x-2 & x-3 \\ x+2 & 0 & x-4 \\ x+3 & x+4 & 0 \end{vmatrix} = 0 \text{ has 0 as a root}$$

27. For what values of x is it true that $\begin{vmatrix} x & 2 \\ 2 & x \end{vmatrix} > 0$?

28. What kind of solutions does $ax^2 + bx + c = 0$, $a \neq 0$, have if:

(a) $\begin{vmatrix} b & 4a \\ c & b \end{vmatrix} > 0$

(b) $\begin{vmatrix} b & 4a \\ c & b \end{vmatrix} = 0$

(c) $\begin{vmatrix} b & 4a \\ c & b \end{vmatrix} < 0$

10.5 Cramer's Rule

As we mentioned in Section 10.4, the solution to the linear system

$$\begin{cases} a_1 x + b_1 y = c_1 \\ a_2 x + b_2 y = c_2 \end{cases}$$

is the ordered pair (x, y), where

$$x = \frac{c_1 b_2 - c_2 b_1}{a_1 b_2 - a_2 b_1} \quad \text{and} \quad y = \frac{a_1 c_2 - a_2 c_1}{a_1 b_2 - a_2 b_1}.$$

We can express the numerators and the denominators of these two equations by using determinants:

$$a_1 b_2 - a_2 b_1 = \begin{vmatrix} a_1 & b_1 \\ a_2 & b_2 \end{vmatrix}.$$

$$c_1 b_2 - c_2 b_1 = \begin{vmatrix} c_1 & b_1 \\ c_2 & b_2 \end{vmatrix} \quad \text{and} \quad a_1 c_2 - a_2 c_1 = \begin{vmatrix} a_1 & c_1 \\ a_2 & c_2 \end{vmatrix}.$$

We may therefore write the solution of the linear system

$$\begin{cases} a_1 x + b_1 y = c_1 \\ a_2 x + b_2 y = c_2 \end{cases}$$

in the form

$$x = \frac{\begin{vmatrix} c_1 & b_1 \\ c_2 & b_2 \end{vmatrix}}{\begin{vmatrix} a_1 & b_1 \\ a_2 & b_2 \end{vmatrix}} \quad \text{and} \quad y = \frac{\begin{vmatrix} a_1 & c_1 \\ a_2 & c_2 \end{vmatrix}}{\begin{vmatrix} a_1 & b_1 \\ a_2 & b_2 \end{vmatrix}}.$$

This is **Cramer's rule** for two linear equations in two unknowns, which is stated formally as follows:

Cramer's Rule for Solving
Linear Systems of Two Equations in Two Unknowns

Suppose that $D \neq 0$, and that D, D_x, and D_y are given by

$$D = \begin{vmatrix} a_1 & b_1 \\ a_2 & b_2 \end{vmatrix}, \qquad D_x = \begin{vmatrix} c_1 & b_1 \\ c_2 & b_2 \end{vmatrix}, \qquad D_y = \begin{vmatrix} a_1 & c_1 \\ a_2 & c_2 \end{vmatrix},$$

then the system

$$\begin{cases} a_1 x + b_1 y = c_1 \\ a_2 x + b_2 y = c_2 \end{cases}$$

has one and only one solution:

$$x = \frac{D_x}{D}, \qquad y = \frac{D_y}{D}.$$

In Cramer's rule, we call D the **coefficient determinant** because the elements of D are the coefficients of the unknowns in the system:

$$\begin{cases} a_1 x + b_1 y = c_1 \\ a_2 x + b_2 y = c_2 \end{cases}, \qquad D = \begin{vmatrix} a_1 & b_1 \\ a_2 & b_2 \end{vmatrix}.$$

You will notice that we obtain D_x by replacing the *first* column of D (the coefficients of x) with the constants on the right in the system of equations:

$$\begin{cases} a_1 x + b_1 y = c_1 \\ a_2 x + b_2 y = c_2 \end{cases}, \qquad D_x = \begin{vmatrix} c_1 & b_1 \\ c_2 & b_2 \end{vmatrix}, \qquad D_y = \begin{vmatrix} a_1 & c_1 \\ a_2 & c_2 \end{vmatrix}.$$

We obtain D_y by replacing the *second* column of D (the coefficients of y) with these same constants.

In Examples 1–3, solve the given systems by using Cramer's rule.

EXAMPLE 1
$$\begin{cases} 2x - 3y = 3 \\ x + 4y = 7 \end{cases}$$

SOLUTION

$$D = \begin{vmatrix} 2 & -3 \\ 1 & 4 \end{vmatrix} = 2(4) - 1(-3) = 11$$

$$D_x = \begin{vmatrix} 3 & -3 \\ 7 & 4 \end{vmatrix} = 3(4) - 7(-3) = 33$$

$$D_y = \begin{vmatrix} 2 & 3 \\ 1 & 7 \end{vmatrix} = 2(7) - 1(3) = 11.$$

Therefore,

$$x = \frac{D_x}{D} = \frac{33}{11} = 3 \quad \text{and} \quad y = \frac{D_y}{D} = \frac{11}{11} = 1.$$

EXAMPLE 2

$$\begin{cases} 3x = 4y - 1 \\ y = 2x + 2 \end{cases}$$

SOLUTION   First we write the system in the form

$$\begin{cases} 3x - 4y = -1 \\ -2x + y = 2. \end{cases}$$

Thus,

$$D = \begin{vmatrix} 3 & -4 \\ -2 & 1 \end{vmatrix} = 3(1) - (-2)(-4) = -5$$

$$D_x = \begin{vmatrix} -1 & -4 \\ 2 & 1 \end{vmatrix} = -1(1) - 2(-4) = 7$$

$$D_y = \begin{vmatrix} 3 & -1 \\ -2 & 2 \end{vmatrix} = 3(2) - (-1)(-2) = 4.$$

The solution is

$$x = \frac{D_x}{D} = -\frac{7}{5} \quad \text{and} \quad y = \frac{D_y}{D} = -\frac{4}{5}.$$

EXAMPLE 3

$$\begin{cases} 2x + y = 1 \\ 4x + 2y = 3 \end{cases}$$

SOLUTION

$$D = \begin{vmatrix} 2 & 1 \\ 4 & 2 \end{vmatrix} = 2(2) - 4(1) = 0$$

$$D_x = \begin{vmatrix} 1 & 1 \\ 3 & 2 \end{vmatrix} = 1(2) - 3(1) = -1$$

$$D_y = \begin{vmatrix} 2 & 1 \\ 4 & 3 \end{vmatrix} = 2(3) - 4(1) = 2.$$

The system is inconsistent because $x = D_x/D = -1/0$ and $y = D_y/D = 2/0$. Thus, there is no solution.

When we solve systems of linear equations by this method, we are using the **Cramer's rule method**. This method can be used to solve any system of n linear equations in n variables (n is a positive integer). Here, however, we only extend this rule to include systems of three equations in three variables.

Consider the following system of linear equations:

$$\begin{cases} a_1x + b_1y + c_1z = d_1 \\ a_2x + b_2y + c_2z = d_2 \\ a_3x + b_3y + c_3z = d_3. \end{cases}$$

We can now state Cramer's rule for solving the above system:

Cramer's Rule for Solving
Linear Systems of Three Equations in Three Unknowns

Suppose that $D \neq 0$ and that D, D_x, D_y, and D_z are given by

$$D = \begin{vmatrix} a_1 & b_1 & c_1 \\ a_2 & b_2 & c_2 \\ a_3 & b_3 & c_3 \end{vmatrix}, \qquad D_x = \begin{vmatrix} d_1 & b_1 & c_1 \\ d_2 & b_2 & c_2 \\ d_3 & b_3 & c_3 \end{vmatrix}, \qquad D_y = \begin{vmatrix} a_1 & d_1 & c_1 \\ a_2 & d_2 & c_2 \\ a_3 & d_3 & c_3 \end{vmatrix},$$

$$D_z = \begin{vmatrix} a_1 & b_1 & d_1 \\ a_2 & b_2 & d_2 \\ a_3 & b_3 & d_3 \end{vmatrix}$$

Then the solution of the above system is given by

$$x = \frac{D_x}{D}, \qquad y = \frac{D_y}{D}, \qquad z = \frac{D_z}{D}.$$

If $D = 0$, the system is inconsistent (or dependent), and Cramer's rule is not applicable.

EXAMPLE 4   Use Cramer's rule to find the solution set of the given system:

$$\begin{cases} x + y + z = 2 \\ 2x - y + z = 0 \\ x + 2y - z = 4 \end{cases}$$

SOLUTION

$$D = \begin{vmatrix} 1 & 1 & 1 \\ 2 & -1 & 1 \\ 1 & 2 & -1 \end{vmatrix}$$

$$= 1\begin{vmatrix} -1 & 1 \\ 2 & -1 \end{vmatrix} - 2\begin{vmatrix} 1 & 1 \\ 2 & -1 \end{vmatrix} + 1\begin{vmatrix} 1 & 1 \\ -1 & 1 \end{vmatrix}$$

$$= 1(-1) - 2(-3) + 1(2)$$
$$= 7.$$

Since $D \neq 0$, we proceed to find D_x, D_y, and D_z as follows:

$$D_x = \begin{vmatrix} 2 & 1 & 1 \\ 0 & -1 & 1 \\ 4 & 2 & -1 \end{vmatrix}$$

$$= 2\begin{vmatrix} -1 & 1 \\ 2 & -1 \end{vmatrix} - 0\begin{vmatrix} 1 & 1 \\ 2 & -1 \end{vmatrix} + 4\begin{vmatrix} 1 & 1 \\ -1 & 1 \end{vmatrix}$$

$$= 2(-1) - 0(-3) + 4(2) = 6$$

$$D_y = \begin{vmatrix} 1 & 2 & 1 \\ 2 & 0 & 1 \\ 1 & 4 & -1 \end{vmatrix}$$

$$= 1\begin{vmatrix} 0 & 1 \\ 4 & -1 \end{vmatrix} - 2\begin{vmatrix} 2 & 1 \\ 4 & -1 \end{vmatrix} + 1\begin{vmatrix} 2 & 1 \\ 0 & 1 \end{vmatrix}$$

$$= 1(-4) - 2(-6) + 1(2) = 10$$

$$D_z = \begin{vmatrix} 1 & 1 & 2 \\ 2 & -1 & 0 \\ 1 & 2 & 4 \end{vmatrix}$$

$$= 1\begin{vmatrix} -1 & 0 \\ 2 & 4 \end{vmatrix} - 2\begin{vmatrix} 1 & 2 \\ 2 & 4 \end{vmatrix} + 1\begin{vmatrix} 1 & 2 \\ -1 & 0 \end{vmatrix}$$

$$= 1(-4) - 2(0) + 1(2) = -2.$$

Then we have

$$x = \frac{D_x}{D} = \frac{6}{7}, \qquad y = \frac{D_y}{D} = \frac{10}{7}, \qquad z = \frac{D_z}{D} = -\frac{2}{7}.$$

PROBLEM SET 10.5

In problems 1–20, use Cramer's rule to solve each system.

1. $\begin{cases} 2x - y = 0 \\ x + y = 1 \end{cases}$
    2. $\begin{cases} -3x + y = 3 \\ -2x - y = -5 \end{cases}$
    3. $\begin{cases} u + v = 0 \\ u - v = 0 \end{cases}$

4. $\begin{cases} 3t + s = 1 \\ 9t + 3s = -4 \end{cases}$
    5. $\begin{cases} p + q = 30 \\ 2p - 2q = 25 \end{cases}$
    6. $\begin{cases} 7x - 9y = 13 \\ 5x + 2y = 10 \end{cases}$

7. $\begin{cases} 3x + 7y = 16 \\ 2x + 5y = 13 \end{cases}$
    8. $\begin{cases} 7m + 4n = 1 \\ 9m + 4n = 3 \end{cases}$
    9. $\begin{cases} 8z - 2w = 52 \\ 3z - 5w = 45 \end{cases}$

10. $\begin{cases} 5x + 11y - 102 = 0 \\ x - 3y + 16 = 0 \end{cases}$
    11. $\begin{cases} x + y + z = 6 \\ 3x - y + 2z = 7 \\ 2x + 3y - z = 5 \end{cases}$
    12. $\begin{cases} u + v + w = 9 \\ 27u + 9v + 3w = 93 \\ 8u + 4v + 2w = 36 \end{cases}$

13. $\begin{cases} 2r - s + t = 3 \\ -r + 2s - t = 1 \\ 3r + s + 2t = -1 \end{cases}$ **14.** $\begin{cases} x + y + 2z = 4 \\ x + y - 2z = 0 \\ x - y = 0 \end{cases}$ **15.** $\begin{cases} 2x - 3y = 4 \\ x + y - 2z = 1 \\ x - y - z = 5 \end{cases}$

16. $\begin{cases} a + b + c = 4 \\ a - b + 2c = 8 \\ 2a + b - c = 3 \end{cases}$ **17.** $\begin{cases} 2u + 3v + w = 6 \\ u - 2v + 3w = -3 \\ 3u + v - w = 8 \end{cases}$ **18.** $\begin{cases} 3r + 2s + 2t = 6 \\ r - 5s + 6t = 2 \\ 6r - 8s = 12 \end{cases}$

19. $\begin{cases} x + y = 1 \\ y + z = 9 \\ x + z = 0 \end{cases}$ **20.** $\begin{cases} x - y + z = 3 \\ 2x + 3y - 2z = 5 \\ 3x + y - 4z = 12 \end{cases}$

REVIEW PROBLEM SET

In problems 1–6, write the first four terms of each sequence.

1. $a_n = 3 + (-1)^{n+1}$ **2.** $b_n = 2^n$ **3.** $c_n = 5 - \dfrac{3}{n}$ **4.** $a_n = \dfrac{2}{n+1}$

5. $a_n = \dfrac{n(4n+1)}{5}$ **6.** $c_n = (-1)^n 2^{n-1}$

In problems 7–12, for each arithmetic progression, find the indicated term and the indicated sum.

7. $4, 9, 14, \ldots$; ninth term and S_9

8. $21, 19, 17, \ldots$; tenth term and S_{10}

9. $42, 39, 36, \ldots$; eleventh term and S_{11}

10. $0.3, 1.2, 2.1, \ldots$; fifteenth term and S_{15}

11. $\frac{1}{6}, \frac{1}{3}, \frac{1}{2}, \ldots$; twenty-fourth term and S_{24}

12. $\frac{1}{6}, \frac{1}{4}, \frac{1}{3}, \ldots$; thirtieth term and S_{30}

In problems 13–16, find the value of x so that each will be the first three terms of an arithmetic progression.

13. $2, 1 + 2x, 21 - 3x, \ldots$ **14.** $2x, \frac{1}{2}x + 3, 3x - 10, \ldots$ **15.** $3x, 2x + 1, x^2 - 4, \ldots$

16. $1, x + 1, 3x - 5, \ldots$

In problems 17–22, for each geometric progression, find the indicated term and the indicated sum.

17. $3, 12, 48, \ldots$; eighth term and S_8

18. $16, 8, 4, \ldots$; ninth term and S_9

19. $81, -27, 9, \ldots$; sixth term and S_6

20. $\sqrt{2}, 2, 2\sqrt{2}, \ldots$; tenth term and S_{10}

21. $3, -3\sqrt{2}, 6, \ldots$; eighteenth term and S_{18}

22. $2, -2\sqrt{2}, 4, \ldots$; twentieth term and S_{20}

In problems 23–26, determine the value of x so that the three terms given will be the first three terms of a geometric progression.

23. $x - 6, x + 6, 2x + 2, \ldots$ **24.** $\frac{1}{2}x, x + 2, 3x + 1, \ldots$ **25.** $x - 7, x + 5, 8x - 5, \ldots$

26. $x + 1, x + 2, x - 3, \ldots$

In problems 27–32, evaluate each sum.

27. $\displaystyle\sum_{k=1}^{5} k(2k - 1)$ **28.** $\displaystyle\sum_{k=5}^{10} (2k - 1)^2$ **29.** $\displaystyle\sum_{k=1}^{4} 2k^2(k - 3)$ **30.** $\displaystyle\sum_{k=1}^{6} 3^{k+1}$

31. $\displaystyle\sum_{k=2}^{6} (k + 1)(k + 2)$ **32.** $\displaystyle\sum_{k=4}^{7} \frac{1}{k(k - 3)}$

In problems 33–44, use the binomial theorem to expand each expression.

33. $(x + 2y)^4$ **34.** $(x - 3y)^4$ **35.** $(1 + x)^5$ **36.** $(2x + 1)^5$

37. $(1 - 2x)^6$ **38.** $(a - b)^6$ **39.** $(3x + y)^4$ **40.** $\left(x - \dfrac{1}{x}\right)^8$

41. $(3x + \sqrt{x})^5$ **42.** $\left(3y + \dfrac{1}{3\sqrt{y}}\right)^6$ **43.** $\left(2x + \dfrac{1}{y}\right)^3$ **44.** $\left(x^3 - \dfrac{1}{\sqrt{x}}\right)^9$

In problems 45–50, find the indicated term in each binomial expansion.

45. fifth term of $(x + y)^{10}$ **46.** sixth term of $(x - y)^{11}$ **47.** fifth term of $(2x + y)^{10}$
48. sixth term of $(x - 3y)^9$ **49.** fourth term of $(3x + y)^{11}$ **50.** third term of $(2x + y)^{20}$

In problems 51–54, evaluate each determinant.

51. $\begin{vmatrix} 3 & 4 \\ 1 & 5 \end{vmatrix}$ **52.** $\begin{vmatrix} 4 & 2 \\ 1 & -1 \end{vmatrix}$ **53.** $\begin{vmatrix} 4 & 2 & 1 \\ 5 & 7 & 1 \\ 6 & 2 & 3 \end{vmatrix}$ **54.** $\begin{vmatrix} -1 & 4 & 3 \\ 7 & 1 & 4 \\ 1 & 3 & 5 \end{vmatrix}$

In problems 55–56, solve for x.

55. $\begin{vmatrix} x - 1 & -3 \\ 2 & x + 3 \end{vmatrix} = 6$ **56.** $\begin{vmatrix} x + 1 & 3 \\ 4 & x \end{vmatrix} < 0$

In problems 57–60, use Cramer's rule to solve each system.

57. $\begin{cases} x - y = 3 \\ 2x + y = 3 \end{cases}$ **58.** $\begin{cases} 5p + 2q = 3 \\ 2p - 3q = 5 \end{cases}$ **59.** $\begin{cases} x - y + 2z = 0 \\ 3x + y + z = 2 \\ 2x - y + 5z = 5 \end{cases}$

60. $\begin{cases} 3r + 2s - t = -4 \\ r - s + 2t = 13 \\ 5r + 3s - 4t = -15 \end{cases}$

Appendices

TABLE I COMMON LOGARITHMS

x	0.00	0.01	0.02	0.03	0.04	0.05	0.06	0.07	0.08	0.09
1.0	0.0000	0.0043	0.0086	0.0128	0.0170	0.0212	0.0253	0.0294	0.0334	0.0374
1.1	0.0414	0.0453	0.0492	0.0531	0.0569	0.0607	0.0645	0.0682	0.0719	0.0755
1.2	0.0792	0.0828	0.0864	0.0899	0.0934	0.0969	0.1004	0.1038	0.1072	0.1106
1.3	0.1139	0.1173	0.1206	0.1239	0.1271	0.1303	0.1335	0.1367	0.1399	0.1430
1.4	0.1461	0.1492	0.1523	0.1553	0.1584	0.1614	0.1644	0.1673	0.1703	0.1732
1.5	0.1761	0.1790	0.1818	0.1847	0.1875	0.1903	0.1931	0.1959	0.1987	0.2014
1.6	0.2041	0.2068	0.2095	0.2122	0.2148	0.2175	0.2201	0.2227	0.2253	0.2279
1.7	0.2304	0.2330	0.2355	0.2380	0.2405	0.2430	0.2455	0.2480	0.2504	0.2529
1.8	0.2553	0.2577	0.2601	0.2625	0.2648	0.2672	0.2695	0.2718	0.2742	0.2765
1.9	0.2788	0.2810	0.2833	0.2856	0.2878	0.2900	0.2923	0.2945	0.2967	0.2989
2.0	0.3010	0.3032	0.3054	0.3075	0.3096	0.3118	0.3139	0.3160	0.3181	0.3201
2.1	0.3222	0.3243	0.3263	0.3284	0.3304	0.3324	0.3345	0.3365	0.3385	0.3404
2.2	0.3424	0.3444	0.3464	0.3483	0.3502	0.3522	0.3541	0.3560	0.3579	0.3598
2.3	0.3617	0.3636	0.3655	0.3674	0.3692	0.3711	0.3729	0.3747	0.3766	0.3784
2.4	0.3802	0.3820	0.3838	0.3856	0.3874	0.3892	0.3909	0.3927	0.3945	0.3962
2.5	0.3979	0.3997	0.4014	0.4031	0.4048	0.4065	0.4082	0.4099	0.4116	0.4133
2.6	0.4150	0.4166	0.4183	0.4200	0.4216	0.4232	0.4249	0.4265	0.4281	0.4298
2.7	0.4314	0.4330	0.4346	0.4362	0.4378	0.4393	0.4409	0.4425	0.4440	0.4456
2.8	0.4472	0.4487	0.4502	0.4518	0.4533	0.4548	0.4564	0.4579	0.4594	0.4609
2.9	0.4624	0.4639	0.4654	0.4669	0.4683	0.4698	0.4713	0.4728	0.4742	0.4757
3.0	0.4771	0.4786	0.4800	0.4814	0.4829	0.4843	0.4857	0.4871	0.4886	0.4900
3.1	0.4914	0.4928	0.4942	0.4955	0.4969	0.4983	0.4997	0.5011	0.5024	0.5038
3.2	0.5051	0.5065	0.5079	0.5092	0.5105	0.5119	0.5132	0.5145	0.5159	0.5172
3.3	0.5185	0.5198	0.5211	0.5224	0.5237	0.5250	0.5263	0.5276	0.5289	0.5302
3.4	0.5315	0.5328	0.5340	0.5353	0.5366	0.5378	0.5391	0.5403	0.5416	0.5428
3.5	0.5441	0.5453	0.5465	0.5478	0.5490	0.5502	0.5514	0.5527	0.5539	0.5551
3.6	0.5563	0.5575	0.5587	0.5599	0.5611	0.5623	0.5635	0.5647	0.5658	0.5670
3.7	0.5682	0.5694	0.5705	0.5717	0.5729	0.5740	0.5752	0.5763	0.5775	0.5786
3.8	0.5798	0.5809	0.5821	0.5832	0.5843	0.5855	0.5866	0.5877	0.5888	0.5899
3.9	0.5911	0.5922	0.5933	0.5944	0.5955	0.5966	0.5977	0.5988	0.5999	0.6010
4.0	0.6021	0.6031	0.6042	0.6053	0.6064	0.6075	0.6085	0.6096	0.6107	0.6117
4.1	0.6128	0.6138	0.6149	0.6160	0.6170	0.6180	0.6191	0.6201	0.6212	0.6222
4.2	0.6232	0.6243	0.6253	0.6263	0.6274	0.6284	0.6294	0.6304	0.6314	0.6325
4.3	0.6335	0.6345	0.6355	0.6365	0.6375	0.6385	0.6395	0.6405	0.6415	0.6425
4.4	0.6435	0.6444	0.6454	0.6464	0.6474	0.6484	0.6493	0.6503	0.6513	0.6522
4.5	0.6532	0.6542	0.6551	0.6561	0.6571	0.6580	0.6590	0.6599	0.6609	0.6618
4.6	0.6628	0.6637	0.6646	0.6656	0.6665	0.6675	0.6684	0.6693	0.6702	0.6712
4.7	0.6721	0.6730	0.6739	0.6749	0.6758	0.6767	0.6776	0.6785	0.6794	0.6803
4.8	0.6812	0.6821	0.6830	0.6839	0.6848	0.6857	0.6866	0.6875	0.6884	0.6893
4.9	0.6902	0.6911	0.6920	0.6928	0.6937	0.6946	0.6955	0.6964	0.6972	0.6981
5.0	0.6990	0.6998	0.7007	0.7016	0.7024	0.7033	0.7042	0.7050	0.7059	0.7067
5.1	0.7076	0.7084	0.7093	0.7101	0.7110	0.7118	0.7126	0.7135	0.7143	0.7152
5.2	0.7160	0.7168	0.7177	0.7185	0.7193	0.7202	0.7210	0.7218	0.7226	0.7235
5.3	0.7243	0.7251	0.7259	0.7267	0.7275	0.7284	0.7292	0.7300	0.7308	0.7316
5.4	0.7324	0.7332	0.7340	0.7348	0.7356	0.7364	0.7372	0.7380	0.7388	0.7396
5.5	0.7404	0.7412	0.7419	0.7427	0.7435	0.7443	0.7451	0.7459	0.7466	0.7474
5.6	0.7482	0.7490	0.7497	0.7505	0.7513	0.7520	0.7528	0.7536	0.7543	0.7551
5.7	0.7559	0.7566	0.7574	0.7582	0.7589	0.7597	0.7604	0.7612	0.7619	0.7627
5.8	0.7634	0.7642	0.7649	0.7657	0.7664	0.7672	0.7679	0.7686	0.7694	0.7701
5.9	0.7709	0.7716	0.7723	0.7731	0.7738	0.7745	0.7752	0.7760	0.7767	0.7774

x	0.00	0.01	0.02	0.03	0.04	0.05	0.06	0.07	0.08	0.09
6.0	0.7782	0.7789	0.7796	0.7803	0.7810	0.7818	0.7825	0.7832	0.7839	0.7846
6.1	0.7853	0.7860	0.7868	0.7875	0.7882	0.7889	0.7896	0.7903	0.7910	0.7917
6.2	0.7924	0.7931	0.7938	0.7945	0.7952	0.7959	0.7966	0.7973	0.7980	0.7987
6.3	0.7993	0.8000	0.8007	0.8014	0.8021	0.8028	0.8035	0.8041	0.8048	0.8055
6.4	0.8062	0.8069	0.8075	0.8082	0.8089	0.8096	0.8102	0.8109	0.8116	0.8122
6.5	0.8129	0.8136	0.8142	0.8149	0.8156	0.8162	0.8169	0.8176	0.8182	0.8189
6.6	0.8195	0.8202	0.8209	0.8215	0.8222	0.8228	0.8235	0.8241	0.8248	0.8254
6.7	0.8261	0.8267	0.8274	0.8280	0.8287	0.8293	0.8299	0.8306	0.8312	0.8319
6.8	0.8325	0.8331	0.8338	0.8344	0.8351	0.8357	0.8363	0.8370	0.8376	0.8382
6.9	0.8388	0.8395	0.8401	0.8407	0.8414	0.8420	0.8426	0.8432	0.8439	0.8445
7.0	0.8451	0.8457	0.8463	0.8470	0.8476	0.8482	0.8488	0.8494	0.8500	0.8506
7.1	0.8513	0.8519	0.8525	0.8531	0.8537	0.8543	0.8549	0.8555	0.8561	0.8567
7.2	0.8573	0.8579	0.8585	0.8591	0.8597	0.8603	0.8609	0.8615	0.8621	0.8627
7.3	0.8633	0.8639	0.8645	0.8651	0.8657	0.8663	0.8669	0.8675	0.8681	0.8686
7.4	0.8692	0.8698	0.8704	0.8710	0.8716	0.8722	0.8727	0.8733	0.8739	0.8745
7.5	0.8751	0.8756	0.8762	0.8768	0.8774	0.8779	0.8785	0.8791	0.8797	0.8802
7.6	0.8808	0.8814	0.8820	0.8825	0.8831	0.8837	0.8842	0.8848	0.8854	0.8859
7.7	0.8865	0.8871	0.8876	0.8882	0.8887	0.8893	0.8899	0.8904	0.8910	0.8915
7.8	0.8921	0.8927	0.8932	0.8938	0.8943	0.8949	0.8954	0.8960	0.8965	0.8971
7.9	0.8976	0.8982	0.8987	0.8993	0.8998	0.9004	0.9009	0.9015	0.9020	0.9025
8.0	0.9031	0.9036	0.9042	0.9047	0.9053	0.9058	0.9063	0.9069	0.9074	0.9079
8.1	0.9085	0.9090	0.9096	0.9101	0.9106	0.9112	0.9117	0.9122	0.9128	0.9133
8.2	0.9138	0.9143	0.9149	0.9154	0.9159	0.9165	0.9170	0.9175	0.9180	0.9186
8.3	0.9191	0.9196	0.9201	0.9206	0.9212	0.9217	0.9222	0.9227	0.9232	0.9238
8.4	0.9243	0.9248	0.9253	0.9258	0.9263	0.9269	0.9274	0.9279	0.9284	0.9289
8.5	0.9294	0.9299	0.9304	0.9309	0.9315	0.9320	0.9325	0.9330	0.9335	0.9340
8.6	0.9345	0.9350	0.9355	0.9360	0.9365	0.9370	0.9375	0.9380	0.9385	0.9390
8.7	0.9395	0.9400	0.9405	0.9410	0.9415	0.9420	0.9425	0.9430	0.9435	0.9440
8.8	0.9445	0.9450	0.9455	0.9460	0.9465	0.9469	0.9474	0.9479	0.9484	0.9489
8.9	0.9494	0.9499	0.9504	0.9509	0.9513	0.9518	0.9523	0.9528	0.9533	0.9538
9.0	0.9542	0.9547	0.9552	0.9557	0.9562	0.9566	0.9571	0.9576	0.9581	0.9586
9.1	0.9590	0.9595	0.9600	0.9605	0.9609	0.9614	0.9619	0.9624	0.9628	0.9633
9.2	0.9638	0.9643	0.9647	0.9652	0.9657	0.9661	0.9666	0.9671	0.9675	0.9680
9.3	0.9685	0.9689	0.9694	0.9699	0.9703	0.9708	0.9713	0.9717	0.9722	0.9727
9.4	0.9731	0.9736	0.9741	0.9745	0.9750	0.9754	0.9759	0.9763	0.9768	0.9773
9.5	0.9777	0.9782	0.9786	0.9791	0.9795	0.9800	0.9805	0.9809	0.9814	0.9818
9.6	0.9823	0.9827	0.9832	0.9836	0.9841	0.9845	0.9850	0.9854	0.9859	0.9863
9.7	0.9868	0.9872	0.9877	0.9881	0.9886	0.9890	0.9894	0.9899	0.9903	0.9908
9.8	0.9912	0.9917	0.9921	0.9926	0.9930	0.9934	0.9939	0.9943	0.9948	0.9952
9.9	0.9956	0.9961	0.9965	0.9969	0.9974	0.9978	0.9983	0.9987	0.9991	0.9996

TABLE II   POWERS AND ROOTS

Number	Square	Square Root	Cube	Cube Root	Number	Square	Square Root	Cube	Cube Root
1	1	1.000	1	1.000	51	2,601	7.141	132,651	3.708
2	4	1.414	8	1.260	52	2,704	7.211	140,608	3.733
3	9	1.732	27	1.442	53	2,809	7.280	148,877	3.756
4	16	2.000	64	1.587	54	2,916	7.348	157,464	3.780
5	25	2.236	125	1.710	55	3,025	7.416	166,375	3.803
6	36	2.449	216	1.817	56	3,136	7.483	175,616	3.826
7	49	2.646	343	1.913	57	3,249	7.550	185,193	3.849
8	64	2.828	512	2.000	58	3,364	7.616	195,112	3.871
9	81	3.000	729	2.080	59	3,481	7.681	205,379	3.893
10	100	3.162	1,000	2.154	60	3,600	7.746	216,000	3.915
11	121	3.317	1,331	2.224	61	3,721	7.810	226,981	3.936
12	144	3.464	1,728	2.289	62	3,844	7.874	238,328	3.958
13	169	3.606	2,197	2.351	63	3,969	7.937	250,047	3.979
14	196	3.742	2,744	2.410	64	4,096	8.000	262,144	4.000
15	225	3.873	3,375	2.466	65	4,225	8.062	274,625	4.021
16	256	4.000	4,096	2.520	66	4,356	8.124	287,496	4.041
17	289	4.123	4,913	2.571	67	4,489	8.185	300,763	4.062
18	324	4.243	5,832	2.621	68	4,624	8.246	314,432	4.082
19	361	4.359	6,859	2.668	69	4,761	8.307	328,509	4.102
20	400	4.472	8,000	2.714	70	4,900	8.367	343,000	4.121
21	441	4.583	9,261	2.759	71	5,041	8.426	357,911	4.141
22	484	4.690	10,648	2.802	72	5,184	8.485	373,248	4.160
23	529	4.796	12,167	2.844	73	5,329	8.544	389,017	4.179
24	576	4.899	13,824	2.884	74	5,476	8.602	405,224	4.198
25	625	5.000	15,625	2.924	75	5,625	8.660	421,875	4.217
26	676	5.099	17,576	2.962	76	5,776	8.718	438,976	4.236
27	729	5.196	19,683	3.000	77	5,929	8.775	456,533	4.254
28	784	5.292	21,952	3.037	78	6,084	8.832	474,552	4.273
29	841	5.385	24,389	3.072	79	6,241	8.888	493,039	4.291
30	900	5.477	27,000	3.107	80	6,400	8.944	512,000	4.309
31	961	5.568	29,791	3.141	81	6,561	9.000	531,441	4.327
32	1,024	5.657	32,768	3.175	82	6,724	9.055	551,368	4.344
33	1,089	5.745	35,937	3.208	83	6,889	9.110	571,787	4.362
34	1,156	5.831	39,304	3.240	84	7,056	9.165	592,704	4.380
35	1,225	5.916	42,875	3.271	85	7,225	9.220	614,125	4.397
36	1,296	6.000	46,656	3.302	86	7,396	9.274	636,056	4.414
37	1,369	6.083	50,653	3.332	87	7,569	9.327	658,503	4.431
38	1,444	6.164	54,872	3.362	88	7,744	9.381	681,472	4.448
39	1,521	6.245	59,319	3.391	89	7,921	9.434	704,969	4.465
40	1,600	6.325	64,000	3.420	90	8,100	9.498	729,000	4.481
41	1,681	6.403	68,921	3.448	91	8,281	9.539	753,571	4.498
42	1,764	6.481	74,088	3.476	92	8,464	9.592	778,688	4.514
43	1,849	6.557	79,507	3.503	93	8,649	9.644	804,357	4.531
44	1,936	6.633	85,184	3.530	94	8,836	9.695	830,584	4.547
45	2,025	6.708	91,125	3.557	95	9,025	9.747	857,375	4.563
46	2,116	6.782	97,336	3.583	96	9,216	9.798	884,736	4.579
47	2,209	6.856	103,823	3.609	97	9,409	9.849	912,673	4.595
48	2,304	6.928	110,592	3.634	98	9,604	9.899	941,192	4.610
49	2,401	7.000	117,649	3.659	99	9,801	9.950	970,299	4.626
50	2,500	7.071	125,000	3.684	100	10,000	10.000	1,000,000	4.642

Answers to Selected Problems

Chapter 1

PROBLEM SET 1.1   page 5

1. 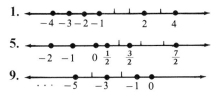   **3.**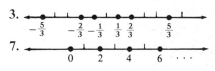

5.   **7.**

9.   **11.** 0.6   **13.** 1.5   **15.** 0.8

17. -1.25 **19.** $-2.\overline{3}$ **21.** $\frac{27}{100}$ **23.** $\frac{66}{25}$ **25.** $-\frac{1}{8}$ **27.** $\frac{527}{10,000}$ **29.** $-\frac{329}{100,000}$ **31.** rational
33. irrational   **35.** rational   **37.** irrational   **39.** rational   **41.** rational   **43.** rational

PROBLEM SET 1.2   page 9

1. 32   **3.** 72   **5.** 1.1   **7.** 0.328   **9.** 15   **11.** 120   **13.** 14.2   **15.** 0.134   **17.** 39   **19.** 48   **21.** 9.45
23. Commutative property for addition   **25.** Associative property for multiplication
27. Distributive property   **29.** Multiplicative inverse   **31.** Negative property (iii)
33. Zero-Factor property (ii)   **35.** Cancellation property for multiplication   **37.** Zero-Factor property (i)
39. Commutative property for addition   **41.** Identity property for addition

PROBLEM SET 1.3   page 18

1. 3   **3.** 11   **5.** -16 **7.** -21 **9.** $-\frac{5}{7}$ **11.** 4   **13.** -14 **15.** -12 **17.** 0   **19.** -7 **21.** -12
23. -53 **25.** 1.5   **27.** -18.2 **29.** 37   **31.** 20   **33.** -15 **35.** -16 **37.** 12   **39.** 0.059   **41.** 8
43. -94 **45.** 154   **47.** -15 **49.** 16   **51.** -2.7 **53.** 0   **55.** -24 **57.** 252   **59.** 30   **61.** 210
63. -2 **65.** -2 **67.** 3   **69.** -13 **71.** 15   **73.** $-1,500$ **75.** 5   **77.** 4   **79.** 5   **81.** 10   **83.** 7
85. 8   **87.** 5.2   **89.** 14   **91.** 9

PROBLEM SET 1.4   page 22

1. 0.13793103   **3.** 0.88235294   **5.** 5.0990195   **7.** 8.4439327   **9.** 0.3   **11.** 5.3   **13.** 8.0   **15.** 15.0
17. 24.1   **19.** 3.19   **21.** 14.36   **23.** 21.00   **25.** 16.51   **27.** 23.70   **29.** 1.73   **31.** 14.3   **33.** 368
35. 5140   **37.** 28.0   **39.** 1.33   **41.** 0.143   **43.** 2.38   **45.** 1.83   **47.** 5.48   **49.** 1.732   **51.** 2.646
53. 4.796   **55.** 6.856   **57.** 10.54

PROBLEM SET 1.5   page 26

1. 5^3 **3.** x^4 **5.** $(-t)^5$ **7.** u^3v^5 **9.** $3^2x^4y^2$ **11.** $8 \cdot 8 \cdot 8 \cdot 8$ **13.** $y \cdot y \cdot y \cdot y \cdot y$ **15.** $5 \cdot 5 \cdot 5 \cdot t \cdot t \cdot t \cdot t$
17. $(-x) \cdot (-x) \cdot (-x) \cdot (-x) \cdot y \cdot y \cdot y$ **19.** $4 \cdot 4 \cdot 4 \cdot u \cdot u \cdot u \cdot u \cdot v$ **21.** $32°$ **23.** $50°$ **25.** $68°$
27. 35 square inches   **29.** 24 square meters   **31.** 150 square feet   **33.** 22 feet   **35.** 52 inches
37. 23.4 meters   **39.** 314 square inches   **41.** 78.5 square feet   **43.** 120.7 square meters   **45.** 150 miles
47. 990 feet   **49.** 7,800 centimeters   **51.** 81 square inches   **53.** 16 square feet   **55.** 225 square inches
57. 27 cubic inches   **59.** 216 cubic feet   **61.** 1,000 cubic centimeters

REVIEW PROBLEM SET   page 28

1.

3.

5. 1.4   **7.** 3.125   **9.** $\frac{17}{50}$ **11.** $\frac{23}{125}$ **13.** $-\frac{3,407}{500}$ **15.** rational   **17.** irrational   **19.** rational
21. irrational   **23.** rational   **25.** rational   **27.** Commutative property for addition
29. Commutative property for multiplication   **31.** Associative property for multiplication
33. Identity property for addition   **35.** Zero-Factor property (i)   **37.** Cancellation property for addition
39. Distributive property   **41.** Identity property for addition   **43.** 13   **45.** -21 **47.** -3 **49.** -14
51. 14   **53.** 30   **55.** -12 **57.** 48   **59.** -40 **61.** -32 **63.** 105   **65.** -9 **67.** 8   **69.** -3
71. 16   **73.** 7   **75.** 14   **77.** 1.08   **79.** 3.82   **81.** 0.04   **83.** 1.035   **85.** 1.013
87. 29.30   **89.** 0.714   **91.** 0.432   **93.** 7.28   **95.** 8.44   **97.** 0.0442   **99.** 0.0177   **101.** $2^2 \cdot x^3$
103. $5u^2v^3$ **105.** $(-2)^2(-t)^2(-s)^2$ **107.** $x \cdot x \cdot y \cdot y \cdot y \cdot y$ **109.** $3 \cdot 3 \cdot 3 \cdot u \cdot u \cdot v \cdot v \cdot v \cdot v$
111. $m \cdot \dot{m} \cdot m \cdot n \cdot n \cdot n \cdot n \cdot p$ **113.** $86°$ **115.** 30 square inches   **117.** 14 feet   **119.** 153.86 square inches

Chapter 2

PROBLEM SET 2.1   page 34

1. Binomial; degree = 1; 3, -2 **3.** Monomial; degree = 2; 4   **5.** Trinomial; degree = 2; 1, $-5, 6$
7. Binomial; degree = 7; 2, -13 **9.** Trinomial; degree = 4; $-1, -1, 13$ **11.** 3   **13.** 2   **15.** 5
17. 6   **19.** 9   **21.** 7   **23.** 19   **25.** -1 **27.** 2   **29.** 20   **31.** 4   **33.** 10   **35.** 0   **37.** No
39. Yes

PROBLEM SET 2.2   page 39

1. $12x^2$ **3.** $11v^3$ **5.** $4t^2$ **7.** $8x + 7$ **9.** $7z^2 + 5z + 5$ **11.** $4x^2 - 2x - 2$ **13.** $5c^4 + c^3 + 4c^2 + c$
15. $4u$ **17.** $2x^2$ **19.** $6v^3$ **21.** $4t^3 - 2$ **23.** $2x^2 + 2x + 4$ **25.** $4s^4 - s^3 + 4s^2 + 2s - 5$
27. $-4t^3 + 5t^2 + 4t - 18$ **29.** $12xy$ **31.** $3uv$ **33.** $2mn^2 + 2m^2n$ **35.** $-2x^2y$ **37.** $4ts - 6$
39. $-6x^2y + 2xy - 4xy^2$ **41.** $5w^2z + 3wz - 10wz^2$ **43.** $5x^2 + 2x - 2$ **45.** $2t^3 + 2t^2 - t + 2$
47. $w^2 + w + 9$ **49.** $10u^3v^2 - 9u^2v - 5w$

PROBLEM SET 2.3 page 45

1. 3^5 **3.** -32 **5.** x^{12} **7.** t^{12} **9.** v^8 **11.** 64 **13.** 64 **15.** x^{35} **17.** t^{22} **19.** w^{12} **21.** $16x^4$

23. u^5v^5 **25.** $x^7y^7z^7$ **27.** $-8w^3$ **29.** $27x^3y^3$ **31.** $\frac{9}{16}$ **33.** $-\frac{8}{27}$ **35.** $\frac{x^4}{y^4}$ **37.** $-\frac{a^5}{b^5}$ **39.** $\frac{t^6}{s^6}$

41. 27 **43.** 64 **45.** x^5 **47.** y^5 **49.** w^3 **51.** $81x^8y^{12}$ **53.** u^3v^4 **55.** $\frac{w^6}{8z^3}$ **57.** $\frac{27a^9}{8b^9}$

59. $-\frac{16y^{10}z^9}{x}$ **61.** 1 **63.** -1

PROBLEM SET 2.4 page 55

1. $6x^6$ **3.** $-30t^7$ **5.** $-28u^5v^7$ **7.** $12x^3y^3z^4$ **9.** $-24a^3b^3c^4$ **11.** $x^2 + x$ **13.** $t^3 + 2t^2$
15. $6w^3 - 12w$ **17.** $-4x^3y^2 + 6x^2y^3 - 10xy^4$ **19.** $12c^5d^2 - 8c^4d^3 + 4c^3d^4$
21. $x^2 + 2xy + y^2 - x - y$ **23.** $2t^3 + 5t^2s + 4ts^2 + s^3$ **25.** $m^5 + 2m^4 + 10m^2 - 9m + 12$
27. $y^4 - y^3 - 10y^2 + 4y + 24$ **29.** $x^5 - x^4y - 2x^3y^2 + 2x^2y^3 + xy^4 - y^5$ **31.** $x^2 + 3x + 2$
33. $u^2 - 9u + 20$ **35.** $t^2 + 3t - 10$ **37.** $y^2 - 3y - 18$ **39.** $6x^2 + x - 1$ **41.** $5w^2 - w - 4$
43. $2x^2 + 7xy + 3y^2$ **45.** $24m^2 - 2mn - 15n^2$ **47.** $28x^2 - 23xy - 15y^2$ **49.** $50v^2 - 115v + 56$
51. $x^2 + 2x + 1$ **53.** $4s^2 + 4st + t^2$ **55.** $u^2 - 6uv + 9v^2$ **57.** $9x^2 - 30x + 25$
59. $16y^2 + 40yz + 25z^2$ **61.** $x^2 - y^2$ **63.** $w^2 - 49$ **65.** $4m^2 - 81$ **67.** $64x^2 - y^2$ **69.** $25u^2 - 36v^2$
75. $x^3 + 3x^2 + 3x + 1$ **77.** $x^3 + 1$ **79.** $c^3 - 6c^2d + 12cd^2 - 8d^3$ **81.** $u^3 - 27v^3$
83. $125t^3 + 150t^2s + 60ts^2 + 8s^3$ **85.** $8x^3 + 27y^3$ **87.** $x^9 - 6x^6 + 12x^3 - 8$ **89.** $u^6 - v^6$

PROBLEM SET 2.5 page 59

1. $x(x - 1)$ **3.** $3x(3x + 1)$ **5.** $x(4x + 7y)$ **7.** $ab(a - b)$ **9.** $6pq(p + 4q)$ **11.** $6ab(b + 5a)$
13. $12x^2y(x - 4y)$ **15.** $2ab(a^2 - 4ab - 3b^2)$ **17.** $xy^2(x^2 + xy + 2y^2)$ **19.** $9mn(m + 2n - 3)$
21. $(3x + 5y)(2a + b)$ **23.** $(5x + 9ay + 9by)(a + b)$ **25.** $(m - 1)(x - y)$
27. $[7x + 14(2a + 7b) + (2a + 7b)^2](2a + 7b)$ **29.** $(xy + 2)[y(xy + 2)^2 - 5x(xy + 2) + 7]$
31. $(a + b)(x + y)$ **33.** $(x^4 + 1)(x + 3)$ **35.** $(y - 1)(z + 2)$ **37.** $(b^2 - d)(a - c)$
39. $(2a - b)(x - y)$ **41.** $(x - a)(x + b)$ **43.** $(a + b)(x + y + 1)$ **45.** $(x^2 + y)(2x + y - 1)$

PROBLEM SET 2.6 page 64

1. $(x - 2)(x + 2)$ **3.** $(1 - 3y)(1 + 3y)$ **5.** $(6 - 5t)(6 + 5t)$ **7.** $(4u - 5v)(4u + 5v)$
9. $(ab - c)(ab + c)$ **11.** $(a - b - 10c)(a - b + 10c)$ **13.** $(3x - 1)(3x + 1)(9x^2 + 1)$
15. $(u - v)(u + v)(u^2 + v^2)(u^4 + v^4)$ **17.** $(x + y - a + b)(x + y + a - b)$
19. $[t - 3(r + s)][t + 3(r + s)][t^2 + 9(r + s)^2]$ **21.** $(x + 1)(x^2 - x + 1)$ **23.** $(4 - t)(16 + 4t + t^2)$
25. $(3w + z)(9w^2 - 3wz + z^2)$ **27.** $(2x - 3y)(4x^2 + 6xy + 9y^2)$ **29.** $(w - 2yz)(w^2 + 2wyz + 4y^2z^2)$
31. $(x + 2 - y)[(x + 2)^2 + (x + 2)y + y^2]$ **33.** $(y + w + 3)[(y + 1)^2 - (y + 1)(w + 2) + (w + 2)^2]$
35. $(w^2 + 2z^2)(w^4 - 2w^2z^2 + 4z^4)$ **37.** $(x - 1)(x^2 + x + 1)(x^6 + x^3 + 1)$ **39.** $2x(2x - y)(2x + y)$

41. $4y(4 - y)(4 + y)$ **43.** $3uv(u - 2)(u^2 + 2u + 4)$ **45.** $7xy(x^2 + y^2)(x^4 - x^2y^2 + y^4)$
47. $(t - 1)(t + 1)(t^2 + t + 1)(t^2 - t + 1)$ **49.** $2u(u - 2)(u + 2)(u^2 + 2u + 4)(u^2 - 2u + 4)$
51. $(y + 4)^2$ **53.** $(3u - 7v)^2$ **55.** $(x + y - z - 3)(x + y + z + 3)$
57. $(w + 1 - y - z)(w + 1 + y + z)$ **59.** $(x^2 + y^2 - xy)(x^2 + y^2 + xy)$
61. $(2m^2 + n^2 - 2mn)(2m^2 + n^2 + 2mn)$

PROBLEM SET 2.7   page 70

1. $(x + 1)(x + 3)$ **3.** $(t - 1)(t - 2)$ **5.** $(y + 3)(y + 12)$ **7.** $(x + 3)(x - 5)$ **9.** $(u - 7)(u - 9)$
11. $(z + 5w)(z + 6w)$ **13.** $(x + 2)(x - 9)$ **15.** $(m - 10n)(m + 12n)$ **17.** $(2 - x)(6 + x)$
19. $(4 - t)(9 + t)$ **21.** $(2w + 1)(w + 3)$ **23.** $(3x - 1)(x + 2)$ **25.** $(5y - 1)(y - 2)$
27. $(3c + d)(c + 2d)$ **29.** $(3x + 2)(2x + 3)$ **31.** $(3z - 2y)(2z + 3y)$ **33.** $(4v - 1)(3v + 5)$
35. $(7x - 6)(8x - 5)$ **37.** $(3 - 2w)(4 + w)$ **39.** $(5r - 4s)(r + 2s)$ **41.** $5x(x - 4)(x - 7)$
43. $2st(8t - s)^2$ **45.** $y^2(x + 3)(x + 7)$ **47.** $4m^2(n - 1)(n + 7)$ **49.** $wy(x - 2)(x - 7)$

PROBLEM SET 2.8   page 78

1. $3x^3$ **3.** $-\dfrac{3x^2}{y^2}$ **5.** $\dfrac{2w^2}{u^2}$ **7.** $3n^2 - 2m$ **9.** $2xy^2 - 8y^2 + 2$ **11.** $2ab^2 + b - \dfrac{4}{ab^2} + \dfrac{3}{a^2b^3}$
13. $x - 2$ **15.** $v - 2; R = -8$ **17.** $w^2 + w - 4; R = 3$ **19.** $2t^3 + t^2 - 6t + 8; R = -9$
21. $x^2 + 3x + 4$ **23.** $y + 2$ **25.** $x + 2$ **27.** $u^2 - 2uv + v^2; R = v^3$ **29.** $2m^2 - mn + 2n^2; R = 2n^3$
31. $x^3 + 2x^2y - 2xy^2 + y^3; R = -6y^4$ **33.** $x^2 + xy + y^2$

REVIEW PROBLEM SET   page 78

1. Trinomial; degree $= 2$; $4, -3, 2$ **3.** Binomial; degree $= 2$; $-7, 3$ **5.** Monomial; degree $= 5$; 10
7. 17 **9.** 46 **11.** 13 **13.** $9w^2$ **15.** $-3x^2 - 1$ **17.** $v^2 + 4v + 6$ **19.** $4x^2 + 6x - 4$ **21.** x^{11}
23. w^{12} **25.** m^{27} **27.** $-v^3u^6$ **29.** $4y^6z^2$ **31.** $\dfrac{9t^2}{s^6}$ **33.** x^3 **35.** $3z^{10}$ **37.** $6t^5 - 12t^3$
39. $6x^3y^4 - 10x^2y^5 + 2xy^4$ **41.** $a^3 - 3a^2b + 3ab^2 - b^3$ **43.** $w^2 + 10w + 21$ **45.** $2x^2 + 7x - 15$
47. $4u^2 - 17uv + 15v^2$ **49.** $t^2 - 64$ **51.** $9x^2 + 42x + 49$ **53.** $t^3 + 8$ **55.** $y^3 + 9y^2 + 27y + 27$
57. $16 - 24s + 9s^2$ **59.** $27w^3 - 54w^2 + 36w - 8$ **61.** $8x^3 - z^3$ **63.** $7xy(x - 3y^2)$
65. $13a^2b^2(2a + 3a^3b^2 - 4b)$ **67.** $2(y + z)(y - 2x)$ **69.** $(x - y)(3 + z)$ **71.** $(x - 3y)(2u + v)$
73. $(5a - b)(m^2 + n)$ **75.** $(5m - 3n)(5m + 3n)$ **77.** $(x + 4)(x^2 - 4x + 16)$
79. $(2t - 3)(2t + 3)(4t^2 + 9)$ **81.** $(x + y - z + 1)(x + y + z - 1)$ **83.** $9u(u - 3v)(u + 3v)$
85. $5s(t + 4)(t^2 - 4t + 16)$ **87.** $(2x - y)(2x + y)(4x^2 + 2xy + y^2)(4x^2 - 2xy + y^2)$
89. $(x + y - w - z)(x + y + w + z)$ **91.** $(t^2 - t + 1)(t^2 + t + 1)$ **93.** $(x - y)(x + 3y)$
95. $(m + 4)(m - 9)$ **97.** $(3u + 2)(u + 5)$ **99.** $(2y + 3)(y - 2)$ **101.** $(5x - 4y)(4x - 3y)$
103. $w(w + 11)(w - 2)$ **105.** $yz(x - 8y)(x + 2y)$ **107.** $6u^2$ **109.** $8w^2z^2 - 4wz$ **111.** $5x - 1$
113. $x - 2$ **115.** $t^2 + 2ts - s^2$ **117.** $w^{10} - w^5 + 1$

Chapter 3

PROBLEM SET 3.1 page 89

1. 3   **3.** -10 **5.** 6; -7 **7.** -2; 8   **9.** -8; -9; 4   **11.** equivalent   **13.** not equivalent

15. equivalent   **17.** not equivalent   **19.** equivalent   **21.** $\frac{5}{6}$ **23.** $\dfrac{5y^3}{9x}$ **25.** $\dfrac{m+1}{m-1}$ **27.** $\dfrac{t-1}{t+2}$

29. $\dfrac{2x+3}{3x}$ **31.** $\dfrac{v+4}{v+7}$ **33.** $\dfrac{x+1}{x-3}$ **35.** $\dfrac{2u+3}{4u^2+6u+9}$ **37.** $-\dfrac{x+y}{2x+y}$ **39.** $\dfrac{z+w}{z+y}$ **41.** 36

43. $20m^7y^4$ **45.** $15u+15v$ **47.** $6x-6$ **49.** $4t^2+2t$ **51.** y^2-2y-3 **53.** $2x^2+x-6$

55. $4y^2-9x^2$ **57.** $\dfrac{7}{x-y}$ **59.** $\dfrac{-xy}{x-y}$ **61.** $\dfrac{x-3}{x-y}$

PROBLEM SET 3.2 page 95

1. $\frac{3}{5}$ **3.** $\dfrac{2x}{3y^3}$ **5.** $\dfrac{y}{t}$ **7.** $\dfrac{5}{3t}$ **9.** $\dfrac{6}{x-8}$ **11.** $-7a-2$ **13.** $\dfrac{(v+1)(3v+1)}{v-9}$ **15.** $(x-12)(x-4)$

17. $\dfrac{(y-1)^2}{(y+1)^2}$ **19.** $\dfrac{(a+2b)(a-b)}{(a+b)(a-2b)}$ **21.** x^2-y^2 **23.** $\frac{5}{9}$ **25.** $6x^2y$ **27.** $5t$ **29.** $-\dfrac{3y+9}{7}$

31. $\dfrac{x(x+1)}{2x-1}$ **33.** $\dfrac{3(a-2)}{a+2}$ **35.** $\dfrac{3(u-3v)}{(u+3v)(u-v)}$ **37.** $\dfrac{x-3}{3x+1}$ **39.** $\dfrac{4y(2y+3)}{(3y-5)(4y+1)}$

41. $(2x-3)(3x+1)$ **43.** $\dfrac{x-2}{(x+1)^2}$ **45.** $\dfrac{a-1}{(3-a)(b+b^2)}$ **47.** $\dfrac{8}{3(v-2)}$ **49.** $\dfrac{y+2}{y+5}$

PROBLEM SET 3.3 page 105

1. $\frac{3}{4}$ **3.** $\dfrac{4}{x}$ **5.** $\frac{1}{3}$ **7.** $\dfrac{t}{2}$ **9.** $\dfrac{2}{x+2}$ **11.** $6+v$ **13.** $\dfrac{3}{m-1}$ **15.** $\dfrac{3}{3x+2}$ **17.** $-\dfrac{3}{2u+1}$

19. $2u-3v$ **21.** $\frac{59}{56}$ **23.** $\frac{23}{60}$ **25.** $\dfrac{47x}{30}$ **27.** $\dfrac{u}{21}$ **29.** $\dfrac{15y-57}{(y-5)(y-3)}$ **31.** $\dfrac{m^2+m+6}{(m-3)(m+3)}$

33. $\dfrac{x^2-x+2}{(3x+2)(x-4)}$ **35.** $\dfrac{10s^2+3t}{2t^2s^3}$ **37.** $\dfrac{36n^4-20m}{15m^3n^5}$ **39.** $\dfrac{5x-3}{x(x+1)(x-1)}$ **41.** $\dfrac{3-4c}{(c-3)(c+3)(c-2)}$

43. $\dfrac{2m^2+4}{m(m+1)(m-6)}$ **45.** $\dfrac{2x^2+14x-6}{(x+8)(x+2)(x+7)}$ **47.** $\dfrac{2y^2-14y}{(y+2)(y+5)(y-4)}$ **49.** $\dfrac{3}{3t-2}$

51. $\dfrac{8x^2+15}{(2x+3)(x-4)(3x-1)}$ **53.** $\dfrac{v^2+4v}{4-v^2}$ **55.** $\dfrac{17t^2+t-9}{27-3t^2}$ **57.** $\dfrac{2x^2+13x-7}{(x-3)^2(x+2)^2}$

59. $\dfrac{4y^2+17y-2}{(3y+2)(y-1)(y+3)}$

PROBLEM SET 3.4 page 109

1. $\frac{5}{6}$ **3.** $\frac{15y}{8}$ **5.** $\frac{3c+1}{2c-3}$ **7.** $\frac{1}{2y-1}$ **9.** $\frac{2}{3v+4}$ **11.** $\frac{m}{m-2}$ **13.** $\frac{x^2-y^2}{x^2+y^2}$ **15.** $-\frac{b}{2}$ **17.** $3-x$

19. $\frac{3t+2}{2t-1}$ **21.** $\frac{-2x}{x^2+1}$ **23.** 1

REVIEW PROBLEM SET page 110

1. 5 **3.** $-2;\ 7$ **5.** equivalent **7.** equivalent **9.** not equivalent **11.** $\frac{2v^3}{3u^4}$ **13.** $\frac{1}{3m-3n}$

15. $\frac{x-4}{x+6}$ **17.** $\frac{t-5}{2t-5}$ **19.** $\frac{x-2}{x}$ **21.** $21x^2y^2$ **23.** $3a(a+b)$ **25.** $(2z+1)(3z+2)$

27. $w(w^2-3w+9)$ **29.** $\frac{-4b}{b-a}$ **31.** $\frac{-3b}{b-a}$ **33.** $\frac{x^2u^2}{10yv^3}$ **35.** $\frac{3}{2(b-5)}$ **37.** $\frac{3-y}{x(x+1)}$ **39.** $\frac{u-v}{2u}$

41. 1 **43.** $\frac{2t+1}{t-3}$ **45.** $\frac{2}{bx^2}$ **47.** $\frac{v-2}{v-7}$ **49.** $\frac{m+1}{m-3}$ **51.** $\frac{x+2y}{x}$ **53.** $\frac{(w-2)(w+2)}{(w-4)(w+1)}$ **55.** $\frac{x-3}{x+2}$

57. $\frac{u+2}{u-2}$ **59.** $\frac{2}{x}$ **61.** $\frac{1}{3+t}$ **63.** $\frac{3}{y-2}$ **65.** $\frac{x^2+y^2}{x^2-y^2}$ **67.** $\frac{7}{3-t}$ **69.** $\frac{5y+16}{(y+4)(y+6)}$

71. $\frac{16n^2+15m}{20m^2n^3}$ **73.** $\frac{v^2+3v}{(2v-3)(2v+3)(v+2)}$ **75.** $\frac{13x-19}{2(x-3)(x-4)(x-1)}$

77. $\frac{9m+15}{(2m+3)(m-1)(m+1)}$ **79.** $\frac{y^2-5xy}{(x-3y)(x+3y)(x+y)}$ **81.** $\frac{1}{a^2x}$ **83.** $\frac{3+x}{4+2y}$ **85.** y **87.** $\frac{1}{t-1}$

89. $\frac{6}{x}$

Chapter 4

PROBLEM SET 4.1 page 119

1. 7 **3.** 12 **5.** 6 **7.** 3 **9.** -3 **11.** $-\frac{9}{5}$ **13.** 5 **15.** $\frac{15}{8}$ **17.** $-\frac{2}{3}$ **19.** 7 **21.** 4 **23.** 3
25. $\frac{25}{3}$ **27.** $\frac{5}{4}$ **29.** $\frac{15}{7}$ **31.** -3 **33.** 2 **35.** 3 **37.** 1 **39.** 0 **41.** 5 **43.** 1 **45.** 3 **47.** 3
49. 9 **51.** $\frac{5}{9}$ **53.** $\frac{4}{3}$ **55.** $\frac{1}{2}$ **57.** $-\frac{3,125}{999}$ **59.** -1.20 **61.** -33.572

PROBLEM SET 4.2 page 124

1. -3 **3.** 7 **5.** 1 **7.** 24 **9.** 2 **11.** $\frac{267}{40}$ **13.** 3 **15.** $-\frac{2}{3}$ **17.** 2 **19.** 5 **21.** 7 **23.** $\frac{25}{16}$
25. 2 **27.** $\frac{1}{4}$ **29.** 2 **31.** $-\frac{1}{3}$ **33.** $-\frac{10}{39}$ **35.** 8 **37.** $-\frac{6}{13}$ **39.** $-\frac{5}{2}$ **41.** No solution

PROBLEM SET 4.3   page 129

1. $5c$ **3.** $\dfrac{c-b}{a}$ **5.** $-\dfrac{b}{2}$ **7.** $\dfrac{9c}{2}$ **9.** $-\dfrac{5a}{3}$ **11.** $\dfrac{h}{4}$ **13.** $-\dfrac{c}{2}$ **15.** $-7b$ **17.** $\dfrac{5+2b}{7}$ **19.** $\dfrac{bc+d}{a}$

21. $\dfrac{30a-35b}{16}$ **23.** $\dfrac{1+2c}{2}$ **25.** $-2a$ **27.** 1 **29.** $\dfrac{9b}{17}$ **31.** $\dfrac{1-3a-a^2}{3-a}$ **33.** $-\dfrac{a}{5}$ **35.** $h=\dfrac{2A}{b}$

37. $l=\dfrac{V}{wh}$ **39.** $r=\dfrac{C}{2\pi}$ **41.** $t=\dfrac{V}{q}$ **43.** $t=\dfrac{I}{Pr}$ **45.** $m=\dfrac{F-b}{x}$ **47.** $w=\dfrac{P-2l}{2}$

49. $a=\dfrac{2S-nl}{n}$ **51.** $d=\dfrac{2(S-na)}{n(n-1)}$ **53.** $d=\dfrac{L-a}{m-1}$ **55.** $t=\dfrac{mpv-km}{k}$

PROBLEM SET 4.4   page 142

1. 27   **3.** 32; 34   **5.** 36   **7.** 12 and 17   **9.** 14 and 11   **11.** 40   **13.** 11 quarters; 15 dimes; 45 nickels
15. 43 nickels; 27 dimes   **17.** 30 quarters; 50 dimes   **19.** $120   **21.** $200   **23.** $8,058   **25.** $16
27. $31   **29.** $12,500   **31.** $28,000 at 13.2%; $34,000 at 13.7%   **33.** $4444.44   **35.** length = 64 meters;
width = 32 meters   **37.** length = 60 meters; width = 15 meters   **39.** 10 feet   **41.** 10 mph; 48 minutes
43. 12 hours   **45.** 15 mph   **47.** 6 milliliters of 10% acid; 4 milliliters of 15% acid   **49.** 200,000 gallons of
9% alcohol; 100,000 gallons of 12% alcohol   **51.** 75 pounds of the $3.80 per pound coffee; 25 pounds of the
$4.00 per pound coffee   **53.** 45 hours   **55.** 4 hours   **57.** $\frac{4}{5}$ hour

PROBLEM SET 4.5   page 153

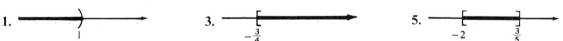

7. Addition property   **9.** Transitive property   **11.** Multiplication property   **13.** Addition property

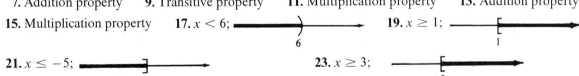

15. Multiplication property   **17.** $x < 6$;   **19.** $x \geq 1$;

21. $x \leq -5$;   **23.** $x \geq 3$;

25. $x > \frac{3}{2}$;   **27.** $x > -3$;

29. $x \leq -\frac{2}{7}$;   **31.** $x \geq 3$;

33. $x < 1$;   **35.** $x < -2$;

37. $x \leq -\frac{4}{3}$;   **39.** $x \leq -\frac{2}{3}$;

41. $x < -1$;   **43.** $x < 3$;

45. $x \le 1$;

47. $x \le \frac{7}{3}$;

49. $x < \frac{7}{5}$;

51. $x \le -2$;

53. $x \le 8$;

55. $x > -3$;

57. $x \ge \frac{9}{2}$;

59. $x \ge 12$;

61. $x > \frac{19}{12}$;

63. $1 \le x \le 4$;

65. 9 years   **67.** \$35,000   **69.** 7 or less

PROBLEM SET 4.6   page 160

1. -3; 3 **3.** -2; 2 **5.** $-\frac{4}{3}$; $\frac{4}{3}$ **7.** -5; 5 **9.** -1; 1 **11.** No solution   **13.** -1; 5 **15.** -1; 11

17. $\frac{1}{3}$; $\frac{8}{3}$ **19.** $-\frac{2}{7}$; $\frac{6}{7}$ **21.** No solution   **23.** $-\frac{2}{3}$ **25.** $-5 < x < 5$;

27. $-4 \le x \le 4$;

29. $-6 \le x \le 6$;

31. $x < -3$ or $x > 3$;

33. $x \le -3$ or $x \ge 3$;

35. $x \le -4$ or $x \ge 4$;

37. $-2 < x < 4$;

39. $-1 \le x \le 2$;

41. $x \le -\frac{5}{2}$ or $x \ge 3$;

43. $\frac{1}{3} \le x \le 1$;

45. $x \le -1$ or $x \ge 5$;

47. $x < -\frac{9}{5}$ or $x > \frac{3}{5}$;

49. $-\frac{11}{2} \le x \le \frac{13}{2}$;

REVIEW PROBLEM SET   page 161

1. 2 **3.** 13 **5.** -3 **7.** -7 **9.** 4 **11.** -7 **13.** -4 **15.** $\frac{7}{4}$ **17.** 4 **19.** $-\frac{2}{7}$ **21.** $\frac{20}{9}$ **23.** 3

25. 6 **27.** 80 **29.** 3 **31.** 8 **33.** No solution   **35.** $\frac{37}{99}$ **37.** $\frac{11}{225}$ **39.** $-2a$ **41.** $-\frac{2a}{b}$ **43.** $\frac{b}{a-c}$

45. $\frac{12a - a^3}{6b}$ **47.** $\frac{2b}{x}$ **49.** $3a - 1$ **51.** $\frac{a^2 + 3a - ac}{a - c}$ **53.** $m = \frac{E}{c^2}$ **55.** $h = \frac{S - 2\pi r^2}{2\pi r}$

57. $m = Gr^3$ **59.** $h = \dfrac{v^2 - gR}{g}$ **61.** Joe, \$15; Dawn, \$30; Mike, \$34   **63.** width $= 42$ meters;

length $= 168$ meters   **65.** \$18,750 and \$6,250   **67.** $1\frac{2}{3}$ hours   **69.** 6.9 miles

71. $x > 2$;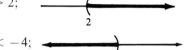
2

73. $x > 6$;
6

75. $x < -4$;
$$$-4$

77. $x \le -10$;
$$$-10$

79. $x \ge 8$;
8

81. $x \ge \frac{1}{18}$;
$$$\frac{1}{18}$

83. Its weight must be less than 52.5 ounces   **85.** $-5, 5$ **87.** $-33;\ 29$ **89.** No solution   **91.** $-4;\ \frac{5}{2}$

93. $-2;\ \frac{36}{11}$

95. $-11 < x < 11$;
$$$-1111$

97. $x \le -4$ or $x \ge 4$
$$$-44$

99. $-83 \le x \le 3$;
$$$-833$

101. $x \le -4$ or $x \ge \frac{1}{2}$;
$$$-4\frac{1}{2}$

103. $-14 < x < 16$;
$$$-1416$

Chapter 5

PROBLEM SET 5.1   page 168

1. $\frac{1}{25}$ **3.** 81   **5.** $\dfrac{1}{p^5}$ **7.** $\dfrac{1}{25x^2}$ **9.** $\dfrac{1}{u^3 v^6}$ **11.** $\dfrac{4b^2}{a^7}$ **13.** 7   **15.** $\dfrac{p^4}{8a^2}$ **17.** $-\dfrac{2x^4 y^5}{3}$ **19.** $\dfrac{q^2}{p^3}$

21. 1, for $y \ne 0$ **23.** 1, for $y \ne 0$ **25.** 29   **27.** $x + y$ **29.** $\frac{144}{25}$ **31.** $\dfrac{u^2 v^2}{v^2 + u^2}$ **33.** $\dfrac{2y}{x}$ **35.** 49

37. $\frac{1}{81}$ **39.** $\dfrac{1}{x^5}$ **41.** $\frac{1}{64}$ **43.** $\frac{1}{256}$ **45.** $\dfrac{1}{81x^4}$ **47.** $\dfrac{25}{p^8}$ **49.** $\dfrac{x^2}{25}$ **51.** $\dfrac{q^4}{p^4}$ **53.** 64   **55.** $\dfrac{1}{x^8}$ **57.** $\dfrac{1}{y^3}$

59. $\frac{1}{9}$ **61.** $\dfrac{x^2}{y^{10}}$ **63.** $\dfrac{c^3}{a^9}$ **65.** $\dfrac{u^{28} z^4}{v^{20}}$ **67.** $\dfrac{1}{a^4 b^4}$

PROBLEM SET 5.2   page 176

1. 4   **3.** 21   **5.** -3 **7.** Undefined   **9.** $13^{1/2}$ **11.** $(3x^2)^{1/5}$ **13.** $(a^3 b^2)^{1/4}$ **15.** $(-3a^2 b^4)^{1/5}$

17. $\sqrt[6]{243}$ **19.** $\sqrt{343 y^3}$ **21.** $\sqrt[7]{1024 x^5 y^5}$ **23.** 4   **25.** 16   **27.** -8 **29.** 27   **31.** Undefined

33. 9   **35.** $\frac{1}{16}$ **37.** $\frac{1}{27}$ **39.** $-\frac{1}{32}$ **41.** 2   **43.** x **45.** 25   **47.** $\dfrac{1}{x^2}$ **49.** $16p^{12}$ **51.** $128 y^3$ **53.** $\dfrac{y}{5}$

55. $5^{17/21}$ **57.** $x^{1/2}$ **59.** $\dfrac{8d}{c}$ **61.** $\dfrac{2304 a^6}{b^5}$ **63.** $\frac{2}{125}$ **65.** q^6 **67.** $-\dfrac{pq^3}{3}$ **69.** 1.592   **71.** 4.743

73. 33.77   **75.** 522.4 square meters

PROBLEM SET 5.3   page 183

1. $3\sqrt{3}$ **3.** $-3\sqrt[3]{2}$ **5.** $12\sqrt{2}$ **7.** $2x\sqrt[3]{x}$ **9.** $-2y^3\sqrt[3]{2y}$ **11.** $7p\sqrt{2p}$ **13.** $\dfrac{\sqrt{7}}{2}$ **15.** $-\dfrac{\sqrt[3]{5}}{2}$

17. $-\dfrac{\sqrt[3]{x^2}}{4}$ **19.** $\dfrac{\sqrt{3}}{2w}$ **21.** $-\dfrac{\sqrt[5]{4}}{2b^4}$ **23.** $-\sqrt[3]{18}$ **25.** $-p$ **27.** 2   **29.** p^4 **31.** 5   **33.** $-p$

35. w **37.** $6a^2b^5$ **39.** $2xy\sqrt[3]{y^2}$ **41.** $x^2\sqrt[8]{x^3y}$ **43.** u^5 **45.** $ab\sqrt[4]{a}$ **47.** $\dfrac{5\sqrt[3]{p^2q}}{2}$ **49.** $\sqrt[6]{x^5}$

51. $y\sqrt[15]{y^7}$ **53.** $b\sqrt[20]{a^{14}b^3}$

PROBLEM SET 5.4   page 190

1. $8\sqrt{7}$ **3.** $8\sqrt{5}$ **5.** $7\sqrt[3]{4}$ **7.** $5\sqrt{2}$ **9.** $3\sqrt{2}$ **11.** $3\sqrt{3}$ **13.** $6\sqrt{5}$ **15.** $13\sqrt[3]{3}$ **17.** $2\sqrt{3x}$
19. $(6p+3)\sqrt{p}$ **21.** 6   **23.** $2\sqrt{6}$ **25.** $4\sqrt{3}-12$ **27.** $2x+15\sqrt{x}+7$ **29.** $33+20\sqrt{2}$
31. $4x-4\sqrt{xy}+y$ **33.** 11   **35.** 42   **37.** $9x-121$ **39.** $x-3$ **41.** $\dfrac{2\sqrt{3}}{3}$ **43.** $\dfrac{8\sqrt{11x}}{77x}$ **45.** $\dfrac{5\sqrt[3]{49}}{7}$
47. $\dfrac{5(\sqrt{5}+1)}{2}$ **49.** $2-\sqrt{2}$ **51.** $\sqrt{7}+\sqrt{5}-\sqrt{21}-\sqrt{15}$ **53.** $\dfrac{3\sqrt{xy}+2y}{9x-4y}$ **55.** $\dfrac{x^2-x\sqrt{y}}{x^2-y}$
57. $\dfrac{37\sqrt{10}+129}{227}$ **59.** $\dfrac{2\sqrt{35}+\sqrt{7}-3\sqrt{5}+8}{38}$

PROBLEM SET 5.5   page 196

1. i **3.** -1 **5.** i **7.** 1   **9.** $15i$ **11.** $5+9i$ **13.** $2-2xi$ **15.** $-1+9i$ **17.** $-6+8i$ **19.** 12
21. $1-i$ **23.** $-16i$ **25.** $1-11i$ **27.** $-9+6i$ **29.** $11-3i$ **31.** $27-5i$ **33.** 41   **35.** 170
37. 53   **39.** $2-4i$ **41.** $-2i$ **43.** 5   **45.** $\frac{4}{25}-\frac{3}{25}i$ **47.** $\frac{35}{41}+\frac{28}{41}i$ **49.** $\frac{7}{13}-\frac{4}{13}i$ **51.** $\frac{14}{29}-\frac{23}{29}i$
53. $\frac{10}{53}+\frac{35}{53}i$ **55.** $-\frac{1}{2}-\frac{1}{2}i$ **57.** $2-5i$ **59.** $\frac{19}{29}-\frac{4}{29}i$ **61.** $-\frac{9}{29}-\frac{21}{29}i$

REVIEW PROBLEM SET   page 197

1. $\frac{1}{9}$ **3.** $\dfrac{1}{y^4}$ **5.** $\dfrac{1}{t^4s^2}$ **7.** $\dfrac{y^3}{x^2}$ **9.** 1   **11.** $\frac{40}{9}$ **13.** $\dfrac{2x^3y^2}{y^2+x^3}$ **15.** 81   **17.** x^7 **19.** $\frac{1}{64}$ **21.** t^{12}

23. $16x^6$ **25.** $\dfrac{1}{8u^6}$ **27.** 25   **29.** $\dfrac{1}{r^3}$ **31.** $4y^5$ **33.** $\dfrac{x^2}{y^{12}}$ **35.** $\dfrac{1}{8u^2v^7}$ **37.** $\dfrac{s^2t^6}{r^8}$ **39.** 9   **41.** -6

43. Undefined   **45.** $15^{1/2}$ **47.** $(a+b)^{3/4}$ **49.** $\sqrt[3]{13^2}$ **51.** $\sqrt[7]{(2x+3y)^2}$ **53.** -8 **55.** $\frac{1}{8}$ **57.** -8

59. $\frac{1}{4}$ **61.** 7   **63.** $x^{1/4}$ **65.** $\frac{1}{8}$ **67.** $\dfrac{1}{x^{1/2}}$ **69.** $-2t^2$ **71.** $4m^2$ **73.** $\dfrac{v^6}{4}$ **75.** $\dfrac{c^{75}}{d^{12}}$ **77.** $-\dfrac{8z^9}{x^3y^6}$

79. $a^{1/21}$ **81.** $r^8y^{16/3}$ **83.** $\dfrac{1}{x^7y^2}$ **85.** 133.3   **87.** 22.36   **89.** $5\sqrt{5}$ **91.** $2t\sqrt[4]{2t}$ **93.** $-2xy^3\sqrt[3]{3xy^2}$

95. $\frac{2}{5}$ **97.** $-\dfrac{t^2}{2}$ **99.** $-\dfrac{\sqrt[3]{5}}{c^2d^3}$ **101.** $5u^3$ **103.** $2x^2$ **105.** v **107.** $4u^2v^3$ **109.** $\sqrt{x}$ **111.** $\sqrt[12]{z^7}$

113. $u\sqrt[6]{72u}$ **117.** $8\sqrt{2}$ **119.** $6\sqrt{x}$ **121.** $10\sqrt{2}$ **123.** $3\sqrt{3}$ **125.** $10\sqrt{2z}$ **127.** $5\sqrt{7u}$
129. $\sqrt{6}+\sqrt{15}$ **131.** $3\sqrt{2}-2\sqrt{6}$ **133.** $12+4\sqrt{5}$ **135.** 6   **137.** $3t-5$ **139.** $3\sqrt{2}$

141. $2x - \sqrt{xy} - y$ **143.** $\dfrac{4\sqrt{3}}{3}$ **145.** $\dfrac{5\sqrt{2}}{8}$ **147.** $\dfrac{2\sqrt{3ts}}{3s}$ **149.** $\dfrac{25 - 5\sqrt{11}}{14}$ **151.** $\dfrac{10\sqrt{x} + 20}{x - 4}$

153. $13 - 2\sqrt{42}$ **155.** $\dfrac{u^2 + u\sqrt{v} - 2v}{u^2 - 4v}$ **157.** i **159.** $-i$ **161.** -1 **163.** $12i$ **165.** $3 + ti$

167. $-1 + 16i$ **169.** $-4 + 4i$ **171.** $-27 + 10i$ **173.** $23 - 10i$ **175.** $12 + 16i$ **177.** $46 + i$

179. $3 - 5i$ **181.** $-4 + 7i$ **183.** $\frac{14}{17} + \frac{5}{17}i$ **185.** $\frac{15}{13} + \frac{23}{13}i$ **187.** $-\frac{11}{26} - \frac{3}{26}i$

Chapter 6

PROBLEM SET 6.1   page 205

1. $1, 2$ **3.** $2, 4$ **5.** $-4, 5$ **7.** $-7, 5$ **9.** $-3, 4$ **11.** $-1, \frac{5}{3}$ **13.** $-\frac{1}{2}, \frac{2}{5}$ **15.** $-\frac{4}{3}, \frac{2}{3}$ **17.** $\frac{1}{2}, \frac{3}{5}$

19. $\frac{5}{4}, 4$ **21.** $-\frac{5}{2}, \frac{4}{3}$ **23.** $-\frac{2}{5}, \frac{7}{2}$ **25.** $0, 7$ **27.** $-7, 0$ **29.** $\frac{1}{7}$ **31.** $-\frac{1}{2}, 1$ **33.** $-5, \frac{4}{3}$ **35.** $-1, \frac{3}{5}$

37. $-\frac{3}{2}, 11$ **39.** $\frac{3}{7}, 4$ **41.** $-1, 7$ **43.** $-\frac{5}{2}, \frac{7}{2}$ **45.** $-3a, 5a$ **47.** $-p - q, p - q$ **49.** $-\dfrac{2b}{3}, \dfrac{b}{2}$

51. $2d, 2d + 1$

PROBLEM SET 6.2   page 212

1. $-8, 8$ **3.** $-3, 3$ **5.** $-\sqrt{15}, \sqrt{15}$ **7.** $-2, 4$ **9.** $-\frac{7}{2}, \frac{5}{2}$ **11.** $\frac{1}{2}, \frac{7}{6}$ **13.** $-\frac{1}{4}, \frac{7}{4}$ **15.** $-9i, 9i$

17. $6 - 5i, 6 + 5i$ **19.** $\frac{2}{3} - \frac{7}{3}i, \frac{2}{3} + \frac{7}{3}i$ **21.** $(x + 3)^2$ **23.** $(y - 5)^2$ **25.** $(m + 10)^2$ **27.** $(x + \frac{7}{2})^2$

29. $(c + \frac{17}{2})^2$ **31.** $-3, 1$ **33.** $6 - \sqrt{53}, 6 + \sqrt{53}$ **35.** $2 - \sqrt{5}, 2 + \sqrt{5}$ **37.** $\dfrac{5 - \sqrt{30}}{5}, \dfrac{5 + \sqrt{30}}{5}$

39. $-\frac{2}{5}, \frac{7}{5}$ **41.** $\dfrac{4}{5} - \dfrac{\sqrt{69}}{5}i, \dfrac{4}{5} + \dfrac{\sqrt{69}}{5}i$ **43.** $-\dfrac{7}{8} - \dfrac{\sqrt{31}}{8}i, -\dfrac{7}{8} + \dfrac{\sqrt{31}}{8}i$ **45.** $\frac{2}{3}, 2$ **47.** $\dfrac{1 - \sqrt{2}}{3}, \dfrac{1 + \sqrt{2}}{3}$

49. $\frac{1}{4}, \frac{5}{4}$ **51.** $\dfrac{-b - \sqrt{b^2 - 12ab}}{2a}, \dfrac{-b + \sqrt{b^2 - 12ab}}{2a}$ **53.** $\dfrac{v - \sqrt{v^2 - 4gs}}{2g}, \dfrac{v + \sqrt{v^2 - 4gs}}{2g}$

PROBLEM SET 6.3   page 217

1. $1, 4$ **3.** $-\frac{1}{3}, \frac{1}{2}$ **5.** $-\frac{1}{2}, 3$ **7.** $-\frac{1}{2}, \frac{5}{3}$ **9.** $\dfrac{4 - \sqrt{34}}{6}, \dfrac{4 + \sqrt{34}}{6}$ **11.** $\dfrac{5 - \sqrt{17}}{4}, \dfrac{5 + \sqrt{17}}{4}$

13. $\dfrac{1}{5} - \dfrac{\sqrt{34}}{5}i, \dfrac{1}{5} + \dfrac{\sqrt{34}}{5}i$ **15.** $\dfrac{4}{7} - \dfrac{\sqrt{5}}{7}i, \dfrac{4}{7} + \dfrac{\sqrt{5}}{7}i$ **17.** $-\frac{4}{5}, \frac{2}{3}$ **19.** $\dfrac{1}{6} - \dfrac{\sqrt{83}}{6}i, \dfrac{1}{6} + \dfrac{\sqrt{83}}{6}i$

21. $\dfrac{1 - 2\sqrt{5}}{2}, \dfrac{1 + 2\sqrt{5}}{2}$ **23.** $\dfrac{5 - \sqrt{21}}{2}, \dfrac{5 + \sqrt{21}}{2}$ **25.** $3 - \sqrt{13}, 3 + \sqrt{13}$ **27.** $-1.45, 1.65$

29. $-1.06, 3.66$ **31.** $-3.91, 1.80$ **33.** $\dfrac{-mn + m\sqrt{n^2 + 4n}}{2n}, \dfrac{-mn - m\sqrt{n^2 + 4n}}{2n}$

35. $\dfrac{-RC - \sqrt{R^2C^2 - 4LC}}{2LC}, \dfrac{-RC + \sqrt{R^2C^2 - 4LC}}{2LC}$ **37.** $D = 64$; 2 unequal real roots

39. $D = 0$; 1 real root   **41.** $D = -11$; 2 unequal complex roots   **43.** $D = 1$; 2 unequal real roots
45. $D = 0$; 1 real root

PROBLEM SET 6.4 page 223

1. 5, 6   **3.** 12, 13   **5.** 6   **7.** 6 feet and 10 feet   **9.** 30 feet and 44 feet   **11.** 30 feet   **13.** 5 feet
15. 5 meters; 12 meters   **17.** 5 mph; 12 mph   **19.** 1 second and 7 seconds   **21.** 5 seconds and 25 seconds
23. 120 mph   **25.** 12 hours   **27.** $540

PROBLEM SET 6.5 page 227

1. $-3, -2, 2, 3$ **3.** $-2, 2, -i, i$ **5.** $-5, -2, 2, 5$ **7.** $-\frac{1}{2}, \frac{1}{4}$ **9.** $-\frac{2}{5}, 3$ **11.** $-\dfrac{\sqrt{5}}{5}, -\frac{1}{2}, \frac{1}{2}, \dfrac{\sqrt{5}}{5}$

13. $-\frac{1}{2}, -\frac{1}{4}, \frac{1}{4}, \frac{1}{2}$ **15.** $-1, 0, 1$ **17.** $-3, -1, 1$ **19.** $-5, -1, 3$ **21.** $-3, 1, -1 - \sqrt{3}i, -1 + \sqrt{3}i$

23. $-1, \frac{1}{2}, \dfrac{-1 - \sqrt{39}i}{4}, \dfrac{-1 + \sqrt{39}i}{4}$ **25.** $-2, -1, \frac{1}{3}, \frac{2}{3}$ **27.** $-4, 3, \dfrac{-1 - \sqrt{13}}{2}, \dfrac{-1 + \sqrt{13}}{2}$ **29.** $-\frac{1}{4}$

31. $-1, -\frac{1}{2}, \dfrac{-1 - i}{2}, \dfrac{-1 + i}{2}$

PROBLEM SET 6.6 page 232

1. 25   **3.** 3   **5.** $\frac{11}{2}$ **7.** 5   **9.** $\frac{5}{2}$ **11.** $\frac{7}{2}$ **13.** 1   **15.** 30   **17.** 4   **19.** 8   **21.** No solution   **23.** 1
25. $\frac{261}{7}$ **27.** 9   **29.** 4   **31.** 4   **33.** 0, 3   **35.** 27   **37.** 7   **39.** 5   **41.** $\frac{1}{4}$ **43.** $-\frac{1}{2}, \frac{3}{2}$

PROBLEM SET 6.7 page 236

1. -243 **3.** -128 **5.** ± 64 **7.** 128   **9.** 0   **11.** 5   **13.** $\frac{8}{3}$ **15.** $\frac{129}{2}$ **17.** $\frac{513}{1024}$ **19.** $\frac{7}{3}$
21. $-64, 8$ **23.** $64, -27$ **25.** 1, 4   **27.** -3 **29.** $-19, 61$ **31.** $-\frac{7}{2}, 3$

PROBLEM SET 6.8 page 245

1. $-2 < x < 1$;

3. $1 \le x \le 2$;

5. $x < -3$ or $x > 1$;

7. $x \le -2$ or $x \ge 6$;

9. $-4 \le x \le 5$;

11. $-1 \le x \le \frac{1}{2}$;

13. $x < -1$ or $x > \frac{5}{3}$

15. $x \le -\frac{1}{4}$ or $x \ge 7$;

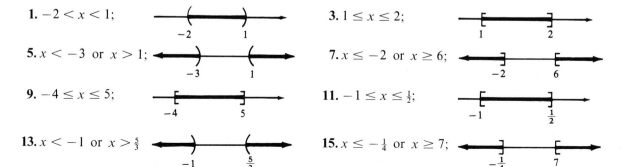

17. All real numbers;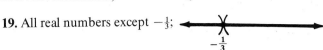

19. All real numbers except $-\frac{1}{3}$;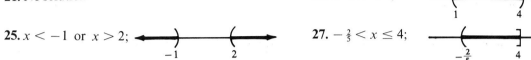

21. No solution

23. $1 < x < 4$;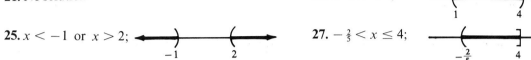

25. $x < -1$ or $x > 2$;

27. $-\frac{2}{5} < x \leq 4$;

29. $x \leq \frac{1}{5}$ or $x > \frac{2}{3}$;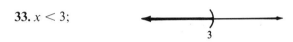

31. $x \leq 1$ or $x > 3$;

33. $x < 3$;

35. $x < 0$ or $x > \frac{1}{2}$;

REVIEW PROBLEM SET   page 246

1. $-1, 15$ **3.** $-4, 3$ **5.** $0, 2$ **7.** $-4, 2$ **9.** $-\frac{4}{9}, 1$ **11.** $-\frac{4}{3}, \frac{5}{2}$ **13.** $-\frac{2}{3}, \frac{3}{4}$ **15.** $-3b, 2b$
17. $-12, 12$ **19.** $-4i, 4i$ **21.** $-\frac{1}{5}, \frac{7}{5}$ **23.** $-5, \frac{17}{3}$ **25.** $(u-4)^2$ **27.** $(x+\frac{5}{2})^2$ **29.** $3 \pm \sqrt{2}$
31. $4 \pm \sqrt{7}$ **33.** $\frac{1}{3} \pm \frac{1}{3}i$ **35.** $\frac{3}{2} \pm i$ **37.** $-\frac{3}{2}, \frac{1}{2}$ **39.** $2c, -3c$ **41.** $-5 \pm \sqrt{38}$ **43.** $-\frac{1}{2} \pm \frac{\sqrt{11}}{2}i$
45. $\frac{1 \pm \sqrt{17}}{8}$ **47.** $\frac{-5 \pm \sqrt{161}}{4}$ **49.** $\frac{-3 \pm \sqrt{65}}{4}$ **51.** $-0.67, 5.22$ **53.** two real and unequal solutions
55. two complex solutions   **57.** one real solution   **59.** $-2, -1, 1, 2$ **61.** $-1, -\frac{1}{4}, \frac{1}{4}, 1$
63. $-2i, 2i, -\frac{\sqrt{3}}{3}, \frac{\sqrt{3}}{3}$ **65.** $-3, -2, 1, 2$ **67.** $-3, 1, -1 \pm \sqrt{2}$ **69.** 17 **71.** -1 **73.** No solution
75. 22 **77.** $\frac{7}{4}$ **79.** 4 **81.** 0 **83.** 4 **85.** -8 **87.** 16 **89.** 29 **91.** $-\frac{5}{8}$ **93.** $-27, 1$

95. $-5 < x < 3$;

97. $t \leq -2$ or $t \geq 5$;

99. $x < -3$ or $x > 2$;

101. $\frac{1}{3} \leq y < \frac{7}{5}$;

103. $6, 7, 8$ **105.** 10 feet   **107.** 89.04 seconds   **109.** 3 mph; 4 mph

Chapter 7

PROBLEM SET 7.1   page 256

1. a) I; b) II; c) III; d) IV; e) y axis; f) x axis; g) x axis; h) y axis   **3.** a) $(2, 3)$; b) $(-4, 2)$; c) $(3, 1)$
d) $(5, -4)$ **5.** 10 **7.** $\sqrt{17}$ **9.** $\sqrt{29}$ **11.** 5 **13.** 4 **15.** 5 **17.** 4 **19.** $\frac{\sqrt{481}}{6}$
21. yes   **23.** yes   **25.** no   **31.** $-2, 3$ **35.** $(-1, 7)$ **37.** $(1, \frac{1}{2})$ **39.** 17.71 **41.** 28.39

PROBLEM SET 7.2   page 264

1.

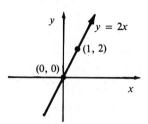

3.

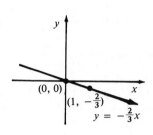

5.

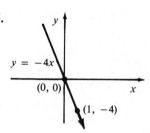

7.

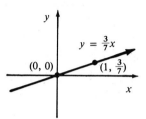

9.

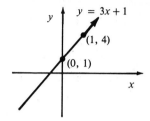

11.

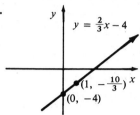

13.

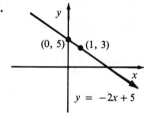

15.

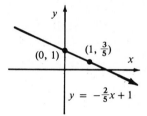

17.

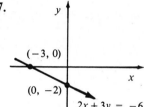

19.

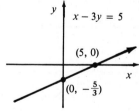

21.

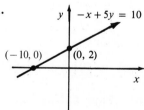

23.

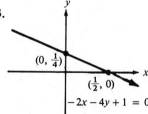

25.

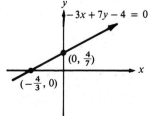

27.

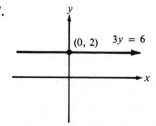

29.

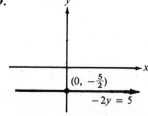

31.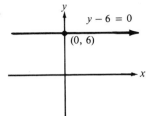

33.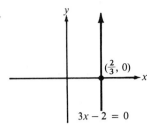

35.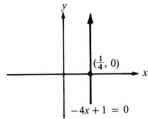

37. 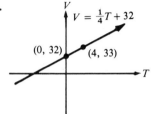 $V = 33\frac{1}{4}$ cubic centimeters when $T = 5°C$

PROBLEM SET 7.3   page 271

1. -1 **3.** $\frac{2}{3}$ **5.** $-\frac{1}{2}$ **7.** 0   **9.** undefined

11.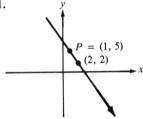

13.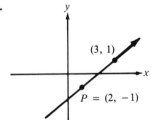

15.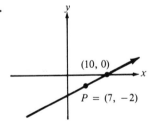

17.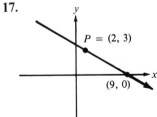

19.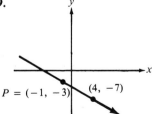

21. parallel   **23.** neither   **25.** perpendicular   **27.** neither   **29.** (a) $\frac{3}{2}$; (b) -1 **31.** (a) -10; (b) 0
33. (a) $-\frac{13}{2}$; (b) $\frac{13}{3}$ **35.** $\overline{AD}$ and $\overline{BC}$ have slope $-\frac{3}{2}$; $\overline{AC}$ and $\overline{BD}$ have slope 3   **37.** $\overline{AB}$ has slope $\frac{2}{5}$; $\overline{BC}$ has slope $-\frac{5}{2}$ **39.** $\overline{AB}$ has slope 1; $\overline{BC}$ has slope -1 **41.** not collinear   **43.** collinear

PROBLEM SET 7.4   page 280

1. $y - 2 = 5(x + 1)$ **3.** $y - 3 = -3(x - 7)$ **5.** $y + 1 = -\frac{3}{7}(x - 5)$ **7.** $y = \frac{3}{8}x$ **9.** $y = -5$
11. $y = \frac{1}{2}x + \frac{7}{2}$ **13.** $y = -3x + 5$ **15.** $y = -\frac{3}{7}x$ **17.** $x = -3$

19. $y = \frac{2}{3}x - \frac{1}{3}$
   $m = \frac{2}{3}$
   x intercept $= \frac{1}{2}$
   y intercept $= -\frac{1}{3}$

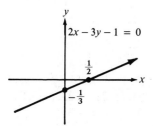

21. $y = -2x + 5$
   $m = -2$
   x intercept $= \frac{5}{2}$
   y intercept $= 5$

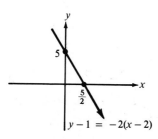

23. $y = 4x + 5$
   $m = 4$
   x intercept $= -\frac{5}{4}$
   y intercept $= 5$

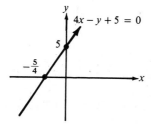

25. $y = 2x$
   $m = 2$
   x intercept $= 0$
   y intercept $= 0$

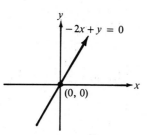

27. a) $y + 1 = -\frac{1}{2}(x - 3)$; b) $y = -\frac{1}{2}x + \frac{1}{2}$; c) $x + 2y - 1 = 0$ **29.** a) $y - 5 = 10(x + \frac{1}{2})$;
b) $y = 10x + 10$; c) $10x - y + 10 = 0$ **31.** a) $y - 5 = \frac{3}{2}(x - 1)$; b) $y = \frac{3}{2}x + \frac{7}{2}$; c) $3x - 2y + 7 = 0$
33. a) $y - 0 = \frac{5}{3}(x - 0)$; b) $y = \frac{5}{3}x$; c) $5x - 3y = 0$ **35.** a) $y - 3 = \frac{1}{2}(x - 0)$; b) $y = \frac{1}{2}x + 3$
c) $x - 2y + 6 = 0$ **37.** a) $y - 0 = -\frac{2}{3}(x + 1)$; b) $y = -\frac{2}{3}x - \frac{2}{3}$; c) $2x + 3y + 2 = 0$

39. a) $y - 6 = 0 (x + 1)$; b) $y = 6$; c) $y - 6 = 0$ **41.** a) $-\frac{5}{2}$; b) $\frac{7}{3}$ **43.** $\dfrac{x}{5} + \dfrac{y}{6} = 1$ **45.** $\dfrac{x}{-3} + \dfrac{y}{-1} = 1$

47. $\dfrac{x}{1} + \dfrac{y}{-\frac{5}{2}} = 1$ **49.** a) $N = \frac{5}{2}V + 50$; b) 16 feet per second

PROBLEM SET 7.5   page 288

1. independent; $(\frac{1}{3}, 0)$ **3.** inconsistent   **5.** independent; $(2, 1)$ **7.** $(4, 3)$ **9.** $(1, 5)$ **11.** $(\frac{7}{3}, -\frac{4}{3})$
13. $(159, -186)$ **15.** $(\frac{7}{11}, -\frac{10}{11})$ **17.** $(3, 1)$ **19.** $(1, 1)$ **21.** $(3, 2)$ **23.** $(-5, 3)$ **25.** $(2, -1)$
27. $(-\frac{3}{2}, \frac{27}{2})$ **29.** $(\frac{2}{5}, \frac{21}{5})$ **31.** $(\frac{22}{25}, \frac{21}{25})$ **33.** $(10, 24)$ **35.** $\left(\dfrac{3a + b}{10}, \dfrac{a - 3b}{10}\right)$ **37.** $(\frac{1}{13}, \frac{1}{17})$ **39.** $(\frac{1}{4}, -\frac{1}{5})$
41. $(-1, \frac{1}{3})$

PROBLEM SET 7.6   page 293

1. $(2, 3, -1)$ **3.** $(4, 3, 2)$ **5.** $(2, -1, 3)$ **7.** $(1, 1, 1)$ **9.** $(-5, \frac{5}{3}, \frac{28}{3})$ **11.** $(2, 2, 1)$ **13.** $(5, 2, 1)$
15. $(\frac{94}{35}, \frac{1}{7}, \frac{1}{5})$ **17.** $(1, -3, -2)$ **19.** $(6, -3, 12)$ **21.** $(3, 4, 6)$

PROBLEM SET 7.7   page 299

1. 5, 7   **3.** 3 and 12   **5.** 4,925 prey; 725 predators   **7.** Dawn is 14; Joe is 17   **9.** 6 dimes; 12 quarters
11. 17 dimes; 28 quarters   **13.** 63 fives; 15 tens   **15.** fixed charge is $30; hourly rate is $20
17. Tom takes 12 hours; John takes 6 hours   **19.** $40; $20   **21.** $25,000 at 10.5%; $15,000 at 13.5%
23. $2,250 at 8.5 %; $1,750 at 7.9%   **25.** $20,000 at 18%; $10,000 at 19.5%   **27.** 210 dimes; 60 quarters;
24 half-dollars   **29.** 45 type I; 20 type II; 15 type III

PROBLEM SET 7.8   page 306

1.

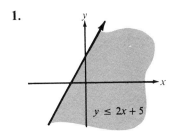

3.

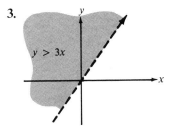

5.

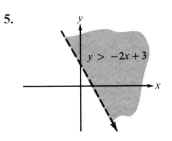

7.

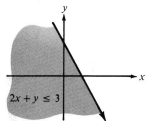

$2x + y \leq 3$

9.

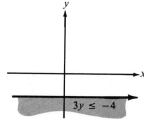

$3y \leq -4$

11.

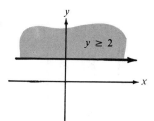

$y \geq 2$

13.

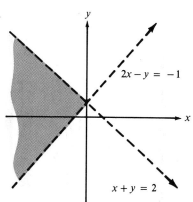

$2x - y = -1$

$x + y = 2$

15.

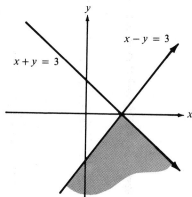

$x - y = 3$

$x + y = 3$

17.

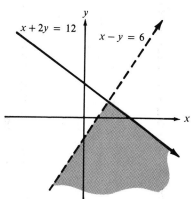

$x + 2y = 12$

$x - y = 6$

19.

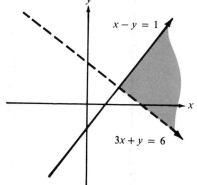

$x - y = 1$

$3x + y = 6$

21.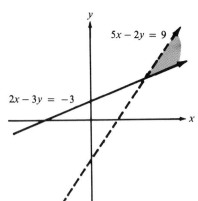

23.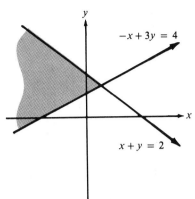

REVIEW PROBLEM SET   page 307

1. I   **3.** II   **5.** y axis   **7.** IV   **9.** 13   **11.** 5   **13.** $|\overline{AB}| = |\overline{AC}| = \sqrt{53}$

15. 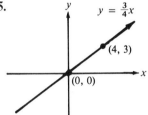   **17.** 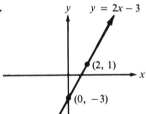   **19.**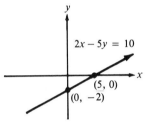

21. 1   **23.** $-\frac{1}{2}$ **25.** 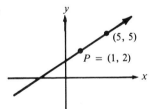   **27.**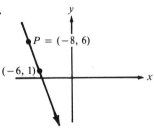

29. parallel   **31.** perpendicular   **33.** $y - 1 = 3(x - 1)$ **35.** $y = -2$ **37.** $y = 2x + 4$ **39.** $x = 3$

41. $y = -2x + 4$
  $m = -2$
  x intercept $= 2$
  y intercept $= 4$
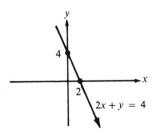

43. $y = \frac{3}{4}x + 3$
  $m = \frac{3}{4}$
  x intercept $= -4$
  y intercept $= 3$
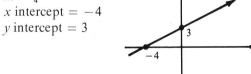

45. a) $y - 3 = \frac{3}{2}(x - 2)$; b) $y = \frac{3}{2}x$; c) $3x - 2y = 0$ **47.** a) $y + 2 = 4(x - 1)$; b) $y = 4x - 6$;
c) $4x - y - 6 = 0$ **49.** independent   **51.** inconsistent   **53.** $(3, 1)$ **55.** $(1, 2, -2)$ **57.** $(2, -1)$
59. $(-1, 3, 2)$ **61.** $(\frac{1}{3}, \frac{1}{2})$

63.

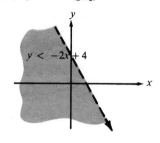

65.

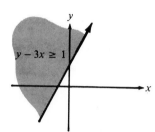

67.

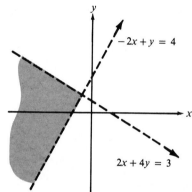

69.

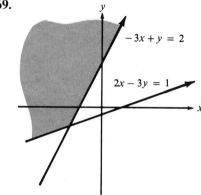

71. $7, -5$ **73.** 37 at 30¢; 63 at 20¢   **75.** 11 nickels; 21 dimes   **77.** $10,000 at 14%; $5,000 at 15%;
$15,000 at 12%

Chapter 8

PROBLEM SET 8.1   page 315

1. 2   **3.** 2   **5.** 8   **7.** -1 **9.** $\frac{9}{4}$ **11.** $-3, 1$ **13.** $3 \pm \sqrt{6}$ **15.** $3 = \log_5 125$ **17.** $5 = \log_{10} 100{,}000$
19. $-2 = \log_4 \frac{1}{16}$ **21.** $-2 = \log_6 \frac{1}{36}$ **23.** $\frac{1}{2} = \log_9 3$ **25.** $-\frac{3}{2} = \log_{100} 0.001$ **27.** $0 = \log_7 1$
29. $3 = \log_x a$ **31.** $9^2 = 81$ **33.** $27^{2/3} = 9$ **35.** $10^{-3} = 0.001$ **37.** $(\frac{1}{3})^{-2} = 9$ **39.** $10^{0.64} = 4.35$
41. $(\sqrt{16})^{1/2} = 2$ **43.** $x^0 = 1$ **45.** 6   **47.** $\frac{1}{2}$ **49.** $-\frac{1}{2}$ **51.** 2   **53.** -3 **55.** $\frac{5}{2}$ **57.** -4 **59.** 6
61. 0   **63.** -3 **65.** 9   **67.** 5   **69.** 25   **71.** 3   **73.** 243   **75.** 0

PROBLEM SET 8.2   page 320

1. $\log_4 5 + \log_4 y$ **3.** $\log_3 u + \log_3 v$ **5.** $\log_5 x - \log_5 3$ **7.** $\log_2 7 - \log_2 15$ **9.** $5 \log_7 3$ **11.** $5 \log_3 c$
13. $\frac{1}{2} \log_4 w$ **15.** $4 \log_7 3 + 2 \log_7 5$ **17.** $4 \log_{11} x + 2 \log_{11} y$ **19.** $\frac{1}{5} \log_2 x + \frac{1}{5} \log_2 y$

21. $2 \log_5 a - 4 \log_5 b$ **23.** $3 \log_7 x + \frac{1}{4} \log_7 y - 3 \log_7 z$ **25.** $4 \log_4 u + 5 \log_4 v - \frac{3}{4} \log_4 z$

27. $\frac{1}{7} \log_3 y - \frac{1}{7} \log_3 (y + 7)$ **29.** $\log_5 \frac{4}{3}$ **31.** $\log_7 \frac{1}{6}$ **33.** $\log_3 \frac{5z}{y}$ **35.** $\log_3 \frac{2}{25}$ **37.** $\log_2 x^3 y^7$

39. $\log_a \frac{y}{3}$ **41.** $\log_4 \frac{\sqrt{a}}{b^3 z^4}$ **43.** $\log_7 (x^2 - 1)$ **45.** $\log_e (y + 5)$ **47.** 4   **49.** $\frac{1}{5}$ **51.** 0.7781

53. 1.2552   **55.** 0.1761   **57.** 0.6990   **59.** 1.9084   **61.** −0.3010   **63.** $\frac{1}{2}$ **65.** 18   **67.** 5   **69.** $\frac{6}{5}$

71. 103   **73.** Let $\log_b b^r = t$, so that $b^t = b^r$, or $t = r$. Therefore, $\log_b b^r = r$

PROBLEM SET 8.3   page 329

1. 3.782×10^3 **3.** 3.81×10^{-3} **5.** 3.75×10^5 **7.** 1.321×10^{-4} **9.** 2.71312×10^{-4} **11.** 210

13. 0.0000314   **15.** 11,300   **17.** 0.00000541   **19.** 0.0003127   **21.** 2.5011   **23.** 4.7275   **25.** 1.2330

27. 0.8331   **29.** 0.0719   **31.** −0.5017   **33.** −1.1469   **35.** −3.7496   **37.** −5.1549   **39.** 3.1889

41. 1.9007   **43.** 0.7253   **45.** −0.2422   **47.** −1.2723   **49.** 2.59   **51.** 19.7   **53.** 553   **55.** 3,560

57. 0.543   **59.** 0.00793   **61.** 0.0235   **63.** 0.000325   **65.** 0.696   **67.** 0.00712   **69.** 1.397   **71.** 34.88

73. 0.01682   **75.** 0.00003448

PROBLEM SET 8.4   page 334

1. 1.518   **3.** 0.7912   **5.** 2.781   **7.** −2.872   **9.** −2.448   **11.** −4.379   **13.** 84.74   **15.** 1.244

17. 29.77   **19.** 25,500   **21.** 0.0328   **23.** 0.001426   **25.** 8.899   **27.** 2.297   **29.** −0.6270   **31.** 8.166

33. 4.113   **35.** 0.8607   **37.** 0.04505   **39.** 41.02   **41.** 0.05388   **43.** 0.04002   **45.** 15.30   **47.** 10.20

49. 1.637   **51.** 2.469   **53.** 0.7357   **55.** 2.807   **57.** −1.850   **59.** 0.5493   **61.** 1.396   **63.** 3.229

65. 0.1989   **67.** a) 7.8; b) 4.2

PROBLEM SET 8.5   page 339

1. $10,100.28   **3.** $20,229.19   **5.** 11%   **7.** 8.243%   **9.** $6,661.68   **11.** $17,131.04   **13.** 2,983,649

15. $2,013.75   **17.** 0.505   **19.** 0.0806

REVIEW PROBLEM SET   page 340

1. $\frac{3}{2}$ **3.** −3   **5.** $\frac{5}{3}$ **7.** 4   **9.** $5 = \log_2 32$ **11.** $\frac{2}{3} = \log_8 4$ **13.** $-\frac{2}{3} = \log_{27} (\frac{1}{9})$ **15.** $n = \log_z w$

17. $2^3 = 8$ **19.** $10^2 = 100$ **21.** $17^0 = 1$ **23.** $10^{-1} = \frac{1}{10}$ **25.** 2   **27.** 0   **29.** $-\frac{3}{2}$ **31.** $-\frac{7}{2}$

33. −4   **35.** $\frac{7}{4}$ **37.** 2   **39.** $\frac{17}{7}$ **41.** 2   **43.** $\log_6 7 + \log_6 u$ **45.** $6 \log_2 3 + 14$ **47.** $\log_4 x + 5 \log_4 y$

49. $2 \log_a w - 4 \log_a z$ **51.** $\log_2 (\frac{2}{9})$ **53.** $\log_5 (\frac{1}{6})$ **55.** $\log_a \frac{x^5}{y^3}$ **57.** $\frac{3}{2}$ **59.** $\frac{9}{4}$ **61.** 15   **63.** $\frac{\sqrt{26}}{2}$

65. 4.68×10 **67.** 1.2×10^{-6} **69.** 5,600   **71.** 0.000192   **73.** 2.9274   **75.** 1.8762   **77.** 0.5030

79. −0.3251   **81.** −2.2316   **83.** 5.58   **85.** 89.8   **87.** 1,728   **89.** 0.0283   **91.** 0.0003912   **93.** 1.168

95. -3.502 **97.** $3,102$ **99.** 0.004822 **101.** 3.814 **103.** 9.356 **105.** 0.1627 **107.** 2.211
109. 2.322 **111.** 0.8510 **113.** 1.315 **115.** $\$11,565.30$ **117.** 14.49% **119.** $\$3,941.04$ **121.** $664,363$

Chapter 9

PROBLEM SET 9.1   page 348

1. 4 **3.** -5 **5.** 1 **7.** $3a + 3z + 1$ **9.** 169 **11.** 4 **13.** 0 **15.** 0 **17.** 3 **19.** 5 **21.** 5
23. 2 **25.** $|-a - 2|$ **27.** all real numbers   **29.** all real numbers except 0 **31.** $x > 1$ **33.** $x \le 2$

35. all real numbers   **37.**   **39.**

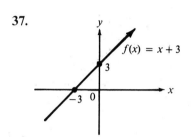

41. function   **43.** function   **45.** not a function

PROBLEM SET 9.2   page 353

1. $f(x) = 2x$; $f(x + 2) = 2x + 4$; $f(2) + f(3) = 10$; $\dfrac{f(x + h) - f(x)}{h} = 2$ **3.** $y = \frac{1}{2}x^3$ **5.** $y = 3x^3$

7. $y = \frac{14}{1331}x^3$ **9.** 4 **11.** $T = \dfrac{0.01x}{y}$ **13.** $y = \dfrac{192}{x^3}$ **15.** $3:2$; $27:8$ **17.** $S = 6x^2$; 864 square inches

PROBLEM SET 9.3   page 360

1. domain = all real numbers; range = all real numbers; slope = -3 **3.** domain = all real numbers;
range = all real numbers; slope = $-\frac{3}{4}$ **5.** domain = all real numbers; range = $\{-1\}$; slope = 0
7. domain = all real numbers; range = all real numbers; slope = 2 **9.** domain = all real numbers;
range = all real numbers; slope = 4 **11.** $f(x) = 2x + 1$ **13.** $f(x) = -\frac{4}{3}x + 4$ **15.** $f(x) = 3$
17. $f(x) = mx$ **19.** $f(x) = \frac{1}{3}x + \frac{14}{5}$ **21.** y intercept = 0; x intercept = 0; vertex = $(0, 0)$;
domain = all real numbers; range = $y \ge 0$ **23.** y intercept = -3; x intercepts = $\pm\dfrac{\sqrt{6}}{2}$; vertex = $(0, -3)$;
domain = all real numbers; range = $y \ge -3$ **25.** y intercept = -1; x intercept = -1; vertex = $(-1, 0)$;
domain = all real numbers; range = $y \le 0$ **27.** y intercept = 6; x intercepts = $-3, -2$;
vertex = $(-\frac{5}{2}, -\frac{1}{4})$; domain = all real numbers; range = $y \ge -\frac{1}{4}$ **29.** y intercept = 0; x intercepts = $0, \frac{3}{2}$;
vertex = $(\frac{3}{4}, -\frac{9}{8})$; domain = all real numbers; range = $y \ge -\frac{9}{8}$ **31.** y intercept = 3;
x intercepts = $-3, -1$; vertex = $(-2, -1)$; domain = all real numbers; range = $y \ge -1$

33.

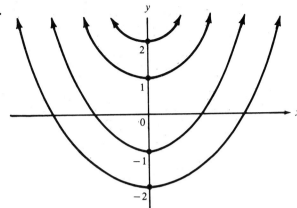

PROBLEM SET 9.4   page 367

1.

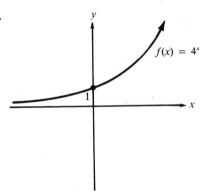

$f(x) = 4^x$

domain = all real numbers;
range = positive real numbers

3.

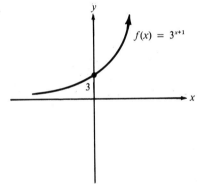

$f(x) = 3^{x+1}$

domain = all real numbers;
range = positive real numbers

5.

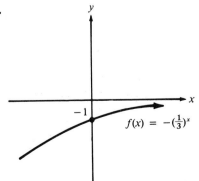

$f(x) = -(\frac{1}{3})^x$

domain = all real numbers;
range = negative real numbers

7.

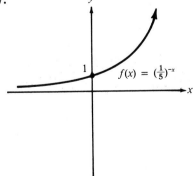

$f(x) = (\frac{1}{5})^{-x}$

domain = all real numbers;
range = positive real numbers

9.

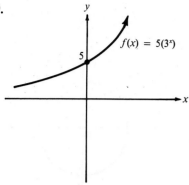

$f(x) = 5(3^x)$

5

domain = all real numbers;
range = positive real numbers

11. 3   **13.** 4   **15.** any number greater than zero

19.

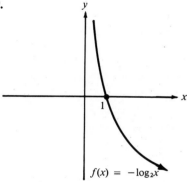

1

$f(x) = -\log_2 x$

domain = positive real numbers;
range = all real numbers

21.

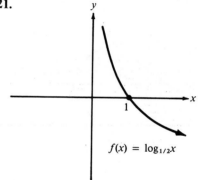

1

$f(x) = \log_{1/2} x$

domain = positive real numbers;
range = all real numbers

23.

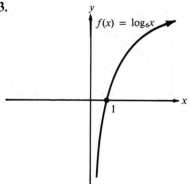

$f(x) = \log_6 x$

1

domain = positive real numbers;
range = all real numbers

25. 2   **27.** 4   **29.** 9

PROBLEM SET 9.5   page 374

1.

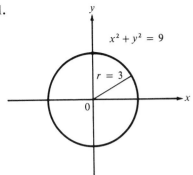

$x^2 + y^2 = 9$

$r = 3$

3.

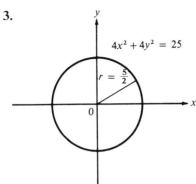

$4x^2 + 4y^2 = 25$

$r = \frac{5}{2}$

5.

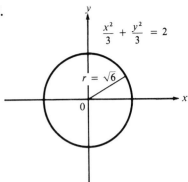

$\frac{x^2}{3} + \frac{y^2}{3} = 2$

$r = \sqrt{6}$

7.

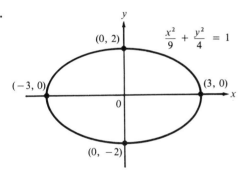

$\frac{x^2}{9} + \frac{y^2}{4} = 1$

$(0, 2)$

$(-3, 0)$ $(3, 0)$

$(0, -2)$

9.

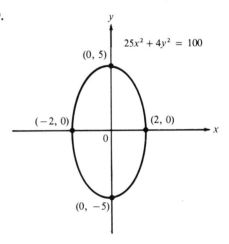

$25x^2 + 4y^2 = 100$

$(0, 5)$

$(-2, 0)$ $(2, 0)$

$(0, -5)$

11.

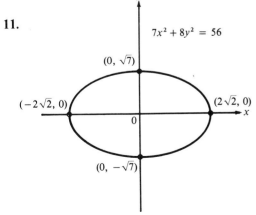

$7x^2 + 8y^2 = 56$

$(0, \sqrt{7})$

$(-2\sqrt{2}, 0)$ $(2\sqrt{2}, 0)$

$(0, -\sqrt{7})$

13.

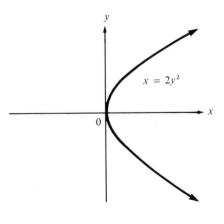

$x = 2y^2$

15.

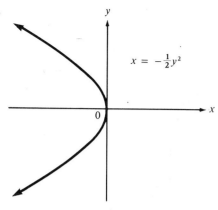

$x = -\frac{1}{2}y^2$

17.

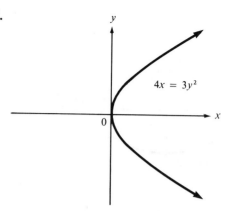

$4x = 3y^2$

19.

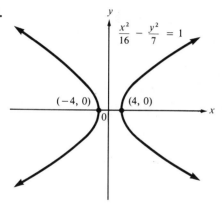

$\frac{x^2}{16} - \frac{y^2}{7} = 1$

$(-4, 0)$ $(4, 0)$

21.

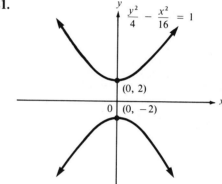

$\frac{y^2}{4} - \frac{x^2}{16} = 1$

$(0, 2)$

$(0, -2)$

23.

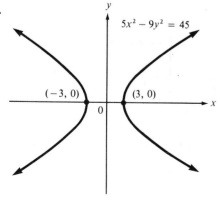

$5x^2 - 9y^2 = 45$

$(-3, 0)$ $(3, 0)$

25.

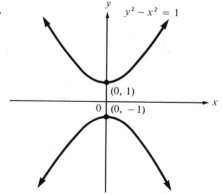

$y^2 - x^2 = 1$

$(0, 1)$

$(0, -1)$

27.

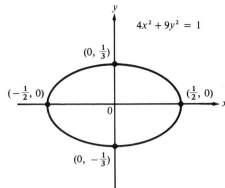

$4x^2 + 9y^2 = 1$

$(0, \frac{1}{3})$

$(-\frac{1}{2}, 0)$ $(\frac{1}{2}, 0)$

$(0, -\frac{1}{3})$

29.

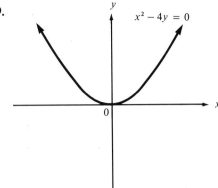

$x^2 - 4y = 0$

31.

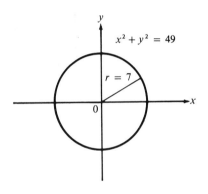

$x^2 + y^2 = 49$

$r = 7$

33.

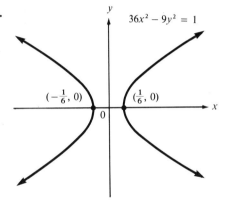

$36x^2 - 9y^2 = 1$

$(-\frac{1}{6}, 0)$ $(\frac{1}{6}, 0)$

35.

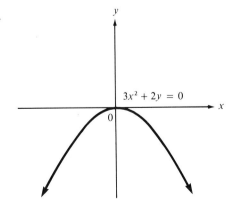

$3x^2 + 2y = 0$

37.

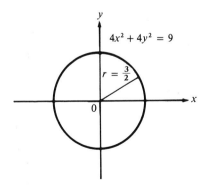

$4x^2 + 4y^2 = 9$

$r = \frac{3}{2}$

39.

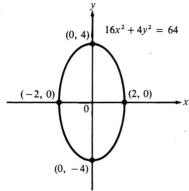

$(0, 4)$ $16x^2 + 4y^2 = 64$

$(-2, 0)$ $(2, 0)$

$(0, -4)$

PROBLEM SET 9.6   page 379

1. $(-1, -2)$; $(2, 1)$ **3.** $(2, 4)$; $(-0.8, -4.4)$ **5.** $(-1 + 2i, 2 - 3i)$; $(-1 - 2i, 2 + 3i)$; $(-6, \sqrt{6}i)$; $(-6, -\sqrt{6}i)$; no point of intersection   **7.** $(\frac{7}{2}, \frac{5}{2})$; $(\frac{11}{4}, \frac{5}{4})$ **9.** $(2 + 3i, 1 - 2i)$; $(2 - 3i, 1 + 2i)$; no point of intersection   **11.** $(5, 4)$ **13.** $(4, 2)$; $(4, -2)$ **15.** $(3, 2)$; $(-3, 2)$; $(3, -2)$; $(-3, -2)$ **17.** $(-2\sqrt{6}, 1)$; $(-2\sqrt{6}, -1)$; $(2\sqrt{6}, 1)$; $(2\sqrt{6}, -1)$ **19.** $(-2, 2\sqrt{3})$; $(-2, -2\sqrt{3})$; $(2, 2\sqrt{3})$; $(2, -2\sqrt{3})$ **21.** $(4, 2)$ $(-4, 2)$; $\left(\frac{4\sqrt{15}}{3}, -\frac{10}{3}\right)$; $\left(\frac{-4\sqrt{15}}{3}, -\frac{10}{3}\right)$ **23.** $(3, 2)$; $(-3, 2)$; $(-3, -2)$; $(3, -2)$ **25.** $(2i, 3)$; $(-2i, 3)$; $(-2, 1)$; $(2, 1)$ **27.** $(3i, 5)$; $(3i, -5)$; $(-3i, 5)$; $(-3i, -5)$; no point of intersection

29. $\left(\frac{41}{10}, \frac{3\sqrt{91}}{10}\right)$; $\left(\frac{41}{10}, -\frac{3\sqrt{91}}{10}\right)$ **31.** 8 centimeters by 12 centimeters   **33.** \$6,800; 2.5%

35. 150 feet by 180 feet

REVIEW PROBLEM SET   page 380

1. 9   **3.** 0   **5.** 3   **7.** 25   **9.** $2u + 5$ **11.** 4   **13.** all real numbers   **15.** all real numbers except -2 and 1   **17.** $x \leq \frac{3}{2}$ **19.** function   **21.** not a function   **23.** $f(x) = \frac{2}{3}x$ **25.** $f(x) = 4\sqrt{x}$ **27.** $f(x) = \dfrac{20}{x}$

29. 48 pounds   **31.** domain = all real numbers; range = all real numbers; slope = 2
33. domain = all real numbers; range = all real numbers; slope = $\frac{3}{2}$ **35.** y intercept = -4;
x intercepts = $-\frac{1}{2}, \frac{4}{3}$; vertex = $(\frac{5}{12}, -\frac{121}{24})$; domain = all real numbers; range = $y \geq -\frac{121}{24}$
37. y intercept = 9; x intercept = -3; vertex = $(-3, 0)$; domain = all real numbers; range = $y \geq 0$
39. y intercept = -3; x intercepts = $-\frac{3}{4}, -\frac{1}{2}$; vertex = $(-\frac{5}{8}, \frac{1}{8})$; domain = all real numbers; range = $y \leq \frac{1}{8}$

41.

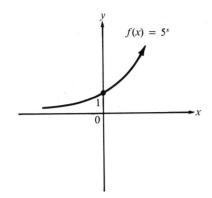

domain = all real numbers;
  range = positive real numbers

43.

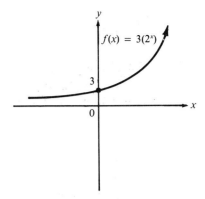

domain = all real numbers;
  range = positive real numbers

45.

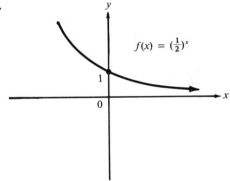

domain = all real numbers;
  range = positive real numbers

47.

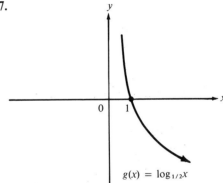

domain = positive real numbers;
  range = all real numbers

49.

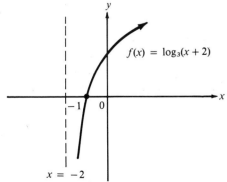

domain = $x > -2$;
  range = all real numbers

51. 1   **53.** $\frac{1}{2}$ **55.** $\frac{5}{4}$ **57.** $\frac{1}{20}$

59. circle

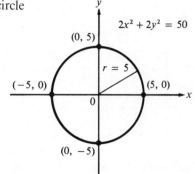

$2x^2 + 2y^2 = 50$
$(0, 5)$
$r = 5$
$(-5, 0)$
$(5, 0)$
$(0, -5)$

61. ellipse

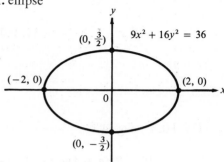

$9x^2 + 16y^2 = 36$
$(0, \frac{3}{2})$
$(-2, 0)$
$(2, 0)$
$(0, -\frac{3}{2})$

63. parabola

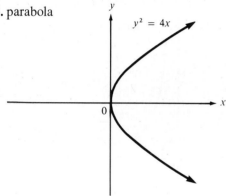

$y^2 = 4x$

65. hyperbola

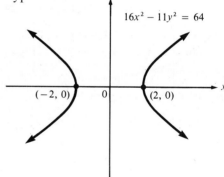

$16x^2 - 11y^2 = 64$
$(-2, 0)$
$(2, 0)$

67. $(3, -4)$ **69.** $(6, 0)$; $(-5, -\sqrt{11})$; $(-5, \sqrt{11})$ **71.** $(-5, -2)$; $(-5, 2)$; $(5, -2)$; $(5, 2)$
73. 16 feet by 30 feet   **75.** 4 centimeters by 5 centimeters

Chapter 10

PROBLEM SET 10.1   page 387

1. $\frac{3}{2}$; 4; $\frac{15}{2}$; 12; $\frac{35}{2}$ **3.** -1; -1; 0; 2; 5   **5.** 2; 4; 2; 4; 2   **7.** A.P.; $d = 3$ **9.** A.P.; $d = 5$
11. A.P.; $d = -13$ **13.** A.P.; $d = 1.2$ **15.** $a_{10} = 50$; $a_{15} = 85$ **17.** $a_6 = 16a + 4b$; $a_9 = 25a - 8b$
19. G.P.; $r = 3$ **21.** G.P.; $r = -2$ **23.** G.P.; $r = \frac{2}{3}$ **25.** G.P.; $r = -\frac{2}{3}$ **27.** $\frac{1}{128}$ **29.** 2
31. $(1 + a)^{n-1}$ **33.** $a_6 = 192$; $a_{10} = 3{,}072$ **35.** 48   **37.** sixth

PROBLEM SET 10.2   page 393

1. 15   **3.** 660   **5.** 15   **7.** 15   **9.** $\frac{163}{60}$ **11.** 500   **13.** $\frac{5}{6}$ **15.** 145   **17.** -28 **19.** $63b - 63$
21. $a_{10} = 33$; $S_{10} = 195$ **23.** $d = 13\frac{5}{51}$; $a_{18} = 239\frac{2}{3}$ **25.** $n = 32$; $d = \frac{21}{31}$ **27.** $\frac{189}{2}$ **29.** 5,460
31. $10(1 + a + a^2 + a^3 + a^4 + a^5)$ **33.** -4 or 3   **35.** -2 or 2   **37.** $a_9 = \frac{1}{128}$; $S_8 = \frac{85}{64}$ **39.** $n = 10$;
$S_{10} = \frac{1,023}{16}$

PROBLEM SET 10.3   page 400

1. $\frac{1}{30}$ **3.** $\frac{1}{9}$ **5.** 2   **7.** 0   **9.** $(n + 1)n(n - 1)(n - 2)$ **11.** $x^5 + 10x^4 + 40x^3 + 80x^2 + 80x + 32$
13. $x^6 + 12x^4y^2 + 48x^2y^4 + 64y^6$ **15.** $a^{18} - 6a^{14} + 15a^{10} - 20a^6 + 15a^2 - 6a^{-2} + a^{-6}$

17. $32 + 80\left(\frac{x}{y}\right) + 80\left(\frac{x}{y}\right)^2 + 40\left(\frac{x}{y}\right)^3 + 10\left(\frac{x}{y}\right)^4 + \left(\frac{x}{y}\right)^5$

19. $256z^8 + 1,024z^7x + 1,792z^6x^2 + 1,792z^5x^3 + 1,120z^4x^4 + 448z^3x^5 + 112z^2x^6 + 16zx^7 + x^8$
21. $y^{10} - 10y^8x + 40y^6x^2 - 80y^4x^3 + 80y^2x^4 - 32x^5$ **23.** $x^{20} - 20x^{18}a + 180x^{16}a^2 - 960x^{14}a^3$

25. $\left(\frac{x}{2}\right)^{7/2} + 14\left(\frac{x}{2}\right)^3 y + 84\left(\frac{x}{2}\right)^{5/2} y^2 + 280\left(\frac{x}{2}\right)^2 y^3$ **27.** $x^{16} + 16x^{15}y + 120x^{14}y^2 + 560x^{13}y^3 + 1,820x^{12}y^4$

29. $a^{11} - 22a^{10}b^2 + 220a^9b^4 - 1,320a^8b^6 + 5,280a^7b^8$ **31.** $x^7 - 14x^6y + 84x^5y^2 - 280x^4y^3 + 560x^3y^4$
33. $a^{27} - 9a^{26} + 36a^{25} - 84a^{24} + 126a^{23}$ **35.** $\frac{455}{4,096}x^{24}a^3$ **37.** $\frac{224}{243}x^6a^{12}$ **39.** $\frac{14}{9}a^5x^{12}$

PROBLEM SET 10.4   page 403

1. 26   **3.** 24   **5.** $-\frac{13}{2}$ **7.** 0   **9.** 1   **11.** 116   **13.** 1   **15.** 18   **17.** -438 **19.** 4   **21.** -1
23. 2 or 8   **25.** $\frac{23}{2}$ **27.** $x < -2$ or $x > 2$

PROBLEM SET 10.5   page 408

1. $\left(\frac{1}{3}, \frac{2}{3}\right)$ **3.** $(0, 0)$ **5.** $\left(\frac{85}{4}, \frac{35}{4}\right)$ **7.** $(-11, 7)$ **9.** $(5, -6)$ **11.** $(1, 2, 3)$ **13.** $\left(\frac{19}{4}, -\frac{3}{4}, -\frac{29}{4}\right)$
15. $\left(-5, -\frac{14}{3}, -\frac{16}{3}\right)$ **17.** $(2, 1, -1)$ **19.** $(-4, 5, 4)$

REVIEW PROBLEM SET   page 409

1. 4, 2, 4, 2   **3.** 2; $\frac{7}{2}$; 4; $\frac{17}{4}$ **5.** 1; $\frac{18}{5}$; $\frac{39}{5}$; $\frac{68}{5}$ **7.** $a_9 = 44$; $s_9 = 216$ **9.** $a_{11} = 12$; $s_{11} = 297$
11. $a_{24} = 4$; $s_{24} = 50$ **13.** 3   **15.** 3 or -2 **17.** $a_8 = 49,152$; $s_8 = 65,535$ **19.** $a_6 = -\frac{1}{3}$; $s_6 = 60\frac{2}{3}$
21. $a_{18} = -768\sqrt{2}$; $s_{18} = 1,533(1 - \sqrt{2})$ **23.** -2 or 24   **25.** $\frac{1}{7}$ or 10   **27.** 95   **29.** 20   **31.** 160
33. $x^4 + 8x^3y + 24x^2y^2 + 32xy^3 + 16y^4$ **35.** $1 + 5x + 10x^2 + 10x^3 + 5x^4 + x^5$
37. $1 - 12x + 60x^2 - 160x^3 + 240x^4 - 192x^5 + 64x^6$ **39.** $81x^4 + 108x^3y + 54x^2y^2 + 12xy^3 + y^4$

41. $243x^5 + 405x^{9/2} + 270x^4 + 90x^{7/2} + 15x^3 + x^{5/2}$ **43.** $8x^3 + \dfrac{12x^2}{y} + \dfrac{6x}{y^2} + \dfrac{1}{y^3}$ **45.** $210x^6y^4$

47. $13,440x^6y^4$ **49.** $1,082,565x^8y^3$ **51.** 11   **53.** 26   **55.** $-3, 1$ **57.** $(2, -1)$ **59.** $(-1, 3, 2)$

Index

Geometry

Assume A = area, C = circumference, V = volume, S = surface area, r = radius, h = altitude, l = length, w = width, b (or a) = length of a base, and s = length of a side.

1 Square $A = s^2; P = 4s$

2 Rectangle $A = lw; P = 2l + 2w$

3 Parallelogram $A = bh$

4 Triangle $A = \frac{1}{2}bh$

5 Circle $A = \pi r^2; C = 2\pi r$

6 Trapezoid $A = \frac{1}{2}(a + b)h$

7 Cube $S = 6s^2; V = s^3$

8 Rectangular Box $S = 2(lw + wh + lh); V = lwh$

9 Cylinder $S = 2\pi rh; V = \pi r^2 h$

10 Sphere $S = 4\pi r^2; V = \frac{4}{3}\pi r^3$

11 Cone $S = \pi r \sqrt{r^2 + h^2}; V = \frac{1}{3}\pi r^2 h$

English–Metric Conversions

Length:

  1 inch = 2.540 centimeters

  1 foot = 30.48 centimeters

  1 yard = 0.9144 meter

  1 mile = 1.609 kilometers

Volume:

  1 pint = 0.4732 liter

  1 quart = 0.9464 liter

  1 gallon = 3.785 liters

Weight:

  1 ounce = 28.35 grams

  1 pound = 453.6 grams

  1 pound = 0.4536 kilogram